Co-simulations of Microwave Circuits and High-Frequency Electromagnetic Fields

Mei Song Tong · Xiao Yu Li

Co-simulations of Microwave Circuits and High-Frequency Electromagnetic Fields

Springer

Mei Song Tong
Department of Electronic Science
and Technology
Tongji University
Shanghai, China

Xiao Yu Li
Department of Electronic Science
and Technology
Tongji University
Shanghai, China

ISBN 978-981-99-8306-3 ISBN 978-981-99-8307-0 (eBook)
https://doi.org/10.1007/978-981-99-8307-0

This Springer imprint is published by the registered company Springer Nature Singapore Pte Ltd.
The registered company address is: 152 Beach Road, #21-01/04 Gateway East, Singapore 189721, Singapore

Preface

With the rapid development of radio frequency (RF) and microwave technologies, there is an increasing need for co-simulations of microwave circuits and high-frequency electromagnetic fields (HFEMFs). This is particularly prevalent in the fields of modern wireless communications and radar systems, where the complexity of microwave devices and antenna structures is continuously rising. Consequently, researchers and designers must not only contemplate the designs of circuit level, but also take into account the impact of electromagnetic fields, and the co-simulations of microwave circuits and HFEMFs have become an inevitable choice. This book aims to offer readers a comprehensive guide to the co-simulations of RF and microwave circuits incorporated with the HFEMFs. Over the past few decades, we have observed the evolution of numerous kinds of powerful simulation software for microwave circuits and HFEMFs. Among these, Advanced Design System (ADS) and High-Frequency Structure Simulator (HFSS) are undoubtedly the two software tools which have been most prevalently used and widely acknowledged. However, these two tools emphasize different functions and application fields, rendering them incapable of fully meeting the needs of some application scenarios. Therefore, it is imperative to develop the co-simulations method based on the ADS and HFSS for microwave circuits and HFEMFs. This book endeavors to present a comprehensive introduction to the concepts, principles, methods, implementations, and applications of co-simulations based on those software tools. It will offer detailed explanations and abundant examples to help readers better understand the related knowledge and provide useful references for practical engineering applications. Also, this book presents some interesting discussions on the further development and applications of co-simulations and relevant software tools in electronic designs, including their features, practical operations, and optimization skills. The book consists of eight chapters and each chapter focuses on an individual topic.

Chapter 1 mainly introduces the background related to the co-simulations of microwave circuits and HFEMFs. We start with presenting an overview of the well-known simulation software tools, ADS and HFSS, facilitating a preliminary understanding on them for readers. This is followed by a detailed discussion on the ADS, highlighting its development history and its significant position in the field of microwave circuit designs. We then explore the main features of ADS, including its robust design functions and extensive library resources. We also present typical application cases to better demonstrate the advantages of ADS in practical engineering applications. Subsequently, we introduce the HFSS by detailing its development route from an inception to becoming an industry leader of electromagnetic simulation software tools. We provide a guidance on using the HFSS to model and simulate complex three-dimensional structures and discuss its key technologies. Also, we showcase typical applications in the area of HFEMFs to illustrate the robust performance of HFSS in practical scenarios. The latter half of this chapter focuses on the software tools for the co-simulations with the ADS and HFSS, and provide researchers and engineers with more efficient ways of addressing practical issues. The advantages of co-simulations for microwave circuits and HFEMFs, such as improved simulation accuracies and reduced design cycles, are discussed. Finally, we offer a prospect for the co-simulations of microwave circuits and HFEMFs and provide useful insights for readers concerning with technological advancements and application expansions.

Chapter 2 addresses the simulation technologies for HFEMFs. We first discuss the definitions, applications, and developments of HFEMF simulations and then delve into their underlying principles, including Maxwell's equations and numerical methods of solving them. We then analyze the potential errors that could occur during the simulations and provide corresponding solutions. Also, we illustrate the applications of ADS and HFSS for HFEMF simulations, including their respective tools, modeling techniques, and simulation steps and guides. Finally, we focus on the definitions, principles, and implementation methods of co-simulations in detail and introduce different kinds of software for the co-simulations with the ADS and HFSS. In order to give readers a better understanding, the specific implementation process of HFSS and co-simulations designs through examples is provided. This will also allow readers to recognize the value and necessity of HFEMF simulations in practical applications. In addition, an overview on the future of HFEMF simulation technologies is given, including the model refinement, computational efficiency, artificial intelligence, interdisciplinary cooperation, and standardization.

In Chap. 3, we discuss the basic principles and applications of simulation technologies for microwave circuits. Starting with an introduction to the fundamental concepts of microwave circuits, we first emphasize the significance and centrality of simulations for optimally designing microwave circuits. We then narrate the involved principles of simulations for microwave circuits, including electromagnetic equations, transmission line theory, equivalent circuit models, finite element method, finite difference time-domain method, and error analysis and compensation methods. We also explore the applications of ADS and HFSS for the low-frequency simulations of microwave circuits and provide detailed insights into their functions and settings. Finally, the co-simulations are discussed and practical examples are provided to demonstrate how to use the Layout and rFactor Pro (rFpro) to accurately simulate microwave integrated circuits.

Chapter 4 concentrates on the basic concepts, principles, and applications of simulation technologies for antenna designs. The importance and role of simulations for antenna designs are first described and the basic simulation workflow is then provided. Also, the involved working principles of antennas in simulations are addressed, including the basic theory, radiation characteristics, electromagnetic fields, optimization algorithms, and error analysis and calibration. In addition, the applications of ADS and HFSS to the simulations of antenna designs are discussed separately and the software tools for the co-simulations of antenna designs with the ADS and HFSS are introduced, including the Computer Simulation Technology (CST) Studio Suite, WIPL-D (WI stands for wires, PL stands for plates, and D stands for dielectrics), X Finite Difference Time Domain (XFdtd), Antenna Magus, and EM Consulting & Software (EMCoS) Antenna VirtualLab. This chapter comes to an end by discussing the prospect of simulation technologies for antenna designs and explaining how to use the HFSS and Ansys Circuit for co-simulating antenna designs.

Chapter 5 highlights the simulation technologies for the analyses and designs of electromagnetic compatibility (EMC), ranging from fundamental principles to practical applications. The definitions and classifications for the EMC are first given so that readers can preliminarily recognize the EMC in simulations. The simulation functions of ADS and HFSS for EMC analyses and designs are then introduced in detail. We also narrate a variety of simulation software that can be incorporated with the ADS and HFSS for the co-simulations of EMC analyses and designs, including CST Studio, Applied Wave Research (AWR) Microwave office, Simbeor, Ansys SIwave, Altium Designer, and HyperLynx. The optimization methods and techniques are very essential for EMC analyses and designs, so we discuss various optimization strategies and involved challenges as well. To help readers better understand, we demonstrate the EMC co-simulations based on the CST Studio Suite and ADS through practical examples. This chapter is finished by discussing the future of simulation technologies on EMC analyses and designs.

In Chap. 6, we focus on the thermal simulations and their influences on electromagnetic simulations. We first introduce the definitions, classifications, principles, implementations, and applications of thermal simulations so as to familiarize readers with this topic. We then delve into the principles of thermal simulations, which include the heat conduction equation, boundary and initial conditions, heat source modeling, meshing, and solution algorithms. With the knowledge, we address the co-simulations or the interaction between thermal simulations and electromagnetic simulations and their mutual impacts. Also, different kinds of software that can co-simulate with the ADS and HFSS are described, respectively, including the FloTHERM and Icepak. Finally, we present a co-simulations example for illustrating the simulation of electromagnetic loss of a circulator in the HFSS and the electro-thermal simulation with a two-way coupling in the Icepak. The illustration can help readers strength their understanding on the thermal-electromagnetic simulations and their influences on analyses and designs.

Chapter 7 presents structural simulations and their incorporations with electromagnetic simulations. A comprehensive overview is provided about the backgrounds, principles, implementations, applications, challenges, and development trends of structural simulations, which can help readers form a holistic understanding on them. The principles of structural simulations are further explained in detail, including stress and strain analyses, material models and mechanical properties, boundary conditions and loads, meshing and solving algorithms, as well as analyses and verifications of results. We also elaborate the requirement, method, framework, process, data transfer, and interface of implementation for structural co-simulations. Furthermore, different kinds of software which can be used to co-simulate with the ADS and HFSS are introduced, including SIMULIA Abaqus and MacNeal-Schwendler Corporation (MSC) Software for the ADS, and Ansys Mechanical, COMSOL Multiphysics, and Ansys Multiphysics for the HFSS. Finally, through the examples of static simulations with the Hypermesh and Ansys Mechanical, we explain the pre-processing, solution, and post-processing stages of the simulations which can help readers understand the incorporation of structural simulations with electromagnetic simulations.

Chapter 8 narrates the automation interfaces and scripting program methods for simulations. It begins with an introduction to the Advanced Extension Language (AEL) interface in the ADS, including the basic concepts, writing, loading, and running of AEL scripts. We then present the Python Data Link (PDL) in the ADS, including its basic concepts and architecture. Also, we illustrate the data transfer method, processing method of ADS simulation results in the MATLAB, and optimization method for the ADS-MATLAB interfaces. For the HFSS, we first discuss the application of VBScript in the HFSS, including the call of VBS, frequency sweep control using the VBS, revision of model parameters, and post-processing operations. We then look into the IronPython scripting method in the HFSS, including the introduction to the IronPython, basic method of writing HFSS automation scripts, and advanced automation design method. Finally, we address the Application Programming Interface (API) based on the MATLAB and HFSS, including the overview of API, framework, and MATLAB-control file writing method. All introductions are accompanied with practical examples to help understand. Readers can well grasp the knowledge and skills of the automation interfaces and script program methods based on those kinds of software by reading this chapter.

Shanghai, China
April 2023

Mei Song Tong

Acknowledgements

The book intends to introduce the co-simulations method for microwave circuits and high-frequency electromagnetic fields which is actually related to our research work. We are highly grateful to the support from the Textbook Construction Project of Tongji University, Shanghai, China, which has greatly helped the publication of this book. We also strongly appreciate our research sponsors for their supports on the relevant research works which have served a basis for publishing the book. In particular, we would like to sincerely appreciate US Air Force Office of Scientific Research (AFOSR), Construction Engineering Research Laboratory (CERL) of US Army, Semiconductor Research Corporation (SRC), Intel Corporation, IBM Corporation, National Natural Science Foundation of China (NSFC), Ministry of Science and Technology (MOST) of China, Ministry of Education (MOE) of China, and Science and Technology Committee of Shanghai Municipality (STCSM) of China. In addition, we would like to truly thank all colleagues and students who have been involved in the related work. Finally, our gratitude would go to the related staffs of Springer Nature who have made possible the publication of this book. Their hard work and tremendous support and help are highly appreciated.

Contents

About the Authors

Prof. Mei Song Tong received the B.S. and M.S. degrees from Huazhong University of Science and Technology, Wuhan, China, respectively, and Ph.D. degree from Arizona State University, Tempe, Arizona, USA, all in Electrical Engineering. He is currently the Distinguished/Permanent Professor; Head of Department of Electronic Science and Technology; and Vice Dean of College of Microelectronics, Tongji University, Shanghai, China. He has also held an Adjunct Professorship at the University of Illinois at Urbana-Champaign, Urbana, Illinois, USA, and an Honorary Professorship at the University of Hong Kong, China. He has published more than 600 papers in refereed journals and conference proceedings and co-authored six books or book chapters. His research interests include electromagnetic field theory, antenna theory and design, simulation and design of RF/microwave circuits and devices, interconnect and packaging analysis, inverse electromagnetic scattering for imaging, and computational electromagnetics. He is a Fellow of the Electromagnetics Academy, Fellow of the Japan Society for the Promotion of Science (JSPS), and Senior Member (Commission B) of the USNC/URSI. He has been the Chair of Shanghai Chapter since 2014 and the Chair of SIGHT committee in 2018, respectively, in IEEE Antennas and Propagation Society. He has served as an associate editor or guest editor for several well-known international journals, including *IEEE Antennas and Propagation Magazine*, *IEEE Transactions on Antennas and Propagation*, *IEEE*

Transactions on Components, Packaging and Manufacturing Technology, *International Journal of Numerical Modeling: Electronic Networks, Devices and Fields*, *Progress in Electromagnetics Research*, and *Journal of Electromagnetic Waves and Applications*. He also frequently served as a session organizer/chair, technical program committee member/chair, and general chair for some prestigious international conferences. He was the recipient of a Visiting Professorship Award from Kyoto University, Japan, in 2012 and from University of Hong Kong, China, in 2013. He advised and co-authored 12 papers that received the Best Student Paper Award from different international conferences. He was the recipient of the Travel Fellowship Award of USNC/URSI for the 31th General Assembly and Scientific Symposium (GASS) in 2014, Award for the Advance of Science and Technology of Shanghai Municipal Government in 2015, Fellowship Award of JSPS in 2016, Innovation Award of Universities' Achievements of Ministry of Education of China in 2017, Innovation Achievement Award of Industry-Academia-Research Collaboration of China in 2019, "Jinqiao" Award of Technology Market Association of China in 2020, Carl Friedrich von Siemens Research Award of the Alexander von Humboldt Foundation of Germany in 2023, and Technical Achievement Award of Applied Computational Electromagnetics Society (USA) in 2024. In 2018, he was selected as the Distinguished Lecturer (DL) of IEEE Antennas and Propagation Society for 2019–2022.

Xiao Yu Li received the B.S. degree in Communication and Information Engineering and the M.S. degree in Electronic and Information Engineering from University of Electronic Science and Technology of China, Chengdu, China, in 2017 and 2020, respectively. He is currently pursuing the Ph.D. degree in Control Science and Engineering with Tongji University, Shanghai, China, and his current research interests include reconfigurable antennas and computational electromagnetics.

Chapter 1
Introduction to Co-simulations of Microwave Circuits and High-Frequency Electromagnetic Fields

1.1 Introduction

With the rapid development in the areas of wireless communications [1–3], radar [4–6], navigation [7–9], and medical imaging [10–12], etc., there is a growing recognition on the significance of simulations for microwave circuits and high-frequency electromagnetic fields (HFEMFs). The simulations of HFEMFs involve the numerical modeling/simulation and analysis of physical wave behaviors such as propagation, scattering, and radiation of high-frequency electromagnetic waves [13]. On the other hand, the simulations of microwave circuits refer to the numerical modeling/simulation and analysis/optimization for the design of microwave circuits [14]. These simulations need to individually meet the requirements on accuracy and efficiency, and more importantly, their co-simulations are also required for many applications in order to realize diverse functions.

The co-simulation is a method that integrates multiple simulation, combining simulation models and tools from different domains to evaluate the behavior and performance of complex systems [15]. Its significance lies not only in improving the accuracy and credibility of simulations but also in fostering cross-fertilization and collaborative innovation across different fields [15, 16]. Co-simulation enables quick assessment of system feasibility and risks, optimization of system design and performance, and enhancement of decision-making based on scientific evidences. Additionally, co-simulation provides a support for system implementation and operation, reducing the risk of failures and errors, and improving system reliability and safety. Hence, co-simulation finds wide-ranging applications and promising prospects in military, aerospace, transportation, and healthcare fields [15–20].

To facilitate the co-simulations for microwave circuits and HFEMFs, the two well-known software tools have been widely used, i.e., Advanced Design System (ADS) [21, 22] and High-Frequency Structure Simulator (HFSS) [23] as shown in Fig. 1.1, and they are the main focus of this book. The ADS is a kind of circuit simulation software based on Electronic Design Automation (EDA) that can enable

M. S. Tong and X. Y. Li, *Co-simulations of Microwave Circuits and High-Frequency Electromagnetic Fields*, https://doi.org/10.1007/978-981-99-8307-0_1

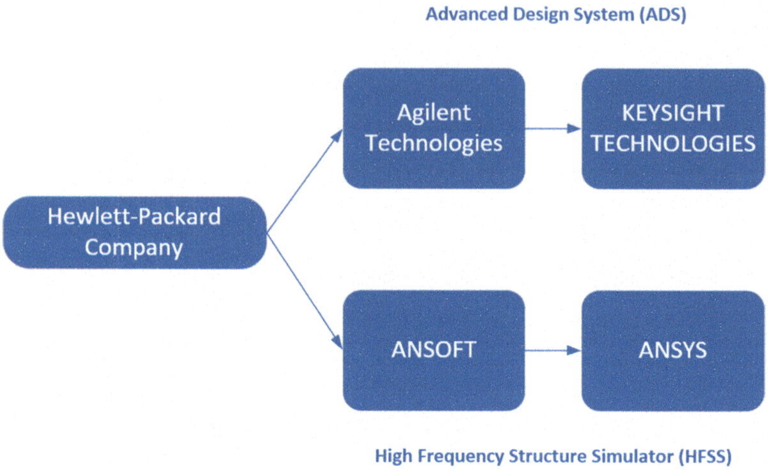

Fig. 1.1 Distinct developmental branches of ADS and HFSS: Tracing the evolution from Hewlett-Packard Company to Agilent Technologies and KEYSIGHT Technologies, and from ANSOFT to ANSYS

the simulations for microwave and RF circuits, among other functions [24]. It offers powerful simulation and analysis capabilities, user-friendly interfaces, and broad application areas. On the other hand, the HFSS is a simulation tool for HFEMFs and has been widely utilized for the modeling/simulations of antennas, electronic packaging, electromagnetic compatibility, radar imaging, etc. [25]. It can provide high-precision simulations of electromagnetic waves, advanced post-processing and analyses, and support co-simulations with other softwares [26, 27].

For co-simulations, both the ADS and HFSS possess unique advantages and application scopes. The ADS excels in microwave circuit simulations, offering diverse simulation functions that enable fast and accurate simulations and optimization for microwave circuits [24, 28]. The HFSS, on the other hand, stands out in electromagnetic field simulations, providing high-precision and high-efficiency simulation capabilities to deliver comprehensive and accurate analyses for complex electromagnetic problems [26, 27, 29, 30].

In addition to the co-simulation capabilities provided by the ADS and HFSS themselves, there are several other kinds of software available that can be integrated with them to expand the range of simulation applications. For instance, the softwares Genesys [31] and CST Studio Suite [32] can be used for the co-simulations of microwave circuits [33]. Similarly, the softwares COMSOL Multiphysics [34] and Designer [35] can be employed for the co-simulations of electromagnetic fields [36, 37]. These software tools can be utilized not only in conjunction with the ADS and HFSS but also with other softwares to cater to the simulation needs of diverse fields.

During the actual co-simulation processes, it is essential to understand and address the data formats and interfaces between different kinds of software. For example, in the co-simulations between the ADS and HFSS, the circuit designs and simulations are conducted using EMPro or Layout software within the ADS. Subsequently, the design results are imported into the HFSS for electromagnetic simulations and analyses. In this process, the data interface between the two kinds of software needs to be properly configured and debugged to ensure accurate data transmission and reliable simulation results. Besides, the interface debugging, optimization, and adjustment of the simulation models and parameters must satisfy the requirements of co-simulations. For instance, in the antenna design and analysis, the parameters such as antenna geometry, material properties, and radiation direction must be finely tuned and optimized to enhance the accuracy and reliability of the simulation results. Finally, the co-simulation results need to be thoroughly analyzed and evaluated to validate the reliability and accuracy of the simulation models and parameters. This involves the comparison and analysis of the simulation results to identify and rectify sources of errors, as well as to determine areas for improvement. Furthermore, the visualization and report of the simulation results are crucial for effectively presenting the findings and conclusions of the analyses.

The co-simulation is an essential part for microwave circuits and HFEMFs, as it can enhance the reliability and accuracy of the circuits with fields. This book aims to introduce the fundamental principles and applications of ADS and HFSS, along with various kinds of software that can facilitate co-simulations. It will also explore the advantages and application scenarios of co-simulations so that readers can gain a better understanding for the applications associated with this approach.

1.2 Overview of Advanced Design System (ADS)

Since the 1970s, the EDA has been extensively utilized in the field of microwave circuit designs [39]. The EDA-based simulation tools have played a crucial role in achieving accurate and efficient designs of microwave circuits [40–42]. Among various EDA software options, the ADS stands out as a highly popular and widely used choice for the simulation and optimization of microwave circuits. Developed by Keysight [43], Inc., the ADS has undergone numerous updates and iterations since its initial release, and continuously enhanced its simulation functions and performances. Over time, it has evolved into a comprehensive and versatile EDA software platform with applications across multiple disciplines. This section provides a comprehensive overview for the ADS, including its historical development, key functions, and notable features.

1.2.1 Development History of ADS

Advanced Design System (ADS) is a kind of premier software developed by PathWave Design, a division of Keysight Technologies, that specializes in electronic circuit design, simulation, and automation. It offers a comprehensive environment for designing RF electronic systems, including wireless networks, mobile phones, satellite communications, high-speed data links, and radar systems. The software

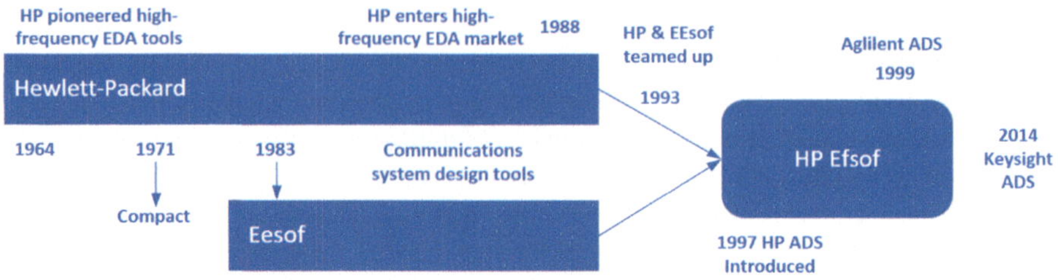

Fig. 1.2 Chronological evolution of Keysight ADS: It begins with Hewlett-Packard in 1964, marking the company's pioneering of high-frequency EDA tools, and follows through various milestones such as the entry into the high-frequency EDA market in 1988, the partnership with EEsof in 1993, and the introduction of HP ADS in 1997. The timeline continues with the evolution to Agilent ADS in 1999 and culminates with Keysight ADS in 2014

initially emerged in 1985 as Microwave Design System (MDS) and was later enhanced and rebranded as ADS in 2016, introducing improvements in speed, performance, and design flexibility [22]. ADS supports designers throughout the entire design process, from schematic capture, design rule checking, and layout, to electromagnetic field (EMF) simulation, and time- and frequency-domain circuit simulation. This enables designers to effortlessly characterize and optimize RF designs without the need to switch between different tools or systems.

In Fig. 1.2, it can be observed that the origins of Advanced Design Software (ADS) can be traced back to the 1970s. During that time, microwave circuit design and simulation heavily relied on manual calculations, experimentation, and hand-drawn diagrams by electronic engineers. However, these methods were time-consuming, labor-intensive, and prone to errors and inaccuracies. Consequently, the exploration of computer-aided design and simulation for microwave circuits began to improve efficiency and accuracy.

Starting in the mid-1970s, several commercial computer programs for microwave circuit design became available. Examples of such software packages include SUPER COMPACT (developed by Compact Software, Inc. in 2000) and TOUCHSTONE (from EEsof, Inc. in 1997). However, due to the limited computational power and storage capacity of computers during that era, these programs could only handle relatively simple circuits and fell short in meeting the simulation demands of complex circuits. As computer technology continued to advance, more microwave circuit simulation software was developed, gradually becoming an indispensable tool for microwave circuit design.

During the early stages of microwave circuit simulation software, numerical methods like the finite element method and finite difference method were primarily utilized for electromagnetic field simulation. Prominent examples of these software packages include HFSS by Ansoft and Sonnet. While these programs offered high-precision simulation capabilities, their efficiency was relatively low and could not satisfy real-time simulation requirements. As computer hardware performance continuously improved, microwave circuit simulation software began to evolve toward high efficiency and multifunctionality. For instance, in the 1990s, the U.S. company Eagleware introduced a type of microwave circuit simulation software called Eagleware Genesys, which boasted high efficiency, precision, and user-friendliness, ultimately becoming a leading product in the market at that time. Furthermore, software offerings such as Mentor Graphics' HyperLynx and Agilent Technologies' ADS gradually gained prominence.

As we enter the twenty-first century, the field of microwave circuit simulation software is witnessing significant advancements in terms of scale, multi-domain capabilities, and intelligent functionalities [37]. One notable example is the evolution of ADS, which has transformed into a versatile EDA software platform catering to various domains such as microwave circuit simulation, RF circuits, digital signal processing, communication systems, and more. ADS now supports co-simulations with other software, enabling users to customize simulation and optimization processes through automated scripts and APIs. Over the years, ADS has undergone numerous updates and iterations to continuously enhance its simulation accuracy and efficiency, gradually evolving into a comprehensive EDA software platform for multiple disciplines. On July 6, 2022, Keysight introduced Keysight PathWave ADS 2023, the latest annual update of ADS, further empowering circuit designers with advanced electromagnetic (EM) simulation capabilities. PathWave ADS 2023 reaffirms its position as the industry's most comprehensive simulation software, offering a wide range of functionalities for RF and microwave, high-speed digital, and power electronics design. The 2023 version introduces new and improved features aimed at enhancing the productivity and usability of RF and microwave circuit and system designers [38].

Circuit Simulations

- Simplified remote and distributed simulation management with support for cloud-based high-performance computing (HPC).
- New Leti-UTSOI 102.6 transistor model for accurate analog and RF circuit design.
- Monte Carlo statistical controller now includes yield optimization.

Electro-thermal Simulations

- Customized multi-technology electro-thermal (ETH) processes are supported even if the layout and schematic levels do not match.
- PathWave ADS provides the industry's most accurate electro-thermal simulation, predicting the location and timing of harmful transient temperature spikes so that they can be fixed before hardware production.
- PathWave ADS' electro-thermal dynamic model generator enables transient electro-thermal simulations to be performed typically 10 times faster and up to 100 times faster to ensure thermal reliability of mission-critical and reputation-critical designs.

High-Performance Computing (HPC)

- EM and Circuit simulation acceleration through parallelized High-Performance Computing (HPC) with cost-effective, powerful, and high-capacity cloud-based hardware resources enable typical speedup from 5x to 20x.
- Parallel EM simulation (RFPro) parameter sweep.
- Even with a modest fivefold increase in speed, this translates to a saving of four days in a typical five-day workweek, allowing for a greater number of simulations in designing high-performance, tolerance-insensitive RF/MW components and substantially reducing the time-to-market.

High-Speed Digital (HSD) Design

Memory Designer

- Supports HBM3, LPDDR5x, IBIS-EMD models.
- IBIS 7.1 parser with custom IBIS file viewer.
- Multi-segment smart bus wire connection.
- Drag-n-drop multi-plots from Expression Manager.
- Improved CA/Data Bus Pre-layout builder editing.
- Cloud based High-Performance Computing (HPC).

SerDes Design

- New Ethernet AMI modeler
- USB AMI model for NZR modulation and Spread Spectrum Clocking (SSC).
- USB reference channel models.
- PCIe AMI model generation for Gen 6 with PAM4 modulation, DFE taps, and 6-lance reference channel with crosstalk.
- PAM6 modulation for Tx in Channel Simulator.
- SmartEye Probe for transient PAM4 simulation.
- Channel Operating Margin (COM) versions 3.4 and 3.7 added.
- Measured Data based model for TX Waveform AMI component in Rx package and external channel models.
- PowerSum crosstalk and ICR/ICN analysis in S-parameter Toolkit.

EM-SIPro/PIpro

- RapidScan support for trapezoidal conductor cross sections and improved coplanar ground detection.
- Python 3.10 update for SIPro, PIPro, RFPro, and PEPro.
- PIPro radiated emission in conducted EMI analysis.
- CEMI analysis on remote, cloud, and cluster compute nodes.
- Power Integrity conducted EMI Picotest SIC531 PCB example.
- Parameter sweeps, e.g., ambient temperature in PIPro-DC simulation.

EM Simulations

RFPro Platform

- OpenAccess component instance swept parameters.
- Opening complex designs, switching component roles and net re-extraction now faster and using less memory.
- Simulations mesh and ports now visible.
- Multi-port excitation in field visualization can be exported/imported via .csv or .json file.
- 3D component editing, measuring, and placement enhancements.
- Improved resource management for remote, cloud, and cluster simulations.
- Improved schematic interoperability between ADS and Virtuoso.

Momentum

- TSMC advanced IC process node features supported including width, thickness, resistivity, bias, multi-patterning colors, and damage K.
- Compliance with foundry recommendations for layout scaling and bias sequence.
- Ground loss correction no longer applied when substrate stack contains a semiconductor layer for better accuracy.

FEM

- Improved meshing for multi-technology designs.
- Improved accuracy for passive device modeling on silicon substrates.
- Gen 2 solver correctly handles temperature dependency of conductor with surface resistance.

HFSS Option

- HFSS simulation support in PathWave Design Cloud simulation service.

For more highlights of the ADS 2023 release update, please visit the official website of Keysight.

1.2.2 Main Functions of ADS

ADS (advanced design system) integrates the benefits of multiple simulation software, offering a diverse range of simulation methods. It encompasses various domains such as time and frequency, digital and analog, linear and nonlinear, high and low frequency, and noise analysis. Its capabilities span from analyzing small components to designing and simulating complex system-level configurations. ADS excels in simulating RF (radio frequency), analog, and digital signal processing (DSP) circuits simultaneously. Moreover, it enables the co-simulations of mixed circuits consisting of both digital and analog components (Fig. 1.3).

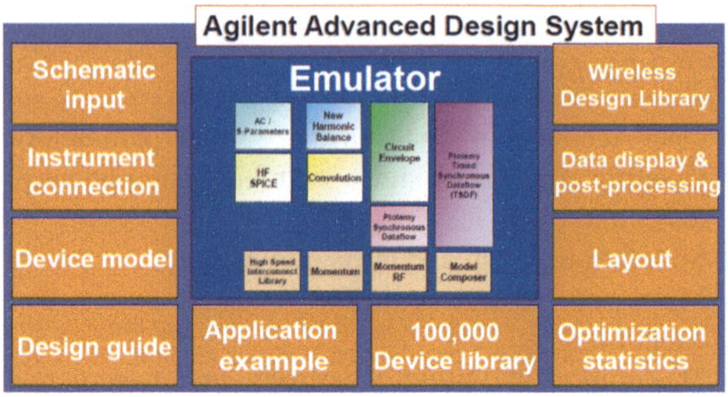

Fig. 1.3 Comprehensive overview of Agilent ADS: Core Components, Emulation Capabilities, and Design Resources

The primary functions of Advanced Design System (ADS) encompass circuit simulation, electromagnetic simulation, system simulation, and optimization design. The circuit simulation capability is fundamental and central to ADS, providing precise simulation and analysis for a wide range of microwave circuits. ADS supports the modeling and simulation of various components like transmission lines, filters, power dividers, mixers, amplifiers, catering to a wide array of circuit simulation needs. The main aspects of the circuit simulation functions in ADS are

- Circuit parameter extraction: ADS can autonomously extract various parameters, such as S-parameters, Y-parameters, and Z-parameters from the circuit's layout diagram and component model for further simulation and analysis.
- Circuit analysis: ADS performs different analyses like DC analysis, AC analysis, frequency-domain analysis, and time-domain analysis on circuits to evaluate their performance and stability.
- Circuit optimization: ADS optimizes various circuit parameters via automatic optimization algorithms, achieving optimal circuit design.
- Circuit layout design: ADS offers a variety of circuit layout design tools to rapidly construct and modify circuit layout diagrams according to varying design requirements.

Beyond circuit simulation, ADS also features robust electromagnetic simulation functions, enabling numerical simulation and analysis of complex electromagnetic field problems. These functions predominantly include

- Electromagnetic field analysis: ADS analyzes various electromagnetic field problems like propagation, radiation, and scattering.
- Electromagnetic field (EMF) optimization: ADS optimizes various EMF parameters through automatic optimization algorithms to achieve optimal EMF design.
- EMF radiation analysis: ADS performs EMF radiation analysis on structures like antennas, arrays to evaluate their radiation characteristics and performance.

- EMF scattering analysis: ADS conducts EMF scattering analysis on various scatterers to assess their scattering characteristics and performance.

Alongside circuit and electromagnetic simulation, ADS also boasts strong system simulation functions. This capability allows the integration of multiple microwave circuit and electromagnetic field simulation models into a system for system-level simulation and analysis. This, in turn, enables optimal design of circuits and systems to meet specific performance specifications. Here's a detailed account of ADS simulators and their functions frequently employed in RF and analog circuit design:

- DC Simulations: DC simulation is the foundation for all simulations, performing circuit topology checks as well as DC operating point scanning and analysis.
- AC Simulations: AC simulation derives small-signal transmission parameters like voltage gain, current gain, linear noise voltage, and current.
- S-parameter Simulations: S-parameter simulation treats the circuit as a four-port network, linearizes the circuit at the operating point, performs linear small-signal analysis, and analyzes various parameter values through its specific algorithm.
- Harmonic Balance Simulations: This simulation is ideal for dealing with the analysis of nonlinear circuits, focusing on the frequency-domain characteristics of the signal.
- Large-Signal S-parameter Simulations (Simulations-LSSP): This type of simulation performs large-signal S-parameter analysis and is instrumental when designing amplifiers.
- Gain Compression Simulations (XDB): Simulations-XDB is used to find user-defined gain compression points.
- Circuit Envelope Simulations: Circuit envelope emulator is a technology for communication systems that processes high-frequency modulated signals in the time and frequency domains.
- Transient Simulations: The Transient Simulator is a kind of traditional SPICE software, which is ideal for all analog and digital circuits but less efficient for high-frequency signals.

Each ADS emulator is designed to simulate a specific metric. Therefore, when simulating a complete circuit, it is crucial to employ multiple emulators to simulate all metrics. For microwave/RF circuit and system design, the most frequently used simulators are S-parameter Simulations, Circuit Envelope Simulations, and Harmonic Balance Simulations.

1.2.3 Characteristics of ADS

In the field of microwave circuit design, Advanced Design System (ADS) stands out as a kind of powerful simulation software. It offers a wide range of features and capabilities that enable engineers to effectively design, optimize, and analyze microwave circuits. This section provides a comprehensive overview of ADS, highlighting its key features and applications.

ADS boasts an extensive library of components and models specifically tailored for microwave circuit simulations. This comprehensive library includes various transmission lines, filters, power dividers, mixers, amplifiers, and more. With such diverse options, ADS can accommodate different simulation requirements. Additionally, ADS incorporates material libraries and device model libraries, such as Smith Chart, S-parameters, and Y-parameters. These libraries simplify the selection of appropriate components and models for specific simulations. Furthermore, ADS supports multiple scripting languages and custom APIs, empowering users to expand and customize the software's functionality. Scripting languages like Python, MATLAB, and Visual Basic enable programming and calculations during simulations, catering to diverse simulation needs. By allowing users to write custom functions and modules, ADS facilitates flexible and efficient simulation and analysis.

ADS offers a broad range of simulation functions, encompassing basic circuit simulation, electromagnetic simulation, system simulation, optimization design, and printed circuit board design, among others. The system simulation function enables the integration of multiple microwave circuit and electromagnetic field simulation models, facilitating system-level analysis. Moreover, ADS is scalable and compatible with other simulation and computer-aided-design (CAD) software, allowing co-simulations and data exchange. It seamlessly interfaces with CAD software such as Cadence Virtuoso, Mentor Graphics, and Altium Designer, enabling circuit and layout design and data transfer.

The graphical user interface (GUI) of ADS is user-friendly, featuring a simple and intuitive design that facilitates simulation and analysis tasks. The GUI adopts a modular approach, organizing different functional modules into categories for easy navigation. Users can swiftly locate and access the required modules and options. Furthermore, the GUI offers customizable toolbars and shortcut keys, empowering users to personalize their workflow according to their preferences.

Performance optimization is another notable aspect of ADS. It includes various techniques to enhance simulation efficiency and accuracy, based on specific requirements and computing resources. ADS supports multi-threaded computing and GPU acceleration, enabling faster and more accurate simulations. Performance monitoring and analysis tools allow users to assess and optimize the performance of simulation processes, ensuring optimal resource allocation and performance.

1.2.4 Application Scenarios of ADS

ADS is a kind of versatile software with a wide range of applications across various fields. It is particularly useful in RF and microwave communications, satellite communications, radar systems, satellite navigation, antenna design, and semiconductor device design and testing, among others. This comprehensive platform allows users to seamlessly integrate design, simulation, and testing processes within a single environment. It offers advanced features such as layout automation, optimization, high-frequency electromagnetic simulation, and system-level simulation. ADS provides diverse modeling and simulation methods, including linear and nonlinear circuit simulation, high-frequency electromagnetic simulation, system simulation, and digital signal processing simulation [44]. Users can create circuits using their own models or utilize the extensive built-in library of components and models. For instance, as the frequencies and speeds of printed circuit boards (PCBs) increase, ensuring signal and power integrity becomes increasingly critical. Transmission line losses can lead to device failures, necessitating accurate simulation of modeled alignments, vias, and interconnects. By employing custom IC designs and electromagnetic simulators, ADS helps improve high-speed link performance in PCB designs and facilitates power and signal integrity analysis. Moreover, ADS offers various automated tools such as PathWave EM Design (EMPro) and PathWave RFIC Design (GoldenGate). EMPro enables the analysis of 3D effects in components and accurately assesses electromagnetic phenomena. GoldenGate, on the other hand, is a kind of simulation software specifically designed for complex integrated circuits. During the early stages of RFIC design, it is crucial to monitor system IC specifications, such as error vector magnitude (EVM), through RF simulations. GoldenGate incorporates layout parasitic effects, complex modulated signals, and digital control circuits into its simulations. Additionally, PathWave RFIC Design allows simulations in both frequency and time domains and facilitates seamless design transfers between Cadence Virtuoso and Synopsys Custom Compiler. The design and optimization of microwave circuits within wireless communication systems directly impact system performance and reliability. ADS offers optimized design capabilities and system-level simulation functions to facilitate performance optimization of wireless communication systems. Radar systems are another major application domain for ADS. The design and optimization of microwave circuits significantly influence the performance and accuracy of radar systems. ADS supports various radar signal processing algorithms and circuit designs. In the field of space technology, microwave circuit design and simulation are crucial for spacecraft communication, navigation, and control. ADS encompasses a rich library of components and models, accommodating a broad range of space applications and standards such as GPS, satellite communications, and solar panels. Beyond these primary application areas, ADS finds utility in other domains as well. For example, the demand to enhance power device efficiency is driven by power supplies, solar inverters, and electric vehicles. ADS models modern materials and switch-mode power supplies to optimize power device designs for maximum efficiency. Moreover, ADS facilitates co-simulations with other CAD software and electromagnetic field (EMF) simulation software. This enables data exchange and collaboration between multiple software tools, enhancing simulation efficiency and accuracy.

1.3 Overview of High Frequency Structure Simulator (HFSS)

The HFSS is a kind of powerful software developed by ANSYS, Inc. specifically designed for high-frequency electromagnetic field simulations. Its primary purpose is to facilitate the design and analysis of various high-frequency electronic products such as antennas, RF and microwave components, high-speed interconnects, filters, connectors, IC components and packages, as well as printed circuit boards [45]. By utilizing the Finite Element Method (FEM), HFSS is capable of accurately solving complex electromagnetic field equations, enabling the simulation of the electromagnetic behavior of a wide range of high-frequency devices and circuits, including antennas, microstrip lines, waveguides, and filters, among others. With its exceptional accuracy, efficient solving speed, and user-friendly interface, HFSS has emerged as a vital tool [13, 45–47] in the field of high-frequency electromagnetic field simulation.

In the subsequent section, a comprehensive overview of HFSS will be presented, encompassing its historical development, key functionalities, and notable features.

1.3.1 Development History of HFSS

HFSS, which stands for High-Frequency Structure Simulator, was originally developed by Prof. Zoltan J. Cendes, at Carnegie Mellon University [48]. Professor Cendes focused on finite element-based electromagnetic calculations and founded Ansoft (short for Analysis Software) in 1984. Ansoft was later acquired by ANSYS in 2011, and HFSS became one of ANSYS' product offerings. Since then, with the support of ANSYS, HFSS has undergone continuous updates and enhancements, incorporating additional functionalities and features. These include advanced parametric modeling, automated design flow, and simulation acceleration, solidifying its position as a kind of leading software in high-frequency electromagnetic field simulation [46].

Initially, HFSS started as a two-dimensional electromagnetic field simulation tool primarily used for solving problems related to antennas and waveguides. As computer performance and software technology advanced, HFSS evolved into a three-dimensional EM field simulation software in the 1990s. Its usage expanded significantly within the microwave and millimeter wave fields. During its early development, HFSS prioritized enhancing simulation accuracy and solution speed to meet the growing demand for high-frequency EMF simulation. In the early 2000s, as the demand for microwave circuit design and optimization increased, HFSS began to incorporate circuit design and optimization capabilities. It introduced various optimization, such as automatic optimization, parameter scanning, and adaptive meshing, enabling the optimization of circuit parameters and achieving optimal design solutions. Additionally, HFSS introduced various component modeling, including microstrip lines, patch antennas, and waveguides, allowing for detailed component modeling and simulation [49].

In the mid-2000s, driven by the development of mobile communication and radar, HFSS expanded its applications to the fields of wireless communication, radar, and aerospace. Its exceptional simulation accuracy, speed, and support for diverse component modeling and optimization led to widespread adoption in these domains. Moreover, HFSS introduced several solvers, such as the Boundary Element Method (BEM), further enhancing simulation efficiency and accuracy [45, 50]. In recent years, with the emergence of new technologies like artificial intelligence and cloud computing, HFSS has ventured into these areas as well. Its integration with machine learning and deep learning technologies enables intelligent optimization design and automated simulation. Furthermore, HFSS can be seamlessly integrated with cloud computing platforms, providing enhanced computational power and improved simulation efficiency. Consequently, HFSS finds extensive applications in high-frequency and high-speed electronic devices for communication systems, advanced driver assistance systems, satellites, and Internet of Things (IoT) products.

1.3.2 Main Functions of HFSS

HFSS is an advanced electromagnetic tool used for research and virtual design prototyping. It offers a wide range of solvers that can effectively handle detailed EM problems, ranging from passive IC components to extensive EM analyses. Unlike other EM simulators, HFSS stands out with its automation capabilities and guaranteed accuracy. It eliminates the requirement for manual user control and multiple solution iterations to generate appropriate and precise meshes. This streamlined process significantly reduces design cycles while enhancing product reliability and performance [45, 51, 52].

EMI/EMC is a common problem in the field of electromagnetic engineering. Ansys Electronics Desktop offers a comprehensive solution by combining an advanced EM field solver with powerful circuit-level and system-level tools in Ansys RF Option. This integrated workflow enables engineers to efficiently diagnose, isolate, and resolve EMI and RF issues at an early stage of the design process. By leveraging Ansys Electronics Desktop, users can accurately predict the EMI/EMC performance of their electrical devices, eliminating the need for repetitive design iterations and costly periodic EMC certification tests [53]. The Ansys HFSS Simulations Suite further enhances the capabilities with its diverse range of EM solvers and circuit simulators. Engineers can evaluate the overall performance of their electrical devices and develop interference-free designs. This comprehensive suite addresses various EMI/EMC challenges such as radiation and conduction, sensitivity issues, crosstalk, RF sensitivity degradation, RF coexistence, co-location concerns, electrostatic discharge, electrical fast transients, burst effects, lightning impacts, high-intensity fields, radiation hazards, electromagnetic environmental effects, electromagnetic pulse, shielding effectiveness, and other EMC applications [54, 55].

Another prevalent electromagnetic issue is the occurrence of radio frequency interference (RFI) in complex environments [56]. By closely collaborating with Ansys HFSS, EMIT (Electromagnetic Interference Toolkit) combines RF system interference analysis with top-notch electromagnetic simulations, allowing for the modeling of inter-antenna coupling in installed configurations. The robust analysis engine of EMIT can compute significant RF interactions, encompassing the effects of nonlinear system components [57]. Through EMIT's dynamic link results view, the origin of any interference can be swiftly identified by visually illustrating the interfering signal's source and its path to the receiver, utilizing graphical signal tracing and diagnostic summary functions. Once the source of interference is determined, EMIT enables rapid evaluation of various RFI mitigation measures to identify optimal solutions. Moreover, the seamless integration between EMIT and Ansys HFSS facilitates an end-to-end workflow through the new HFSS/EMIT data link. This allows the creation of RFI analysis models directly from physical 3D models of HFSS installed antennas, addressing challenges such as large platform co-location interference and meeting RF environment requirements, including receiver sensitivity degradation of electronic devices.

In Ansys HFSS, engineers have the capability to simulate both infinite and finite phased array antennas, taking into account all electromagnetic effects through advanced cell simulation [58]. The versatile array design feature enables the examination of the input impedance of individual elements under various beam scan conditions. Phased array antennas can be optimized for performance at the element, subarray, or complete array level by analyzing the behavior of element-matched (passive or driven) far-field and near-field patterns under different target scan conditions. When it comes to finite array modeling, one or more antenna elements are placed within the cell, allowing for the calculation of element scan impedance and embedded element radiation direction maps, which consider the mutual coupling effects among elements. The finite array simulation employs domain decomposition of the cell to rapidly solve for large finite arrays. This provides a comprehensive array analysis that predicts mutual coupling, scanning impedance, cell orientation maps, array orientation maps, and array edge effects.

Ansys HFSS has the capability to integrate circuit and RF system simulation, enabling the creation of comprehensive, high-performance workflows. Additionally, it offers EMIT, a distinctive multi-fidelity approach that accurately predicts RF system performance in intricate RF environments characterized by multiple sources of interference. EMIT also encompasses diagnostic tools that expedite the identification of root causes for radio frequency interference issues, facilitating their early resolution during the design phase.

Finally, when utilized alongside HFSS, SI Circuits enable the analysis of signal integrity, power integrity, and EMI issues arising from timing and noise tolerance reductions in PCBs, electronic packages, connectors, and other intricate electronic interconnects. HFSS with SI Circuits effectively manages the intricacies of contemporary interconnect designs, encompassing ICs, packages, connectors, and PCBs in die-to-die configurations. By harnessing HFSS's advanced EMF simulation capabilities, seamlessly integrated with robust circuit and system simulation, engineers can gain comprehensive insights into the performance of high-speed electronics well in advance of hardware prototype development.

1.3.3 Characteristics of HFSS

- Basic Principle and Accuracy Guarantee
  High-Frequency Structure Simulations Software (HFSS) is a powerful tool used to solve complex three-dimensional electromagnetic field problems. The functionalities are illustrated in Fig. 1.4 as depicted. Based on the frequency and wavelength of the problem at hand, the physical prototype is divided into a series of initial tetrahedral meshes. These meshes are further discretized to generate a matrix, which is then solved to derive results such as the distribution of the electromagnetic field and port parameters. The solution process involves iterative calculations, wherein the problem is solved, and adaptive grid refinement is performed until the error between the current and previous iterations falls below a user-defined threshold. This ensures the attainment of a converged calculation result. To obtain a com-

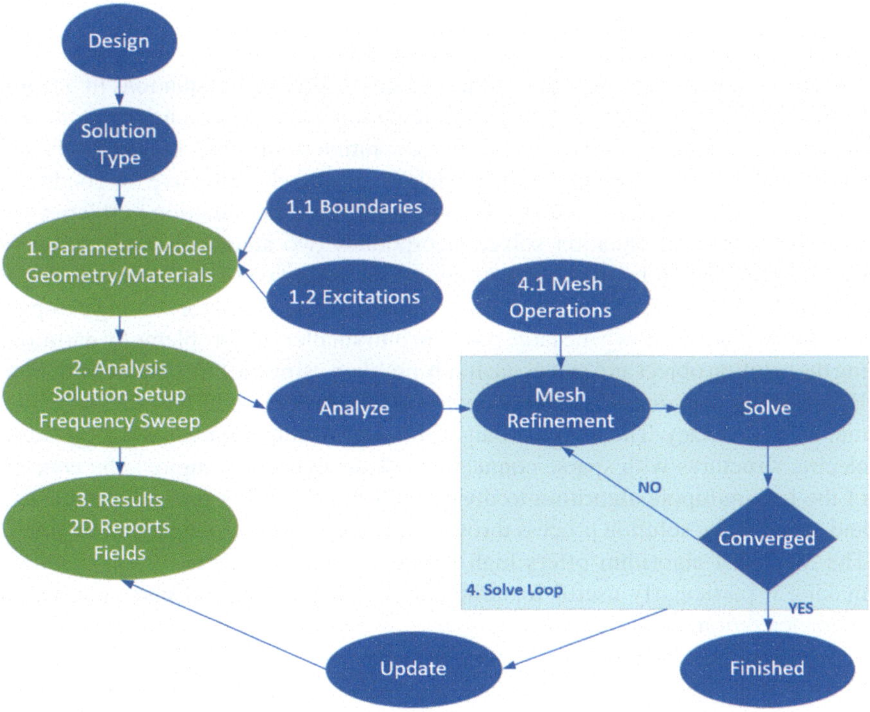

Fig. 1.4 The workflow diagram of HFSS: From design concept to converged solution

prehensive understanding of the electromagnetic behavior, a broadband sweep calculation is carried out across the entire frequency band, yielding a broadband frequency response. It is worth noting that the initial grid generation and grid refinement procedures are automated within the HFSS software. Consequently, no human intervention is required during the simulation process. This automation contributes to the accuracy, reliability, and repeatability of the simulation results.

- Finite Element Method (FEM)
  HFSS employs the hybrid order basis function, which enhances solution efficiency by reducing the number of meshes and unknowns. To handle intricate surface structures, HFSS incorporates curved cell technology for improved mesh segmentation. Furthermore, HFSS employs adaptive mesh technology, automating the entire mesh segmentation process. This includes physical and electrical mesh dissection, initial mesh generation, and iterative calculation for automatic mesh refinement. These advancements greatly ensure the accuracy of finite element solutions. In addition, HFSS introduces Broadband Mesh to address ultra-broadband problems. BAM leverages multiple frequency points for parallel computation, generating an adaptive mesh that covers the entire frequency band. This approach guarantees reliable and accurate mesh computation. A groundbreaking addition to HFSS is the Mesh Fusion technology, which is based on 3D component. Mesh Fusion effectively resolves mesh scaling issues in large scenes, further improving the software's capabilities.
- Integral Equation (IE) Method (ACA and MLFMM)
  The HFSS integral equation algorithm is based on Maxwell's equations in integral form, allowing it to automatically satisfy radiation boundary conditions. It utilizes the integral equation to solve the full-wave solution of the object under study and calculate the surface currents of the model. This method is effective for conductor and dielectric models, particularly for simple models and radiation problems.
  The HFSS integral equation solver incorporates two algorithms: the Adaptive Cross-Approximation (ACA) algorithm and the Multilayer Fast Multipole Algorithm (MLFMM). The ACA algorithm is specifically designed for solving complex structures. It breaks down the large problem into smaller sub-problems by partitioning the solution object and solves each sub-problem using the cross-approximation. By doing so, it reduces computational effort, enhances solution efficiency, and maintains accuracy. The MLFMM algorithm is well suited for solving large-scale electric structures with sparse connectivity. This algorithm employs the concept of the fast multipole algorithm to divide the solution region into multiple layers and expedite the solution process through interactions between near and far fields. The MLFMM algorithm offers high efficiency and low memory consumption, making it particularly useful for problems like antenna layout and radar target characterization on platforms.
- Shooting and Bouncing Rays (SBR+)
  The Enhanced Shooting and Bouncing Rays (SBR+) for HFSS incorporates a high-frequency ray algorithm that considers the impacts of multiple ray reflections, edge consistency, bypassing consistency, and surface creeping waves. This solver offers exceptional computational speed, accuracy, and effectiveness in high-frequency

applications and antenna layout design for extensive platforms. Moreover, the SBR+ solver includes ray-tracing simulations, enabling rapid exploration of ray trajectories and propagation patterns in complex environments.

- Two Complementary Time-Domain Method
  HFSS offers two time-domain solution algorithms: The Discontinuous Galerkin in Time-Domain algorithm and the Implicit Finite Element Time-Domain (FETD) algorithm. Both methods utilize a conformal non-uniform tetrahedral mesh and ensure model accuracy through adaptive mesh refinement.
  The DGTD algorithm, based on the interrupted Galerkin method, efficiently solves time-domain electromagnetic field problems by dividing the solution region into multiple cells. It is well suited for solving various problems such as ground-penetrating radar, ultra-wideband antennas, transient RCS, lightning strikes, electrostatic discharges, and time-domain reflected impedance. The algorithm incorporates adaptive grid refinement and time-step control.
  The FETD algorithm, based on the finite element method, accurately solves electromagnetic field problems and is particularly suitable for determining the self-resonant frequency and resonant Q of closed cavities. When combined with the Perfectly Matched Layer (PML), it can further obtain the on-load Q of the cavity. In electronic device design, the FETD solver, along with periodic boundary conditions, accurately simulates dispersion characteristics, phase velocity versus frequency, and electromagnetic field distribution of the device.
- Multipaction Function
  HFSS provides the Multipaction Solver for analyzing the secondary electron multiplication effect in RF microwave devices. The multipaction solver allows direct import of frequency-domain electromagnetic field results calculated by HFSS. It calculates the multipaction effect of charged particles and predicts the multipaction power threshold by defining the secondary electron emission yield (SEY) of metals and dielectric surfaces using the tetrahedral grid-based Particle in Cell (PIC) method.
- Eigenmode Solver Function
  The Eigenmode Solver (ESS) in HFSS, based on the finite element method, accurately determines the self-resonant frequency and resonant Q of closed cavities. By incorporating the Perfectly Matched Layer (PML), it can also obtain the on-load Q of the cavity. The eigenmode solver, combined with periodic boundary conditions, enables precise simulation of dispersion characteristics, phase velocity versus frequency, and electromagnetic field distribution of the device in electromagnetic device design.
- Characteristic Mode Solver Function
  The characteristic mode analysis in HFSS calculates a set of orthogonal current modes supported by the structure itself using the integral equation method. It provides eigenvalues, eigenangles, mode weighting coefficients for each mode, as well as mode current weighting coefficients corresponding to the excitation current. Structure-based characteristic mode analysis aids in deepening the understanding of antenna radiation mechanisms, optimizing antenna design and layout,

and improving isolation among multiple antennas in MIMO applications through inherent orthogonal modes of the structure.

- Hybrid Algorithm
  HFSS software offers mature finite element method FEM, integral equation method IE, optical algorithm SBR (including PO), and more. Hybrid algorithm leverages the strengths of each algorithm to select the most suitable approach for solving different parts of the same problem. The Region Decomposition Method (DDM) forms the basis of the hybrid algorithm, automatically partitioning the solution domain based on the grid size. Different algorithm types are assigned to different domains, which can be further divided into sub-domains as needed. DDM is crucial for implementing the hybrid algorithm and facilitates large array simulation and 3D component array simulation.
  The Finite Array + DDM (FA-DDM) function maximizes the periodic repetitive characteristics of array antennas, quickly generating a full mesh model of the entire array using a cell model mesh with periodic boundaries. This allows for parallel computation through DDM domain decomposition, rapidly obtaining simulation results for the entire array. The 3D Component Array Simulation (3D Component Array + DDM) is a method for efficiently modeling and simulating non-periodic arrays. Each unique 3D component is solved independently until convergence is achieved. Then, the DDM domain decomposition computes the final mesh for all cells, providing results for the entire array. The 3D component array simulation differs from finite large array simulation in terms of the number of cell models, mesh acquisition method, and the ability to handle non-periodic arrays. It also considers various cell coupling effects and edge effects in order to obtain efficient and accurate simulation results.
  Finite element-boundary integral (FE-BI) boundaries are employed to establish air boxes, serving as radiation boundaries, akin to conventional radiation boundaries and perfect matching layer (PML) boundaries. FE-BI boundaries, such as IE-Region/SBR-Region/PO-Region, represent typical applications of hybrid algorithms. These boundaries are utilized for setting up air boxes, whereas Region boundaries are utilized for establishing other independent metal bodies, material surfaces, or media bodies. The model may consist of multiple FE-BI solved regions and Region solved regions, all of which necessitate the employment of DDM domain decomposition methods for their resolution.
- High-Performance Computing
  High-Performance Computing (HPC) refers to the use of parallel accelerated computing in engineering simulation software. It encompasses various methods such as multi-threaded shared memory, distributed matrix solutions, DDM domain decomposition, GPU acceleration, spectral domain decomposition, and distributed parameter scanning. The Spectral Domain Decomposition Method (SDM) allows for parallel computation of multiple frequency points by distributing broadband frequency scan points across multiple CPUs. This method significantly reduces simulation time and provides precise broadband scan results. Distributed parameter solving involves multifaceted parallel computation in parameter scanning, automatically assigning predefined parameter combinations to different CPUs or

computers for parallel solving. This approach enhances computational efficiency in design exploration, considering model geometry, materials, boundary conditions, and excitation.

- Periodic Structure Simulation Function
  HFSS offers the Floquet port specifically designed for solving periodic structures such as planar phased arrays and frequency selective surface structures. The Floquet port simplifies the setup of incident wave sweeps and provides faster solving speed compared to traditional incident wave excitation methods, especially in broadband sweep cases. This improvement in simulation efficiency is more than tenfold.
- Dynamic Linking Capabilities
  HFSS incorporates a dynamic linking feature that enables dynamic field-to-field linking between different designs and projects. This capability allows complex systems to be decomposed into components for simulation. Examples of its applications include reflecting surface antennas, antenna and radome simulation, and electromagnetic interference and compatibility analysis of complex systems.
- Powerful Post-processing Capabilities
  HFSS provides powerful post-processing capabilities to efficiently and accurately obtain various results required by designers. These capabilities include matrix parameters, electromagnetic field distribution, far-field and near-field radiation characteristics, electromagnetic interference and compatibility analysis, and more. By changing the power and phase of the port input signal, internal field strength, radiation characteristic variations of multi-port antennas, and other results can be easily obtained without pre-setting or re-solving. The field calculator function in HFSS further allows for the calculation of various results such as local voltage, current, conductor loss, and dielectric loss. For port parameter post-processing, calculations can be performed for port modes, impedances in the 2D eigenmode solver, as well as S-, Y-, and Z-parameters (single-ended, differential, de-embedding, and renormalization). Field post-processing enables 2D and 3D mapping of far-field and near-field radiation characteristics, including gain and flap width. Three-dimensional static and animated mapping of electromagnetic field distribution, vector, amplitude, and streamline mapping on any surface or in space within the solution space is also possible. Additional capabilities include obtaining field distributions with loaded intrinsic resonant frequencies and resonant modes, performing specific absorption rate (SAR) calculations for cell phones or biomedical applications, and conducting post-processing analyses such as single- and dual-station RCS and radiation tests. The field calculator post-processing function offers enhanced functionality (ACT) for radar pre-processing and post-processing.
- Comprehensive In-depth Optimization
  HFSS offers a concomitant derivation for performing partial derivative analysis of changes in structure, material, and boundary conditions. This analysis involves calculating S-parameters and far-field data to reveal the impact of design sensitivity and processing errors on device performance. The concomitant derivation provides a real-time parameter tuning function, enabling quick optimization of S-parameters and directional maps. The automatic optimization technology is the

core of HFSS optimization function, incorporating various advanced optimization algorithms. Designers can easily perform parametric scanning, optimization, sensitivity analysis, and statistical analysis through a user-friendly interface. HFSS also integrates the Design of Experiments and Response Surface analysis from Ansys DesignXplorer software, known as DX-Optimization. In HFSS, designers can directly perform design space exploration and response surface optimization, or pass data to DesignXplorer for multidisciplinary collaborative design and optimization. Additionally, HFSS can be combined with optiSLang optimization technology for more efficient optimization and robust design. Ansys optiSLang is a simulation flow management and multidisciplinary optimization tool from Ansys that incorporates advanced optimal predictive metamodeling (MoP) technology and parameter sensitivity analysis.

- All-in-one Electro-thermal Coupling Design Environment
  Starting from version 19, Ansys eDesktop integrates the Icepak solver, creating an integrated electric-thermal coupled design environment called AEDT-Icepak. This environment focuses on electric-thermal coupling and is tailored to the operating habits of electrical engineers. It allows for considering thermal design aspects during the electric design phase, thereby shortening the overall R& D process.
- Interdisciplinary Multiphysics Field Coupling Simulations Capability
  HFSS software is integrated with Ansys' other physical field simulation software within a unified Workbench environment, providing a comprehensive multiphysics field coupling simulation capability. By simulating electromagnetic fields with HFSS and seamlessly linking them to Ansys Mechanical or Fluent/CFX/Icepak for thermal and stress simulation, it becomes possible to import thermal loads and external loads into the structural solver to calculate deformations. This bidirectional coupling enables HFSS simulations of electrical properties with temperature variation characteristics based on thermal simulation results. Furthermore, the deformation mesh results from structural analysis can be fed back to HFSS

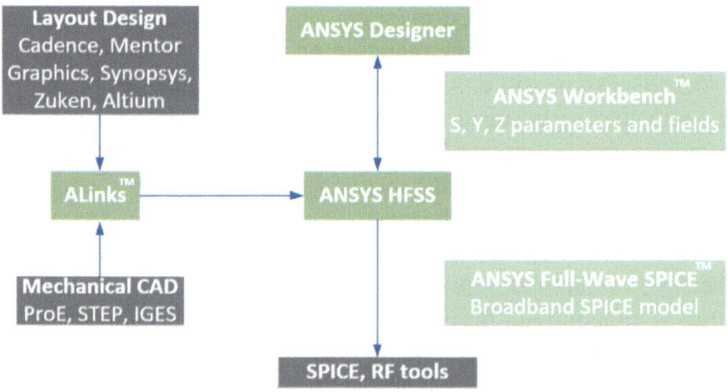

Fig. 1.5 The detailed connectivity framework of ANSYS HFSS with multidisciplinary design tools and Workbench for comprehensive high-frequency electromagnetic simulations

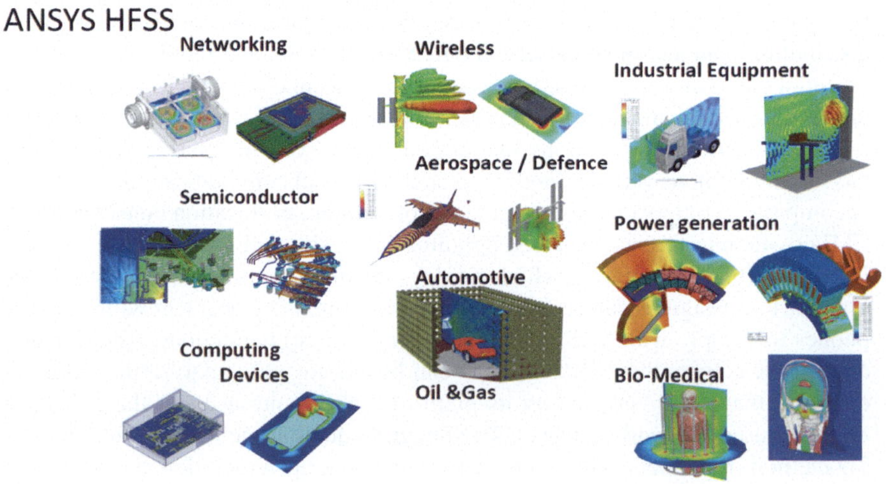

Fig. 1.6 Expansive application spectrum of ANSYS HFSS across industry verticals: Networking to Bio-Medical simulations

for further analysis, allowing for iterative simulation processes until steady-state characteristics are achieved. Additionally, tools like DesignXplorer and optiS-Lang facilitate interdisciplinary design space exploration, significantly enhancing the smoothness and efficiency of the simulation process while uncovering the potential of product design (Figs. 1.5 and 1.6).

The following are the key features and main contents of HFSS software, a comprehensive and powerful electromagnetic (EM) field simulation tool. With HFSS, engineers can accurately analyze and simulate a wide range of EM field problems. This enables them to optimize EM designs, enhance product performance, and provide reliable solutions for signal integrity simulation in various applications such as antennas, waveguides, PCB/packages, and more. Additionally, HFSS offers extensive optimization and post-processing capabilities that empower engineers to gain deeper insights into their designs and optimize them effectively. Whether it's academic research, product development, or engineering design, HFSS plays a crucial role by providing users with robust simulation and analysis tools.

1.3.4 Application Scenarios of HFSS

- Antenna Design and Layout
  HFSS (high-frequency structure simulator) is an advanced electromagnetic simulation tool that enables efficient and accurate calculations of various antenna properties. These properties include 2D and 3D far-field and near-field radiation direc-

tion maps, antenna directivity, gain, axis ratio, half-power flap width, electric field distribution, antenna impedance, VSWR, S-parameters, and more. Antennas are essential components in modern communication technology and find widespread use in commercial applications such as smartphones, RFID tags, wireless printers, as well as defense applications like aircraft radar systems, phased array antennas for autonomous vehicles, and integrated terrestrial communication systems. It encompasses component design, antenna placement, co-location considerations, EMI/co-site interference analysis, lightning strike simulations, and more. By utilizing the HFSS FEM solver, which employs the finite element method, and HFSS optimetrics, designers can accurately determine antenna placement within smartphones to comply with SAR (specific absorption rate) regulations. Additionally, asymptotic solvers like SBR+ and PO can be employed to analyze the radiation direction map of an antenna on its platform, taking into account the reflection effects caused by nearby objects. The integral equation method (IE) utilizes the 3D method of moments (MoM) to efficiently solve open radiation and scattering problems. Furthermore, the Finite Element Transient (FEM) transient solver combines HFSS adaptive meshing with the analysis of transient applications such as ground-penetrating radar (GPR), electrostatic discharge, time-domain reflectometry (TDR), and identification of radar cross-section (RCS) scattering centers.

In technology areas that demand increasingly advanced antenna capabilities, such as 5G and IoT, simulation has become an essential solution to achieve high-performance antenna and platform designs while minimizing wasted effort and costs. Antenna beamforming and beam control are crucial aspects of 5G technology for increasing capacity and data rates, and large-scale MIMO requires phased array antennas to optimize links and ensure targeted coverage. In the IoT domain, advanced 3D simulation technology enables customers to assemble different IoT components into a comprehensive virtual prototype of a software-controlled multi-domain system. HFSS empowers designers with powerful simulation tools for efficient and enhanced antenna design in various application scenarios. As communication technology continues to evolve, antenna design faces increasing challenges in meeting performance requirements, cost control, and product size limitations. By leveraging HFSS simulation technology, engineers can effectively evaluate various real-life "what-if" scenarios during the design phase, enabling them to optimize antenna performance and system integration.

- Advancements in Autonomous Sensor Development

  The development and application of high-performance sensors are crucial for the success of self-driving vehicles [15]. These sensors must consistently deliver reliable and efficient functionality in various challenging environments, such as rain, ice, and snow, to facilitate intelligent and safe decision-making. Simulation technology plays a pivotal role in this process, empowering engineers to enhance sensor performance, expedite sensor-vehicle integration, and conduct rapid performance analysis under diverse operating conditions. To fulfill the sensor requirements of self-driving vehicles across different scenarios, Ansys offers a comprehensive suite of sensor development tools. These tools encompass three key sensors: LiDAR, camera, and radar. For LiDAR, Ansys provides a multiphysics field solution that enables swift design and optimization of emitters and receivers. This solution incorporates optomechanical and thermal interactions, as well as optical material property libraries. On the camera front, Ansys leverages its multiphysics field simulation solution to facilitate visual performance analysis and high-fidelity lens transfer functions. This, in turn, allows for efficient optomechanical optimization, layout, and verification during camera design and layout studies. Furthermore, Ansys equips engineers with electromagnetic simulation capabilities, empowering them to analyze the performance of radar systems in complex dynamic scenarios. By doing so, engineers can improve the accuracy of radar system responses to potential hazards. Ansys also offers SCADE Vision for perception software testing. This tool automatically identifies potential vulnerabilities in self-driving vehicle perception systems and helps identify edge cases where these systems may exhibit unsafe behavior, enabling effective classification of such cases.

- Electromagnetic Interference and Compatibility (EMI/EMC)
  Electromagnetic Interference and Compatibility (EMI/EMC) poses significant challenges in the realm of modern electronic devices. With the widespread adoption of technology, ranging from 5G-connected devices to self-driving cars and the Internet of Things, electronic systems have become the backbone of the global technological revolution. In order to ensure the delivery of high-performance, compliant, and safe electronic systems, it is crucial to minimize EMI at every level, down to the microchip, by employing effective simulation. During the production of electronic products, various aspects such as PCB and component design, antenna design and layout, wire harness design, and system integration play a critical role in addressing EMI and compatibility issues. Advanced simulation tools like Ansys HFSS enable engineers to identify and tackle these issues at the design phase, thus reducing the need for extensive physical testing and enabling the delivery of high-performance, safe, and compliant designs. Ansys offers a range of capabilities, including near-/far-field EMI simulations, parasitic extraction, and circuit simulations, which effectively simulate EMC compliance during the PCB and component design phases. Additionally, in the wire harness design phase, Ansys provides a simplified cable simulation workflow that aids in designing layouts with appropriate shielding to minimize crosstalk, radiation, and susceptibility, even for complex structures and cable routing. In the domain of antenna design and layout, Ansys' full-wave electromagnetic field solver and electromagnetic interference toolkit prove invaluable in determining radiation patterns of installed antennas, enabling optimal placement to minimize co-location interference. For system integration, cable and circuit simulations, along with techniques like beam and reflected ray enhancements and full-wave electromagnetic solvers, facilitate the examination of electronic subsystem performance within electrically large environments. Simulations also play a pivotal role in diagnosing and mitigating electromagnetic interference sources in 5G devices, ensuring optimal wireless performance. As wireless systems continue to evolve, the number of interference and performance degradation issues increases. In this context, simulation empowers engineers to proactively identify potential interference problems early on and address issues related to interference and efficiency. Therefore, in the realm of HFSS applica-

tions, solutions for electromagnetic interference and compatibility (EMI/EMC) hold critical importance.

- PCB, IC, and IC Packaging
The field of software solutions for PCB design and simulation has witnessed rapid advancements in recent years [59]. Ansys, a kind of prominent software company, has emerged as a leader in offering a comprehensive and powerful simulation solution for printed circuit boards (PCBs), integrated circuits (ICs), and IC packages. This solution enables accurate evaluation of entire systems, catering to diverse industries such as automotive, aerospace, consumer electronics, healthcare, and energy. As electronic devices continue to decrease in size, engineers face the challenge of designing smaller boards while incorporating essential functionality. Consequently, precise modeling and simulation of these components become crucial in ensuring the reliability of the final product. Ansys solutions encompass signal integrity (SI), power integrity (PI), and electromagnetic interference (EMI) analyses, ensuring proper operation of electronic devices across various application scenarios. Additionally, features like electrical and thermal analysis, shock and vibration analysis, and reliability prediction contribute to evaluating system performance under diverse environmental conditions. Serving as a comprehensive solution for PCBs, ICs, and IC packaging, Ansys also provides a foundation for semiconductor power integrity analysis and simulation, voltage (IR) and dynamic voltage drop (DVD) analysis, as well as system-on-chip (SoC) designs. Given the increasing power dissipation and shrinking board sizes, thermal and mechanical loads exerted on PCBs can lead to system failures. Therefore, preventing PCB failures caused by thermal and mechanical stress has become a critical task. Furthermore, it is essential to pay attention to PCB design rules for electromagnetic compatibility, especially regarding signal references. Ansys SIwave facilitates automatic and customizable signal reference rule checking for PCBs through its built-in EMI scanner, effectively identifying potential problematic areas. Reliability assurance for electronic products involves a multifaceted effort encompassing selection, optimization, and adherence to design rules. At its core, every organization needs to conduct reliability predictions to gain insights into product life cycles and potential failure modes. Ansys simulation solutions empower engineers to identify and address potential reliability issues, thereby enhancing product performance and quality. To fully leverage the capabilities of Ansys' simulation solutions, engineers must possess a comprehensive understanding of the PCB, IC, and IC packaging design and manufacturing processes. This includes familiarity with the properties of various materials such as substrates, conductors, and insulation, as well as their behavior under diverse environmental conditions. By gaining insights into these parameters, engineers can optimize their designs more effectively, leading to improved system performance and reliability.
- Comprehensive and Advanced Optimization
ADS and HFSS are widely used software tools for microwave circuit simulation and electromagnetic field analysis, each offering distinct advantages and applicability in co-simulations [60]. Although they differ in principles, calculation methods, and solutions, they also share similarities and connections. Both software

packages are specialized in electromagnetic simulations, providing highly accurate results. They offer extensive libraries of simulation models for microwave circuits and electromagnetic fields, and both support co-simulations with other softwares, enabling the optimization and design of circuits and electromagnetic fields. While HFSS excels in complex electromagnetic field simulations, such as microwave antennas and devices, in certain scenarios, the two types of software packages can be used in combination for co-simulations and optimization. For instance, when designing intricate wireless communication devices, a co-simulation of the circuit and electromagnetic field is necessary to evaluate performance and optimize the entire system. In such cases, ADS can be used to simulate and optimize the circuit, with the circuit model then imported into HFSS for electromagnetic field simulation. Finally, the simulation results can be fed back into ADS for comprehensive analysis and optimal design. Proficiency in ADS and HFSS simulation software is highly beneficial for microwave circuit and antenna design engineers, as it enhances work efficiency and simulation accuracy. These tools provide robust support for achieving high-quality and high-performance microwave circuit and antenna designs. Looking ahead, as microwave circuit and antenna technology continues to advance, it is expected that both ADS and HFSS simulation softwares will undergo constant updates and improvements, offering new opportunities for innovation and development in microwave circuit and antenna design.

In addition to the co-simulation capabilities of ADS and HFSS, there are numerous third-party software options available for achieving even more comprehensive and accurate simulation analysis. EMF simulation software:

- CST Studio Suite.
- ANSYS Maxwell.
- Feldberechnung für Körper mit beliebiger Oberfläche (FEKO).
- EMCoS Studio.
- EMPIRE XPU.
- Sonnet.
- SEMCAD X.
- Remcom XFDTD.
- QuickWave.

Microwave circuit simulation software:

- AWR Microwave Office.
- National Instruments (NI) AWR Design Environment.
- CST Microwave Studio.
- Keysight Momentum.
- Agilent Genesys.
- ANSYS Circuit. (ANSYS Designer)

Antenna design software:

- Antenna Magus.
- ANSYS HFSS Antenna Design Kit.
- HFWorks.

- Antenna Magus.
- Delcross Electromagnetic Interference Toolkit (EMIT).

Electromagnetic compatibility (EMC) analysis software:
- EMIT.
- Ansys quasi-static 3D (Q3D) Extractor.
- PathWave EM Design (EMPro).
- Simbeor.
- ANSYS SIwave.

Thermal simulation software:
- ANSYS Fluent.
- COMSOL Multiphysics.
- SolidWorks Flow Simulations.
- Simcenter Flotherm.
- CST MPHYSICS STUDIO.

Structural simulation software:
- ANSYS Mechanical.
- Abaqus.
- MSC Nastran.
- Simulia.
- (LS-DYNA is an advanced general-purpose multiphysics simulation software package developed by the former Livermore Software Technology Corporation (LSTC), which was acquired by Ansys in 2019).
- Système pour l'Analyse des Milieux Continus par Eléments Finis (SAMCEF).

Other simulation software:
- MATLAB.
- Simulink.
- LabVIEW.
- Altair HyperWorks.
- DIgSILENT PowerFactory.
- PSCAD (Power Systems Computer Aided Design).
- Dassault Systemes Computer-aided three-dimensional interactive application (CATIA).
- Autodesk Inventor.
- SolidWorks.

It is important to acknowledge that the list provided is not exhaustive, as there are numerous other software options available in various fields that offer specialized solutions for different simulation scenarios. By employing co-simulations with these software tools, it becomes possible to achieve a more comprehensive and accurate analysis of circuit and electromagnetic field simulations, thereby enhancing design efficiency, optimizing performance, and expediting product development and promotion. The incorporation of co-simulation capabilities within these software packages reflects the growing trend of interconnection and interoperability among different software platforms. In the future, we can expect more software to adopt co-simulation functionality to better cater to diverse application scenarios. To support co-simulations effectively, many kinds of software vendors also provide a range of interfaces and standardized formats that facilitate seamless data transfer and sharing between different software applications. For instance, Ansys offers the Ansys Electronics Desktop (AEDT) platform, which enables data sharing and interaction between multiple simulation software, including HFSS, Maxwell, and Q3D Extractor. In addition to AEDT, other software vendors have developed their own interfaces and standardized formats to facilitate data exchange. For example, Simulink, developed by MathWorks, supports co-simulations with other simulation software and provides interfaces like Simulink Design Verifier, Simulink Verification and Validation, and Simulink Code Inspector. OpenModelica supports standardized formats such as Modelica and FMI, while LabVIEW offers interfaces like NI TestStand, NI VeriStand, and NI DIAdem. COMSOL Multiphysics supports a variety of interfaces, including LiveLink for MATLAB, LiveLink for Excel, and LiveLink for SolidWorks. These interfaces and standardized formats simplify data transfer and sharing between different simulation software platforms, enabling co-simulations to be more easily realized. Furthermore, several standardization organizations have introduced standardized formats that facilitate data exchange and sharing between different simulation softwares. Examples of these standardized formats include VHDL-AMS, which is a hardware description and simulation language for analog circuits and systems, and SPICE, an analog circuit simulation language. Modelica, on the other hand, is a language for multi-domain system modeling and simulation that can describe electrical, mechanical, thermodynamic, and other systems. FMI (functional mock-up interface) is a standardized format for model exchange and co-simulations, designed to facilitate data transfer and sharing between different simulation softwares. These standardized formats not only simplify data transfer and sharing between different simulation softwares, but also enhance the portability and reusability of simulation models, thereby reducing design costs and improving overall efficiency. Additionally, there are other standardized formats available, such as HDF5, NetCDF, and MATLAB. HDF5 is a versatile data storage format capable of storing various types of data, including numerical data, images, and sound. NetCDF, on the other hand, is a data storage format specifically designed for multi-dimensional numerical data, such as weather and ocean data. MATLAB, a type of powerful mathematical computing software, provides a variety of toolboxes and functions that facilitate data transfer and sharing between different simulation softwares. The emergence of these standardized formats and interfaces has made data transfer and sharing between different simulation softwares more convenient and efficient, thereby enabling co-simulations to become a reality.

1.4 Advantages of Co-simulations

Co-simulation plays a crucial role as an essential tool for combining microwave circuit simulations and high-frequency electromagnetic field analyses. By integrating these two domains, co-simulation enables a more comprehensive and precise representation of their interactions, resulting in improved design efficiency and quality. This approach has a significant impact on circuit performance and the distribution of electromagnetic fields. In comparison to standalone microwave circuit simulation or high-frequency EMF simulation, co-simulation offers more accurate results by considering the effects of circuit component location and shape on EMF distribution. By simultaneously conducting microwave circuit and high-frequency EMF simulations, co-simulation reduces simulation time and computational costs while providing more intuitive and visual outcomes. For instance, it can generate images depicting the EMF distribution of circuit components, facilitating a better understanding of circuit performance and characteristics. Co-simulation is particularly beneficial in microwave filter design, where it enables the simultaneous consideration of transmission characteristics, attenuation characteristics, and electromagnetic compatibility. This comprehensive approach leads to more effective design solutions. Additionally, co-simulation is instrumental in antenna design to meet diverse requirements such as frequency, orientation, and gain. It empowers rapid design iteration and optimization and finds applications in RF integrated circuits, antennas, radars, wireless communications, and other areas supporting design and research. Co-simulation offers several advantages in microwave circuit and high-frequency electromagnetic field design and research. It delivers more accurate, comprehensive, and efficient simulation results, benefiting the design of microwave amplifiers, filters, antennas, and other related aspects.

1.5 Outlook

Co-simulation is an indispensable tool with diverse applications in microwave circuit design and high-frequency electromagnetic field research. As technology advances, co-simulation faces both challenges and opportunities for future applications. While some progress has been made in current co-simulations, there are still certain shortcomings that need to be addressed, including

- Lack of standardized interfaces
  The absence of unified standardized interfaces and data formats between different softwares makes seamless integration and data exchange difficult.
- Improvement in simulation accuracy and efficiency
  Different simulation softwares employ distinct algorithms and calculation methods, resulting in accuracy errors and low simulation efficiency.
- Insufficient support for complex scenarios

Current co-simulation primarily focuses on simple scene simulations and lacks adequate support for complex scenarios. To overcome these challenges, future co-simulations should prioritize the standardization of interfaces and data formats across various softwares. It should also develop more accurate and efficient algorithms and calculation methods while expanding support for complex scenarios such as nonlinear and multiphysics fields. These advancements will cater to practical engineering needs. Co-simulations will continue to play a vital role in various domains. For instance, in communication, it can aid in designing and optimizing antenna arrays and RF integrated circuits. In radar applications, co-simulations can assist in the optimization of antennas, antenna arrays, and front-end circuits. In the future, co-simulations will face new challenges and opportunities, particularly with the integration of technologies like artificial intelligence and machine learning. These advancements will expand the prospects of co-simulations. For example, machine learning can be leveraged to analyze and process simulation data, improving accuracy, efficiency, and accelerating the design and optimization process. Additionally, virtual reality technology can present co-simulation results in a three-dimensional form, enhancing engineers' understanding and analysis of the simulation outcomes.

To achieve these goals, future co-simulation requires continuous technological innovation and close integration with practical engineering applications. In antenna design, co-simulations can be combined with actual measurement results for verification and correction to ensure accuracy and reliability. In RF integrated circuit design, co-simulation enables more precise analysis of electromagnetic field distribution and signal transmission characteristics, facilitating optimized circuit design and layout. Likewise, co-simulation finds application in wireless communication system-level design and optimization, aerospace electromagnetic compatibility analysis, and design optimization for equipment such as aircraft, missiles, and satellites. Furthermore, co-simulations should prioritize integration and collaboration across multiple disciplines, considering the interdisciplinary nature of microwave circuit and high-frequency electromagnetic field design involving electronics, electromagnetism, and computer science.

Additionally, future co-simulations should focus on the following areas of development:

- Co-simulations for large-scale complex systems

  As wireless and satellite communications advance, associated systems grow in size and complexity. Thus, future co-simulations should support the simulation and optimization of large-scale, complex systems. Methods such as parallel computing and distributed simulation can be employed to accelerate simulation and improve efficiency by distributing computation and simulation across multiple modules and subsystems.
- Support for multiphysics field coupling simulation

  To realistically simulate operating conditions, future co-simulations need to support the coupling of multiple physical fields involved in microwave circuit and high-frequency electromagnetic field design, such as electromagnetic, thermal, and mechanical fields. Multiphysics field simulation software can model and simulate the coupling relationship between different physical fields.

- Focus on system-level design and optimization
  Future co-simulations should prioritize system-level design and optimization rather than limiting itself to circuit-level design and simulation. For example, in wireless communication system design, co-simulations can aid in system-level optimization, including aspects such as antennas, RF front ends, and signal processing. This comprehensive analysis of system performance enables system-level optimization and improvement.
- Support for artificial intelligence and machine learning technology
  To leverage advancements in artificial intelligence and machine learning, future co-simulations should fully utilize these technologies for more intelligent simulation and optimization. Machine learning algorithms can be employed to process and analyze simulation data, leading to more accurate simulation results and optimization solutions.
- Support for rapid prototyping and testing
  As product development cycles shorten, engineers require the ability to design and test products within shorter timeframes. Co-simulation allows for early-stage simulation and optimization, reducing product development time and costs. Additionally, co-simulations can be combined with rapid prototyping techniques to facilitate rapid prototyping and testing.
- Support for visual simulation and interactive design
  Engineers can benefit from virtual simulation environments that enable circuit and scenario simulation and testing. Interactive design capabilities empower engineers to design and optimize more efficiently, enhancing both accuracy and efficiency.
- Enhanced standardized interfaces and data formats
  Future co-simulations should prioritize the development and improvement of standardized interfaces and data formats. While some standardized interfaces and data formats, such as HDF5 and VTK, already exist, further refinement is necessary. These standardized formats facilitate seamless data transfer and sharing between different simulation softwares, thereby improving simulation efficiency and accuracy.

Future co-simulation demands continuous innovation and development, both technically and in terms of practical application. It should strive for more efficient, accurate, and reliable simulation and optimization. Co-simulation needs to align with engineering applications and practices to meet the diverse needs and challenges of various fields. With collaborative efforts, future co-simulations will provide comprehensive and efficient design and optimization solutions for engineers and researchers.

Problems

P 1.1 Low-Noise Amplifier Design Using ADS

Design a low-noise amplifier using the ADS with following specifications and components:

Consider designing a low-noise amplifier (LNA) using ADS for an application operating in the frequency range of 1.8–2.2 GHz. The design should meet the following specifications:

- Gain: Greater than or equal to 10 dB.
- Input return loss: Greater than or equal to 10 dB.
- Output return loss: Greater than or equal to 10 dB.
- Noise figure: Less than or equal to 2 dB.

For this design, you are required to use the HP-AT41511 transistor. The AT-41511 is a general-purpose NPN bipolar transistor manufactured by Hewlett-Packard. The transistor is characterized for use at 3, 5, and 8 V and has a performance of 1 dB noise figure and 15.5 dB gain at 900 MHz. The AT-41511 is available in SOT-23 and SOT-143 surface-mount plastic packages (Fig. 1.7).

P 1.2 Design and analyze an air-filled rectangular waveguide (WR-90) using HFSS

Objective:

1. Perform a frequency sweep from 4 to 20 GHz.
2. Analysis must include the first five modes (TE10, TE20, TE01, TE11, TM11).
3. Generate graphs for wave number (β), wavelength (γ), and intrinsic impedance (η) as functions of frequency for each mode using HFSS. Compare the results for wave number (β), wavelength (γ), and intrinsic impedance (η) with values obtained using corresponding theoretical expressions.

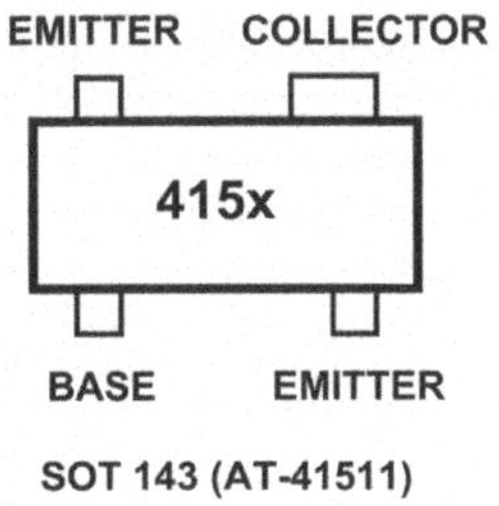

Fig. 1.7 Outline drawing of AT-41511

References

1. Sennur U et al (2015) Energy harvesting wireless communications: a review of recent advances. IEEE J Sel Areas Commun 33(3):360–381
2. Choongil Y et al (2023) Perspectives on 6G wireless communications. ICT Express 9(1):82–91
3. Muntadher A et al (2021) 6G wireless communications networks: a comprehensive survey. IEEE Access 9:148191–148243
4. Hamed SJ, Farina A (2020) Radar networks: a review of features and challenges. Inf Fusion 61:48–55
5. Susan CSD et al (2017) Radar remote sensing of agricultural canopies: a review. IEEE J Selected Topics Appl Earth Obs Remote Sens 10(5):2249–2273
6. Le Z et al (2019) Radar and communication coexistence: an overview: a review of recent methods. IEEE Signal Process Mag 36(5):85–99
7. Guoquan H (2019) Visual-inertial navigation: a concise review. In: 2019 international conference on robotics and automation (ICRA). IEEE
8. B. K P, et al (2019) A review: on path planning strategies for navigation of mobile robot. Defence Technol 15(4):582–606
9. Josué GG et al (2020) Autonomous underwater vehicles: localization, navigation, and communication for collaborative missions. Appl Sci 10(4):1256
10. Zhou SK et al (2021) A review of deep learning in medical imaging: imaging traits, technology trends, case studies with progress highlights, and future promises. Proc IEEE 109(5):820–838
11. Martin JW et al (2020) Preparing medical imaging data for machine learning. Radiology 295(1):4–15
12. Yabo F et al (2020) Deep learning in medical image registration: a review. Phys Med & Biol 65(20):20TR01
13. Ansys HFSS (2019) High frequency electromagnetic field simulation
14. Daniel GS, Hoefer WJR (2003) Microwave circuit modeling using electromagnetic field simulation. Artech House
15. Cláudio G et al (2018) Co-simulations: a survey. ACM Comput Surv (CSUR) 51(3):1–33
16. Cláudio G et al (2017) Co-simulations: state of the art. arXiv:1702.00686
17. Puppala R, Jeyaraj BE (2019) Application of DR and co-simulation approach for renewable integrated HEMS: a review. IET Gener Trans & Distrib 13(16):3501–3512
18. Tim G et al (2010) Modeling smart grid applications with Co-simulations. In: 2010 first IEEE international conference on smart grid communications. IEEE
19. Christian A (2013) A software framework for implementation and evaluation of co-simulations algorithms. Centre for Mathematical Sciences, Numerical Analysis, Lund University
20. Robert S, Ratajczyk R, Stępien J (2022) High isolation X-band and S-band diplexers in economical technologies for space applications. In; 2022 24th international microwave and radar conference (MIKON). IEEE
21. Agilent EEsof EDA (2009) Advanced design system. The industry's leading RF, microwave and high-speed design platform
22. Agilent EEsof EDA (2000) Advanced design system overview
23. Otman EM (2006) High frequency structure simulator (HFSS) tutorial. IETR, UMR CNRS 6164(2006): 2005–2006
24. Arpitha B, Jabeen F (2015) Design of microwave modules for microstrip antenna using Advanced Design System (ADS). In: 2015 international conference on trends in automation, communications and computing technology (I-TACT-15). IEEE
25. Zoltan C (2016) The development of HFSS. In: 2016 USNC-URSI radio science meeting. IEEE
26. Mikhail K, Turner R (2010) A comparison of Ansoft HFSS and CST microwave studio simulation software for multi-channel coil design and SAR estimation at 7T MRI. Piers online 6(4):395–399

27. Yangping Z et al (2011) 8th-order cavity filter design with Co-simulations by HFSS and designer. In: 2011 international conference on electronics, communications and control (ICECC). IEEE
28. Kyung-Whan Y (2015) Microwave circuit design: a practical approach using ADS. Prentice Hall Press
29. Otman EM (2006) High frequency structure simulator (HFSS) tutorial. IETR, UMR CNRS 6164 (2006): 2005–2006
30. Zoltan C (2016) The development of HFSS. In: 2016 USNC-URSI radio science meeting. IEEE
31. Neil T (2019) Keysight Unveils PathWave Design 2020:7–7
32. Electromagnetic Simulations Solvers, C. Compatibility. CST Studio Suite. https://www.3ds.com/ko/products-services/simulia/products/cst-studio-suite/solvers/. Accessed 23 Nov 2022. DOI 10 (2020): s19051212
33. Albert Kh S et al (2016) IR-UWB radio-over-fiber system components development. Optical Technologies in Telecommunications 2015, vol. 9807. SPIE
34. C. O. M. S. O. L. Multiphysics. Introduction to comsol multiphysics®. COMSOL Multiphysics, Burlington, MA. Accessed 9 Feb 9 (2018) (1998): 32
35. Ansoft Designer (2005) User manual. Ansoft Corporate
36. Aamir HH, Mian A (2019) Radio frequency response of flexible microstrip patch antennas under compressive and bending loads using multiphysics modeling approach. Int J RF Microwave Comput Aided Eng 29(3):e21649
37. Moradi M, Shirazi RS, Abdipour A (2016) Design and non-linear modeling of a wide tuning range four-plate MEMS varactor with high Q-factor for RF application. Appl Comput Electromagn Soc J (ACES)
38. Keysight (2023) https://www.keysight.com/us/en/lib/resources/software-releases/pathwave-advanced-design-system-ads-2023.html
39. Don M et al (2000) An industrial view of electronic design automation. IEEE Trans Comput Aided Des Integr Circuits Syst 19(12):1428–1448
40. Laung-Terng W, Yao-Wen C, Kwang-Ting Tim C (eds) (2009) Electronic design automation: synthesis, verification, and test. Morgan Kaufmann
41. Don M et al (2000) An industrial view of electronic design automation. IEEE Trans Comput Aided Des Integr Circuits Syst 19(12):1428–1448
42. Dirk J (ed) (2003) The electronic design automation handbook. Kluwer Academic Publishers, Norwell, MA
43. Keysight X, Anayzers SS (2014) Keysight technologies. Santa Rosa, CA, USA
44. George D V et al (2021) Microwave circuit design using linear and nonlinear techniques. Wiley
45. Zoltan C (2016) The development of HFSS. In: 2016 USNC-URSI radio science meeting. IEEE
46. Stefano S (2022) A brief history of finite element method and its applications to computational electromagnetics. Appl Comput Electromagn Soc J (ACES) 517–525
47. Antonino S (2000) Passive component modelling with HFSS. In: Advanced topics in microelectronics and system design, pp 139–154
48. Lee J-F, Sun D-K, Cendes ZJ (1991) Full-wave analysis of dielectric waveguides using tangential vector finite elements. IEEE Trans Microwave Theory Tech 39(8):1262–1271
49. Daniel GS, Hoefer WJR (2003) Microwave circuit modeling using electromagnetic field simulation. Artech House
50. Wenjian Y, Yan C, Wang Z (2008) Fast multi-frequency extraction of 3D impedance based on boundary element method. Microw Opt Technol Lett 50(8):2191–2197
51. Antonino S (2000) Passive component modelling with HFSS. In: Advanced topics in microelectronics and system design, pp 139–154
52. https://www.ansys.com/products/electronics/ansys-hfss
53. Marcelo PM et al (2016) A simulation technique for RF amplifier circuits using ANSYS electronics desktop. In: 2016 IEEE international symposium on antennas and propagation (APSURSI). IEEE

54. Bruce RA, Ramahi OM, Brench C (2012) EMI/EMC computational modeling handbook, vol. 630. Springer Science & Business Media
55. Bruce A, Brench C, Connor S (2010) Review of printed-circuit-board level EMI/EMC issues and tools. IEEE Trans Electromagn Compat 52(2):455–461
56. Mingliang T et al (2019) Mitigation of radio frequency interference in synthetic aperture radar data: current status and future trends. Remote Sens 11(20):2438
57. Manivannan S et al (2010) HFSS simulation, experimental investigation and optimisation of heat sink EMI. IET Power Electron 3(6):881–891
58. Thomas PK et al (2015) A 2.45 GHz phased array antenna unit cell fabricated using 3-D multi-layer direct digital manufacturing. IEEE Trans Microwave Theory Techn 63(12):4382–4394
59. Yogesh Dilip S et al (2013) Oscad: an open source EDA tool for circuit design, simulation, analysis and PCB design. In: 2013 IEEE 20th international conference on electronics, circuits, and systems (ICECS). IEEE
60. Mikhail K, Turner R (2009) Fast MRI coil analysis based on 3-D electromagnetic and RF circuit co-simulations. J Magn Reson 200(1):147–152

Chapter 2
Co-simulations of High-Frequency Electromagnetic Fields (HFEMFs)

2.1 Introduction

Scientific research aims to reveal the secrets of nature and pursue a better understanding on the laws of governing them [1]. Scientists or researchers will employ various methods to achieve this purpose. Basically, there are three primary methods, i.e., theoretical method, experimental method, and numerical method [2–4]. The theoretical method serves as the foundation of scientific research and it mainly uses various mathematical methods or tools to solve the physical equations of describing problems in an analytical approach with exact solutions, so that the physical mechanisms hidden in the problems can be accurately disclosed. The physical mechanisms are usually represented through various laws and principles, constituting a basis of theoretical research methods. The results from theoretical studies are instructive and can serve as a guideline for experimental research methods and numerical simulation methods. The experimental research methods are indispensable for scientific investigations in the early stage, especially in some areas like chemistry and biology [5, 6]. Through practical measurements and observations, researchers can not only verify the findings of theoretical research but also look into other new findings by the experimental methods themselves. The experimental results can enable researchers to comprehend and grasp the inherent laws of governing physical phenomena to a certain extent while also providing practical evidences for proving the effectiveness of scientific theories. However, the experimental research methods have been used less and less in recent years as the computer technologies have developed rapidly. The numerical simulation methods have been thought of as the third method of scientific research and have gradually replaced the expensive experimental methods which require special laboratories, instruments, and many professionals [7, 8]. The numerical simulation methods are very flexible since they do not have a limitation on time and space in use, and also very cheap because they do not require many resources except computers. Therefore, the methods have been very welcome by researchers since they appeared and have been widely used in various disciplinaries.

M. S. Tong and X. Y. Li, *Co-simulations of Microwave Circuits and High-Frequency Electromagnetic Fields*, https://doi.org/10.1007/978-981-99-8307-0_2

By simulating practical problems using the methods, researchers can obtain effective and accurate results much faster. Also, the methods are so powerful that almost all problems can be solved by using them. As the computer technologies continue to advance, it can be expected that the numerical simulation methods will become much more powerful and play an irreplaceable role in scientific research.

The numerical simulation methods can be programmed into computer codes and built into various kinds of software, called computer-aided design (CAD) software or computer-aided engineering (CAE) software [9]. Simulation technologies usually refer to those using various sorts of simulation software to conduct scientific research works or solve problems and have become the first choice in modern scientific research and product designs. By utilizing the simulation software, researchers can quickly analyze and make clear the involved mechanisms of physical phenomena or design optimal products. The simulation technologies represented by CAD software or CAE software offer numerous advantages, including the significant reduction of costs and economic losses associated with trial-error processes. Furthermore, the simulation technologies can mitigate potential risks, improve the reliability of products, and foster more innovative designs, providing researchers and engineers with additional inspirations to pursue the perfection of products and technologies. In various domains, such as automobile manufacture, aerospace electronics, and medical equipment, the simulation technologies have effectively helped enhance the performances and quality of products due to involved simulation optimizations.

Electromagnetic simulations aiming to solve electromagnetic problems by simulations constitute an important branch of simulation technologies and have been very essential in predicting and controlling the performances of electromagnetic products and systems in order to optimally design them [10, 11]. By accurately establishing simulation models and setting up related parameters based on the property of solved problems, electromagnetic simulations can rapidly give rise to simulation results which can be fully used to replace actual measurement results. Similar to other simulation technologies, the electromagnetic simulation technologies are cost-effective, flexible, accurate, and robust, and have been widely applied in the fields like wireless communications, electronic packaging, radar navigation, geophysical exploration, medical imaging, etc. For instance, in the realm of wireless communications, the electromagnetic simulations are indispensable for studying the characteristics of used antennas and optimizing their design [12]. In the medical field, the electromagnetic simulations can be employed to investigate the impact of electromagnetic radiation on human body, thereby ensuring the safe use of electronic devices with strong electromagnetic radiation [13]. As the science and technology and society progress, the electromagnetic simulation technologies will continue to develop and further expand their scope of applications. This chapter will focus on the introduction to the simulation technologies and their applications of HFEMFs.

2.1.1 Definition and Application of HFEMFs

The high-frequency electromagnetic simulation refers to the utilization of computer simulation methods to analyze and investigate the properties and behavior of high-frequency electromagnetic fields. This approach aims to optimize the design of various high-frequency electronic devices and systems [14]. In the realm of microwave circuit and antenna design, simulation of high-frequency electromagnetic fields finds extensive application in enhancing the performance of microwave devices and systems, including power amplifiers, filters, mixers, and antennas [15–19]. By simulating high-frequency electromagnetic fields, it becomes possible to predict parameters such as electromagnetic field distribution, power loss, and noise within the circuit. This information can then be utilized to optimize the design and achieve superior performance. Figure 2.1 demonstrates the capability of high-frequency electromagnetic simulation in obtaining the electromagnetic field distribution and radiation pattern of an antenna.

In communication and radar systems, high-frequency electromagnetic simulation plays a crucial role in predicting and analyzing various influential factors, including signal transmission, interference, and attenuation. By optimizing the system's design and performance, this simulation aids in achieving enhanced signal transmission, detection, and localization effects [20]. Moreover, high-frequency electromagnetic simulation finds extensive applications in medical imaging and bioelectromagnetics. It is widely employed to study and analyze the response and interaction

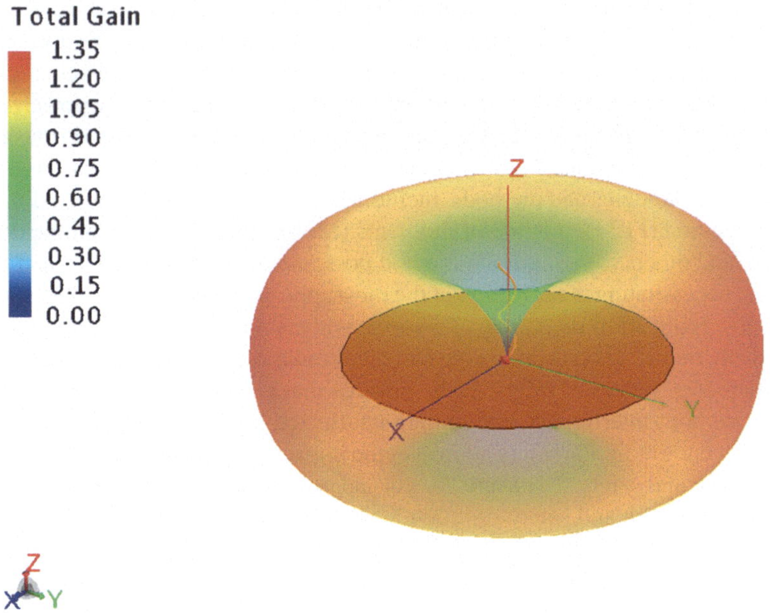

Fig. 2.1 Electromagnetic field distribution and radiation pattern of antenna

of human tissues with high-frequency electromagnetic fields, thereby optimizing medical imaging and bioelectromagnetics applications [21]. Through the simulation of high-frequency electromagnetic fields, we can accurately predict and simulate phenomena such as absorption, scattering, and reflection of these fields in human tissues. Additionally, we can assess the thermal and biological effects on human tissues, ultimately assisting in the design of safer and more effective medical equipment and treatment methods. For instance, in magnetic resonance imaging (MRI), high-frequency electromagnetic simulation is instrumental in optimizing the design of gradient coils to enhance imaging quality and spatial resolution [22]. Similarly, in ultrasound imaging, high-frequency electromagnetic simulation enables the simulation and prediction of ultrasonic wave propagation and reflection in different tissues, thus optimizing imaging and diagnostic outcomes. In the field of bioelectromagnetics, simulation of high-frequency electromagnetic fields aids in studying and analyzing the effects of these fields on physiological systems such as human nerves and the heart. This knowledge facilitates the optimization of treatment methods and medical device designs.

With the rapid development of simulation algorithms, the application of high-frequency electromagnetic simulation is becoming increasingly extensive and profound across various fields. High-frequency electromagnetic simulation offers numerous benefits, including the optimization and design of high-frequency electronic devices and systems, as well as the resolution of complex electromagnetic problems and engineering challenges such as electromagnetic compatibility (EMC), RF radiation safety, and electromagnetic stealth [23–25]. Furthermore, high-frequency electromagnetic simulation can be combined with other simulations like thermal, mechanical, and hydrodynamic simulations to model more intricate systems and phenomena [26–28]. The co-simulations of these multi-physics simulation enables more accurate predictions and analysis of electronic devices and systems' performance and behavior in real-world working environments. It provides a powerful tool for optimal design, performance analysis, and technical support across various fields.

In practical applications, high-frequency electromagnetic field simulations involve a range of technologies and methods, including computer-aided design (CAD), finite element method (FEM), method of moments (MoM), finite-difference time domain (FDTD), and others [29–32]. Each method possesses its own advantages and disadvantages, necessitating the selection of the most appropriate simulation method based on specific application scenarios. Moreover, when simulating high-frequency electromagnetic fields, it is crucial to consider electromagnetic compatibility (EMC) and radio frequency radiation safety. EMC ensures that electronic devices and systems function without interference and remain unaffected by other devices and systems. Radio frequency radiation safety addresses the impact of electromagnetic radiation on human health and safety. To ensure the normal operation of equipment and systems, as well as the health and safety of individuals, high-frequency electromagnetic field simulations need to account for these factors and conduct corresponding optimization and analysis. The simulation technology for high-frequency electromagnetic fields holds immense potential for applications in electronic engineering, medicine, bioelectromagnetics, and other domains. Through ongoing research and development, high-frequency electromagnetic field simulations will be able to more accurately and efficiently predict and analyze the performance and behavior of diverse electronic devices and systems, while providing stronger technical support and a theoretical foundation for research and application in other fields.

Looking towards the future, high-frequency electromagnetic simulation will encounter numerous challenges and opportunities [33–35]. Some potential trends include:

- Enhanced simulation accuracy and efficiency
  As electronic devices and systems grow increasingly complex, high-frequency electromagnetic simulation requires higher levels of accuracy and efficiency to meet practical demands. To achieve this, advanced algorithms and computational methods need to be developed, incorporating new hardware and software to improve simulation efficiency and accuracy.
- Development of multi-physics co-simulations
  Multi-physics co-simulations will become a crucial trend. For instance, in high-speed communication systems, electromagnetic simulations must consider the influence of signal transmission, thermal effects, mechanical stress, and other factors. Therefore, the development of more comprehensive multi-physics simulation is necessary to achieve holistic performance analysis and optimization.
- Cloud computing-based simulation platforms
  Simulations platforms based on cloud computing will emerge as a new trend. These platforms enable efficient sharing and allocation of computing resources, online storage and analysis of simulation results, and facilitate intelligent simulation analysis and optimization.
- Interdisciplinary collaboration
  High-frequency electromagnetic field simulations encompass knowledge from various disciplines, such as electronic engineering, electromagnetics, computer science, and physics. Therefore, interdisciplinary collaboration will become an important direction for development, fostering comprehensive and in-depth research and development through interdisciplinary exchanges and cooperation.
  High-frequency electromagnetic simulation will continue to play a crucial role in the future, providing robust support for the optimization, design, and performance analysis of electronic equipment and systems. Additionally, as science and technology progress, high-frequency electromagnetic simulation will continue to innovate and evolve, achieving higher levels of accuracy and efficiency. This will present more opportunities and challenges for scientific research and engineering applications. The development of high-frequency electromagnetic simulation has revolutionized various fields by enabling the optimization and design of electronic devices and systems, as well as addressing complex electromagnetic problems. By integrating with other simulations, high-frequency electromagnetic simulation can tackle more intricate systems and phenomena. The selection of appropriate simulation methods and considerations of electromagnetic compatibility and radiation safety are essential for accurate and effective simulations. Looking ahead, advancements in algorithms, computational methods, and hardware/software will drive higher accuracy and efficiency in simulations. The development of multi-physics co-simulations, cloud computing-based simulation platforms, and inter-

disciplinary collaborations will further enhance the capabilities of high-frequency electromagnetic simulation. Ultimately, this technology will continue to play a pivotal role, supporting the design, analysis, and optimization of electronic equipment and systems, and driving further advancements in research and engineering applications.

2.1.2 Development History of HFEMFs

Traditional numerical methods, namely the finite-difference method (FDM) and finite element time-domain method (FETD), were primarily employed for simulating high-frequency electromagnetic fields [36, 37]. The realm of high-frequency electromagnetic simulation encompassed various traditional methods, including the Finite-Difference Method (FDM), Finite Element Method (FEM), and Vector Potential Method (VPM) [38, 39]. These methods initially found application in antenna design and microwave circuit simulation, laying the groundwork for the advancement of high-frequency electromagnetic simulation. The finite-difference method, one of the earliest approaches for simulating high-frequency electromagnetic fields, discretizes a continuous domain into a finite number of elements. It then utilizes distinct equations within each element to solve for the components of the electric and magnetic fields. The finite-difference method offers fast computation speeds; however, it is subject to limitations arising from element size and numerical calculation accuracy. The time-domain finite element method is a high-frequency electromagnetic simulation method that evolved from the finite element method. It subdivides a continuous domain into a finite number of triangular or quadrilateral mesh units. By considering the temporal variation of the electromagnetic field, the time-domain finite element method solves the time-domain finite element equations to obtain the distribution of the electromagnetic field. Compared to the finite-difference method, this approach offers higher numerical accuracy and broader applicability. The vector potential method is yet another traditional high-frequency electromagnetic simulation. It is a static electromagnetic field solution method based on Maxwell's equations. By solving the vector potential function and the magnetic field divergence, the vector potential method determines the distribution of the electromagnetic field. Notably, the vector potential method exhibits advantages such as rapid computation and high numerical accuracy. However, when the electric field and magnetic field are coupled, the computational complexity and solution accuracy of this method can be affected. These methods find extensive employment in the domain of electromagnetic simulation, effectively addressing numerous electromagnetic field problems. Nonetheless, they possess certain limitations, such as constraints on accuracy and computational efficiency. Consequently, obtaining precise solutions for complex electromagnetic field problems remains challenging using these approaches [40]. The finite element method is a numerical approach that effectively solves electromagnetic field problems involving irregular and complex geometries. Compared to traditional methods like finite-difference and time-domain finite element methods, the finite element method offers superior computational accuracy and efficiency, better satisfying the demands of practical engineering. As computer hardware and software technology continually advance, the finite element method has progressively become one of the dominant methodologies in high-frequency electromagnetic simulation. The finite element method is a numerical solution method based on differential equations. It involves discretizing the simulation domain into small finite elements, establishing appropriate equations for each element, and solving these equations using numerical methods to obtain simulation results. The key advantage of the finite element method lies in its broad applicability, enabling the handling of complex geometric shapes and material properties, as well as effectively addressing nonlinear problems. In the realm of high-frequency electromagnetic simulation, the finite element method finds applications in electric fields, magnetic fields, and electromagnetic waves, such as antenna design, microwave devices, and electromagnetic shielding. Typically, the simulation of high-frequency electromagnetic fields using the finite element method employs either a time-domain or frequency-domain approach. The time-domain finite element method is suitable for transient problems like impulse excitation and transient response, while the frequency-domain finite element method is adept at handling periodic problems like antenna radiation and filtering. These methods exhibit high accuracy and efficiency in analyzing the distribution, field intensity, and propagation characteristics of electric fields, magnetic fields, and electromagnetic waves. Furthermore, the finite element method has given rise to several improved methods, such as the adaptive finite element method, mixed finite element method, and boundary element method, which further enhance simulation accuracy and efficiency. These methods have found extensive use and research in the domain of high-frequency electromagnetic simulation. Additionally, specialized simulation software based on the finite element method, such as Ansys HFSS and COMSOL Multiphysics, has emerged, thereby advancing the development of high-frequency electromagnetic simulation (Fig. 2.2).

In the mid-2000s, a revolutionary method to simulating high-frequency electromagnetic fields emerged, known as the hybrid simulation method. This method combines the traditional finite-difference method with the finite element method, harnessing the strengths of both to overcome their respective limitations. Consequently, the accuracy and efficiency of high-frequency electromagnetic field simulations have been significantly improved. Currently, the hybrid method, based on high-frequency electromagnetic field simulation, has gained widespread adoption and is considered a crucial area of research in this field.

1. Hybrid Method: Finite Element and Time-Domain Integration

The hybrid method integrates two powerful computational methods, namely the Finite Element Method (FEM) and the Time-Domain Integral Equation Method (TDIE), to solve electromagnetic field problems [41, 42]. This approach involves dividing the problem domain into two distinct regions. The first region employs a finite element mesh to handle geometric and material complexities. The second region, referred to as the time-domain integration region, focuses on boundary conditions and scattering phenomena. By establishing a suitable interface condition between these regions, the finite element region and the time-domain integration region are coupled to form a comprehensive solution method. The hybrid method

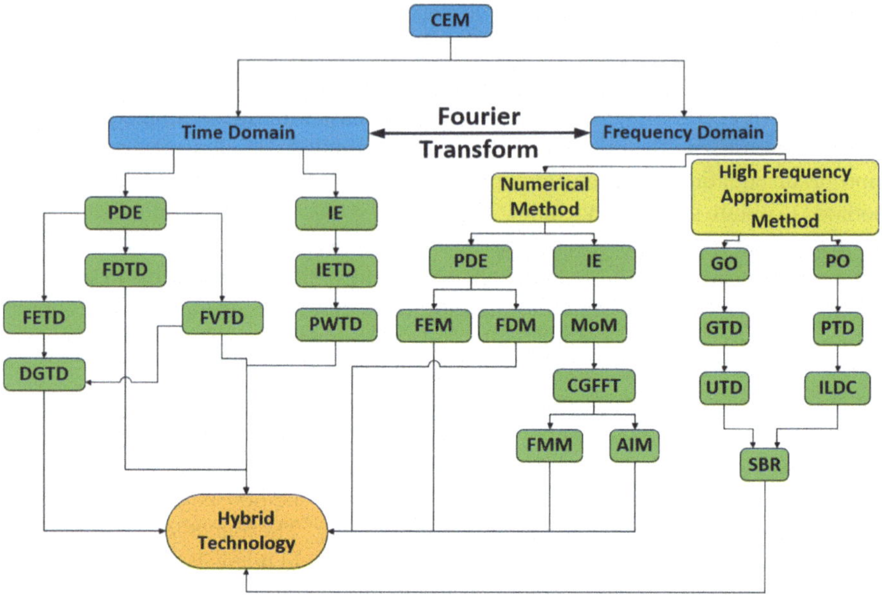

Fig. 2.2 Classification of electromagnetic computational methods

based on the finite element method and the time-domain integral method offers the flexibility of the finite element method and the efficiency of the time-domain integral method. It excels in addressing electromagnetic field problems involving intricate geometries, varying material properties, and accounting for boundary conditions and scattering phenomena. This versatile approach finds extensive applications in electromagnetic compatibility, antenna design, scattering analysis, and other related fields.

2. Hybrid Method: Finite Element and Methods of Moment

The Finite Element Method (FEM) and the Moment Method (MM) are two widely used numerical methods. In certain scenarios, these methods can be combined to create a hybrid approach that leverages the strengths of both FEM and MM [43, 44]. This hybrid method aims to enhance the accuracy and efficiency of calculations. The Finite Element Method is a versatile numerical approach commonly employed in engineering. It discretizes the solution space into finite elements and defines a local approximate function for each element. The global approximate solution is then obtained by minimizing the energy function or satisfying the variational principle. FEM is renowned for its adaptability, enabling it to handle complex geometries and boundary conditions. Its successful applications span various fields, including structural mechanics, fluid dynamics, heat conduction, and electromagnetic fields. The Method of Moments (MoM) is a numerical method based on integral equations, primarily utilized in electromagnetic and acoustic fields. By transforming the problem into an integral equation and discretizing it using basis functions and weight functions, MoM solves for the unknowns. This method exhibits superior convergence performance when dealing with open-domain and infinite-domain problems. By combining the finite element method and the moment method, the hybrid approach capitalizes on the advantages of both methods. FEM excels in handling complex geometric structures and boundary conditions, while MoM demonstrates excellent convergence performance for open-domain problems. In this hybrid method, the finite element method is generally employed to solve the interior domain of the problem, while the moment method is utilized for the boundary and open domain. This synergistic utilization allows us to harness the benefits of both methods, enhancing solution accuracy and reducing computational requirements. The hybrid method has proven successful in solving problems within electromagnetic fields, acoustic fields, structural mechanics, and other disciplines.

3. Hybrid Method: Finite Element and Harmonic Balance

The hybrid method that combines finite element and harmonic balance is a powerful numerical analysis approach widely employed in solving engineering problems, particularly in the areas of structural mechanics and vibration analysis. By integrating the strengths of both the finite element method (FEM) and the harmonic balance method (HB), this method significantly enhances the accuracy and efficiency of calculations. The harmonic balance method, a frequency-domain approach for addressing nonlinear vibration problems, assumes that the vibration response can be expressed as the sum of multiple harmonic components. By transforming the nonlinear equations into a linear system, the vibration response can be efficiently solved in the frequency domain. This method offers computational efficiency; however, its application may be limited when dealing with structures possessing complex geometries. In contrast, the hybrid method, which combines the finite element method with the harmonic balance method, excels in handling intricate geometries while efficiently solving nonlinear vibration problems in the frequency domain. Consequently, this hybrid approach has gained extensive usage in the analysis of nonlinear vibrations in fields such as structural engineering, automotive engineering, and aerospace engineering. The evolution of high-frequency electromagnetic simulation can be summarized as a progression from traditional methods like the finite-difference method and time-domain finite element method to the simulation method based on the finite element methods, and subsequently to the simulation method incorporating hybrid approaches. The emergence and advancement of these methods have not only propelled the development of electronic engineering and communications but also imposed higher demands on the application of electromagnetic simulation.

2.2 Principle of HFEMFs

High-frequency electromagnetic simulation employs numerical methods to model and forecast the behavior and characteristics of electromagnetic fields. It finds extensive application in wireless communications, radar systems, antenna design, electromagnetic compatibility, and various other fields. The fundamental principle underlying this method involves formulating Maxwell's equations and solving them using numerical solution methods to obtain the numerical representation of electromagnetic field distribution. This paper aims to provide an overview of Maxwell's equations and the numerical solution methods employed, including time-domain and frequency-domain analysis methods, boundary element method, finite element method, as well as error analysis in the context of high-frequency electromagnetic simulation.

2.2.1 Maxwell's Equations and Numerical Solution Method

Maxwell's equations are the basic set of equations that describe the behavior of electromagnetic fields and include four equations [45], respectively:

Gauss's law of electrostatic field:

$$\oint_S \boldsymbol{D} \cdot \mathrm{d}\boldsymbol{S} = \int_V \rho \mathrm{d}V \tag{2.1}$$

Loop theorem of electrostatic field:

$$\oint_C \mathbf{E} \cdot \mathrm{d}\mathbf{l} = -\int_S \frac{\partial \mathbf{B}}{\partial t} \cdot \mathrm{d}\mathbf{S} \tag{2.2}$$

Gauss theorem of magnetic field:

$$\oint_S \mathbf{B} \cdot \mathrm{d}\mathbf{S} = 0 \tag{2.3}$$

Ampere's theorem of magnetic field:

$$\oint_C \mathbf{H} \cdot \mathrm{d}\mathbf{l} = \int_S \mathbf{J} \cdot \mathrm{d}\mathbf{S} + \int_S \frac{\partial \mathbf{D}}{\partial t} \cdot \mathrm{d}\mathbf{S} \tag{2.4}$$

In this set of Maxwell's equations, ($\mathbf{E}$) represents the electric field intensity, characterizing the force of the electric field; ($\mathbf{B}$) is the magnetic flux density, representing the strength of the magnetic field; ($\mathbf{H}$) is the magnetic field strength, similar to the magnetic flux density but taking into account the material's magnetic permeability; ($\mathbf{D}$) represents the electric displacement vector, related to the electric field intensity but also including the effect of the dielectric. ($\mathbf{J}$) denotes the current density, which is the amount of current per unit area; (ρ) is the charge density, indicating the amount of charge per unit volume. ($d\mathbf{l}$), ($d\mathbf{S}$), and (dV) are differential elements of line, surface, and volume, respectively, used for integral operations.

Below, we will present an overview of the finite-difference method, finite element method, and spectral method, which are numerical methods employed to solve Maxwell's equations. The underlying principles, mathematical formulations, solution procedures, and derivation processes of these three methods will be elaborated upon.

- Finite-Difference Method
  The finite-difference method [46] is a discretization method that converts derivatives into difference quotients and discretizes continuous problems into a set of algebraic equations, allowing for numerical solutions. In the case of Maxwell's equations, it is possible to discretize them into a system of different equations, leading to numerical solutions [47]. Let's take Gauss's law of the electrostatic field as an example. The space is divided into N grid points, and the coordinates of the ith grid point are denoted as $\mathbf{r}_i = (x_i, y_i, z_i)$, with a grid spacing of Δx. At this grid point, the electric field is represented by $\mathbf{E}_i = (E_{xi}, E_{yi}, E_{zi})$, and the charge density is denoted as ρ_i. Then there are:

$$\nabla \cdot \mathbf{E} = \frac{\rho}{\varepsilon_0} \tag{2.5}$$

$$\frac{E_{x,i+1,j,k} - E_{x,i,j,k}}{\Delta x} + \frac{E_{y,i,j+1,k} - E_{y,i,j,k}}{\Delta y} + \frac{E_{z,i,j,k+1} - E_{z,i,j,k}}{\Delta z} = \frac{\rho_i}{\varepsilon_0} \tag{2.6}$$

  In a similar vein, the remaining Maxwell's equations can be discretized, yielding a set of discrete equations. Subsequently, various numerical methods, including iteration, direct solution, or relaxation methods, can be employed to solve these differential equations and acquire the numerical approximation of the electric field.
- Finite Element Method
  The finite element method [48, 49] is a discretization method based on the weak form. It transforms a continuous problem into a collection of discrete problems by approximating the solution using the solutions of these discrete problems [50]. In this method, the computational domain is initially divided into small regions called finite elements. Then, suitable functions are chosen within each element as trial functions, and the continuous problem is discretized using the weighted residual method [51]. Subsequently, numerical methods are employed to solve the discrete problem, and the resulting solution fragments are assembled to obtain the solution for the entire computational domain. To illustrate this, let's consider the Gauss law of the electrostatic field. The calculation region is divided into N triangular elements, and within each triangular element, a set of linear trial functions, such as linear triangular functions, are selected.

$$\varphi_i(\mathbf{r}) = a_i + b_i x + c_i y \tag{2.7}$$

  Next, we can derive the discretized system of equations:

$$\sum_{i=1}^{N}\sum_{j=1}^{N}\left(\int_{\Omega_i}(\nabla\varphi_i\cdot\nabla\varphi_j)\,\mathrm{d}\Omega\right)E_j=\int_{\Omega}\varphi_i\rho\mathrm{d}\Omega,\quad\forall i=1,2,\ldots,N \tag{2.8}$$

Let Ω_I represents the I-th triangular element, E_J denotes the electric field strength at the J-th node, and ρ symbolizes the charge density. To discretize the above equation, a numerical integration method is employed, resulting in a system of linear equations. This system is subsequently solved using either a direct or an iterative method, yielding a numerical solution for the electric field.

- Spectral Method

  The spectral method [52, 53] is a numerical method that utilizes the expansion of a specific function to provide an approximate solution to a differential equation. In the case of Maxwell's equations, the electric and magnetic field solutions can be estimated by employing expansions of particular basis functions [54]. One such approach is to expand the electric and magnetic fields using the Fourier series:

$$\mathbf{E}(\mathbf{r},t)=\sum_{n=-\infty}^{\infty}\sum_{m=-\infty}^{\infty}\sum_{p=-\infty}^{\infty}\mathbf{E}_{nmp}(t)e^{i\left(nk_xx+mk_yy+pk_zz\right)} \tag{2.9}$$

$$\mathbf{B}(\mathbf{r},t)=\sum_{n=-\infty}^{\infty}\sum_{m=-\infty}^{\infty}\sum_{p=-\infty}^{\infty}\mathbf{B}_{nmp}(t)e^{i\left(nk_xx+mk_yy+pk_zz\right)} \tag{2.10}$$

where k_x, k_y, andk_z represent the wave numbers in space, respectively, and $\mathbf{E}_{nmp}$ and $\mathbf{B}_{nmp}$ denote the expansion coefficients of the electric and magnetic fields, respectively. By substituting the above equation into Maxwell's equations, we can derive a set of coupled ordinary differential equations for $\mathbf{E}_{nmp}$ and $\mathbf{B}_{nmp}$. The system of equations is then solved using numerical methods, and the expansion coefficients are reinserted into the original equations to obtain the numerical solutions for the electric and magnetic fields.

2.2.2 *Time-Domain and Frequency-Domain Analysis Methods*

The time-domain analysis method involves numerically simulating the changes in an electromagnetic field over time. Commonly used methods for this purpose include the finite-difference method and the time-domain finite element method [51]. In this discussion, we will focus on the finite-difference time-domain (FDTD) method as an example. The fundamental concept behind the FDTD method is to discretize Maxwell's equations in the time domain, resulting in a set of difference equations. These equations can then be solved using numerical methods such as the iterative method to obtain a numerical solution for the electromagnetic field in the time domain. Let's consider the two-dimensional case, assuming the calculation area is a rectangular region with a boundary denoted by Γ, and the variable representing time is denoted as t. Therefore, the time-domain Maxwell equations governing the behavior of the electromagnetic field can be expressed as follows:

$$\nabla \times \mathbf{E} = -\frac{\partial \mathbf{B}}{\partial t} \tag{2.11}$$

$$\nabla \times \mathbf{H} = \frac{\partial \mathbf{D}}{\partial t} + \mathbf{J} \tag{2.12}$$

Let **E** and **H** represent the electric and magnetic fields, respectively. Similarly, **B** and **D** represent the magnetic induction and electric displacement vector, respectively, while **J** denotes the current density. In the numerical method, space is divided into a set of finite-sized grids, and time is discretized into a sequence of time steps. Consequently, the time-domain Maxwell equations are discretized to derive differential equations. For instance, discretizing the time-domain Maxwell equations for the electric field results in the following discrete equation:

$$\frac{E^{n+1}_{x,i,j} - E^{n}_{x,i,j}}{\Delta t} = -\frac{1}{\mu}\left(\frac{H^{n+1}_{z,i,j+1} - H^{n+1}_{z,i,j}}{\Delta y} - \frac{H^{n+1}_{z,i+1,j} - H^{n+1}_{z,i,j}}{\Delta x}\right) \tag{2.13}$$

Similarly, we can derive the difference equations for other electric and magnetic fields. Subsequently, the iterative method is employed to obtain the numerical solution of the electromagnetic field in the time domain. Frequency-domain analysis provides a means to simulate the variations of the electromagnetic field with respect to frequency. Commonly used methods in this domain include the spectral method and the finite element method [54]. For illustration purposes, we will focus on the spectral method. The fundamental concept behind the spectral method is to represent the solution of the electromagnetic field as an expansion involving specific basis functions. By leveraging the properties of these special functions, we can determine the solution of the electromagnetic field. Let's consider the two-dimensional scenario, where the calculation area is assumed to be rectangular with a boundary denoted as Γ and a frequency of Ω. In this case, we express the electromagnetic field as a spectral expansion in the following form:

$$\mathbf{E}(\mathbf{r}, t) = \sum_{m,n} \mathbf{E}_{mn}(t) e^{i(mk_x x + nk_y y)} \tag{2.14}$$

$$\mathbf{B}(\mathbf{r}, t) = \sum_{m,n} \mathbf{B}_{mn}(t) e^{i(mk_x x + nk_y y)} \tag{2.15}$$

Let k_x and k_y represent the wave numbers of the space, while $\mathbf{E}_{mn}$ and $\mathbf{B}mn$ denote the expansion coefficients for the electric and magnetic fields, respectively. By substituting the aforementioned equation into Maxwell's equations, we can derive a set of interconnected ordinary differential equations for $\mathbf{E}_{mn}$ and $\mathbf{B}_{mn}$. Specifically, when considering the expansion coefficients $\mathbf{E}_{mn}$ for the electric field, we obtain a differential equation in the following form:

$$\frac{d\mathbf{E}_{mn}}{dt} + i\omega\mu_0\mathbf{B}_{mn} = 0 \tag{2.16}$$

Similarly, we can establish a differential equation that describes the expansion coefficients of the magnetic field. Then, we utilize a numerical method to solve the system of equations and reintroduce the obtained expansion coefficients back into the original equations. This process yields the numerical solution of the electromagnetic field in the frequency domain. The basic principles of the Finite-Difference Time-Domain (FDTD) method are summarized as follows:

$$\frac{E^{n+1}_{x,i,j} - E^{n}_{x,i,j}}{\Delta t} = -\frac{1}{\epsilon}\left(\frac{H^{n}_{z,i,j+1} - H^{n}_{z,i,j}}{\Delta y} - \frac{H^{n}_{z,i+1,j} - H^{n}_{z,i,j}}{\Delta x}\right), \tag{2.17}$$

$$\frac{H^{n+1}_{z,i,j} - H^{n}_{z,i,j}}{\Delta t} = \frac{1}{\mu}\left(\frac{E^{n+1}_{x,i,j+1} - E^{n+1}_{x,i,j}}{\Delta y} - \frac{E^{n+1}_{y,i+1,j} - E^{n+1}_{y,i,j}}{\Delta x}\right). \tag{2.18}$$

The basic formulation of the spectral method is as follows:

$$\begin{aligned}
\mathbf{E}(\mathbf{r}, t) &= \sum_{m,n} \mathbf{E}_{mn}(t)e^{i(mk_x x + nk_y y)}, \\
\mathbf{B}(\mathbf{r}, t) &= \sum_{m,n} \mathbf{B}_{mn}(t)e^{i(mk_x x + nk_y y)}, \\
\frac{d\mathbf{E}_{mn}}{dt} &+ i\omega\mu_0\mathbf{B}_{mn} = 0, \\
\frac{d\mathbf{B}_{mn}}{dt} &- i\omega\epsilon_0\epsilon_r\mathbf{E}_{mn} = 0.
\end{aligned} \tag{2.19}$$

where μ_0 and ϵ_0 represent the permeability and permittivity in free space, respectively, while ϵ_R denotes the relative permittivity. The values of k_x and k_y can be determined using the expressions $k_x = \frac{2\pi m}{L_x}$ and $k_y = \frac{2\pi n}{L_y}$, where L_x and L_y refer to the length and width of the rectangular region, respectively. Both time-domain and frequency-domain analysis methods are widely used in electromagnetic simulation. The time-domain analysis method is suitable for analyzing the time-varying characteristics of electromagnetic fields, such as transient electromagnetic fields, pulsed electromagnetic fields, and so on. The frequency-domain analysis method is suitable for analyzing the characteristics of electromagnetic fields at different frequencies, such as impedance matching, resonance, and so on. In practical applications, we usually choose the appropriate method to simulate according to the needs and characteristics of the specific problem.

2.2.3 Boundary Element Method and Finite Element Method

- Boundary Element Method
  Boundary Element Method (BEM), also known as the boundary integral equation method, is a numerical method based on the integral equation on the boundary to solve physical problems [55]. It is similar to the finite element method, but only needs to be discretized on the boundary, so there is no need to discretize the whole computational domain. The boundary element method is mainly divided into the iterative method and the direct method. Iterative methods usually use iterative methods to solve matrix equations, such as the iterative method, GMRES algorithm, and so on. The direct method usually uses the method of matrix decomposition, such as LU decomposition, Cholesky decomposition, and so on. The basic idea of the boundary element method is to transform the integral equation on the boundary of the physical problem into a matrix form, and then solve the matrix equation.
- Finite Element Method
  The Finite Element Method (FEM) is a powerful method that discretizes a continuous domain of a physical problem into a finite number of simple shapes [50]. It involves dividing the entire computational domain of the physical problem into numerous small regions and selecting appropriate basis functions within each region. These basis functions are utilized to represent the physical quantity as a linear combination, thus enabling the derivation of a discretized numerical solution. The fundamental concept underlying the finite element method is to transform the partial differential equation of the physical problem into a weak form. Subsequently, the discretized equation is solved through the processes of discretization and approximation. For instance, in the context of electromagnetic field problems, Maxwell's equations can be represented as follows:

$$\nabla \times \nabla \times \mathbf{E} + k^2\mathbf{E} = -\nabla \times \mathbf{J} \tag{2.20}$$

where $\mathbf{E}$ is the electric field, $\mathbf{J}$ is the current density, and k is the wave number. Multiplying this equation by a test function $\mathbf{v}$ and integrating over the entire computational domain, one obtains the weak form:

$$\int_{\Omega} \left(\nabla \times \nabla \times \mathbf{E} + k^2\mathbf{E}\right) \cdot \mathbf{v} dV = -\int_{\Omega} (\nabla \times \mathbf{J}) \cdot \mathbf{v} dV \tag{2.21}$$

where Ω is the calculation area. The electric field and the test function are represented by selecting appropriate basis functions in the calculation region, and the electric field and the test function are expanded into finite element functions to obtain a discretized equation:

$$\mathbf{K} \cdot \mathbf{E} = \mathbf{F} \tag{2.22}$$

where **K** and **F** are the stiffness matrix and the load vector, respectively, obtained by numerically evaluating the integrals. The solution process of the finite element method typically involves the following steps, which can be further enhanced as follows [56]:

1. Discretization: The computational domain is divided into smaller regions, and a suitable basis function is selected for each region. The physical quantity is expressed as a linear combination of these basis functions, enabling the derivation of a discretized numerical solution.
2. Solve the stiffness matrix and load vector: The stiffness matrix and load vector are determined by numerically evaluating the discretized weak form. This involves calculating the contributions of each element in the discretized domain.
3. Solve a matrix equation: Utilizing the stiffness matrix and load vector, a matrix equation is formulated and solved to obtain the numerical solution. This step involves solving a system of equations representing the relationships between the unknowns.
4. Post-processing: After obtaining the numerical solution, further analysis is performed to derive additional information. This includes calculating field quantities, field strengths, or other relevant parameters to gain deeper insights into the problem.

Both the boundary element method and finite element method are extensively employed in electromagnetic simulations. The selection of the appropriate method depends on the specific problem's requirements and characteristics to achieve accurate and reliable simulation results.

2.3 Error Analysis in HFEMFs

Error analysis plays a crucial role in simulating high-frequency electromagnetic fields [57, 58]. This is due to the unique characteristics of such fields, namely their short wavelengths and high spatial resolution. Errors in these simulations can be greatly amplified, making error analysis an essential tool. By conducting error analysis, we gain insights into the accuracy and reliability of simulation results, and it offers guidance for optimizing calculation parameters. The most common sources of errors include numerical errors, boundary condition errors, and model errors.

Numerical errors arise from the inherent limitations of numerical methods, such as discretization errors and rounding errors. Generally, numerical errors decrease as calculation parameters increase. However, excessively high calculation parameters result in a significant computational burden. Hence, striking a balance between computational accuracy and efficiency becomes imperative. Boundary condition errors occur when inaccuracies or incompleteness exist in the specified boundary conditions. For instance, in the finite element method, boundary conditions are typically calculated using the boundary integral equation, which inherently carries its own errors. The accuracy of the boundary condition is influenced by the boundary shape and the precision of its formulation. On the other hand, model errors emerge from approximations and simplifications made in the simulation model. For instance, when simulating antenna radiation, the shape and material parameters of the antenna are often approximated, leading to model errors. Although model errors are challenging to completely eliminate, they can be mitigated by optimizing the simulation model and employing more accurate material parameters. Below are some common error estimation methods:

1. Manual adjustment method: Minimize the error between the numerical solution and the exact solution or experimental result by manually adjusting calculation parameters (e.g., grid density, time step, etc.).
2. Richardson extrapolation method: Estimate the error of the numerical solution by analyzing the relationship between numerical solutions with different mesh sizes.
3. Repeated calculation method: Estimate the error of the numerical solution by performing repeated calculations and averaging the results.
4. Error propagation method: Estimate the size of the numerical solution error by calculating and superimposing the contributions of various error sources.
5. Corrected error estimation method: Improve the accuracy of error estimation by applying error correction.

In addition to error analysis, other methods can enhance the accuracy and reliability of high-frequency electromagnetic simulation. Adaptive mesh, for example, can improve calculation accuracy by dynamically adjusting the mesh size and density based on simulation results while reducing computing time and storage space. Additionally, multi-physical field coupling can be utilized to couple different physical fields (e.g., electric field, thermal field, fluid field) and enhance the accuracy and credibility of the simulation. When simulating high-frequency electromagnetic fields, it is crucial to select appropriate numerical methods based on specific problem requirements and characteristics. Analyzing and optimizing errors helps improve the accuracy and reliability of simulation results and provides support for practical applications.

2.4 Application of ADS to HFEMFs

2.4.1 Simulations Tools in ADS

The Advanced Design System (ADS) offers a range of powerful electromagnetic simulation tools that are invaluable for designing and analyzing microwave and millimeter-wave circuits, antennas, filters, RF integrated circuits, and more. One notable example is Key sight's EMPro, which stands out as a versatile 3D electromagnetic simulation software. Additionally, ADS itself is widely recognized as the industry's leading circuit simulation software. In addition to EMPro, there are other electromagnetic simulation software options available, such as Momentum, RFPro, CILD, Coilsys, Via Designer, SIPro, PIPro, and PEPro. Keysight provides an extensive selection of electromagnetic (EM) simulation methods, ensuring users have a wide array of options integrated directly into the ADS design process. This seamless integration facilitates co-simulations between schematics, layout, and 3D components, eliminating the need to navigate outside the circuit design flow. This capability proves particularly beneficial for RF module designs, where simulating 3D interconnects and packaging alongside the circuit is essential.

1. Momentum Simulator is an industry-leading 3D planar electromagnetic (EM) simulator renowned for its exceptional capability in passive circuit modeling and analysis. By utilizing the frequency-domain method of moments (MoM) method, it accurately captures coupling and parasitic effects in intricate multilayer designs. Momentum seamlessly integrates with ADS, Genesys, and GoldenGate and also offers compatibility with third-party tools like Cadence and Mentor.

2. The Keysight Finite Element Method (FEM) simulator relies on an industry-proven finite element method to tackle the simulation of 3D structures with arbitrary shapes. Unlike 3D planar simulators, this provides the flexibility to handle various structural complexities. Integrated with ADS, the Keysight FEM eliminates the need to export the layout design to a third-party EM simulator. Furthermore, Keysight FEM can be used within EMPro to simulate 3D models imported from other CAD tools or created natively.

3. SIPro simulation excels in achieving high-frequency accuracy while catering to the speed and capacity requirements of advanced printed circuit board designs with dense wiring. Featuring an automatic schematic generation function, SIPro enables the direct integration of the extracted EM model into ADS transient simulation and ADS channel simulation, enabling comprehensive channel analysis.

4. Keysight's finite-difference time-domain (FDTD) simulator operates on the finite-difference time-domain methods, much like FEM. Both methods handle arbitrarily shaped 3D structures. However, FEM generates a large matrix during the solution process, whereas FDTD employs an iterative process to update field values at each time step. Keysight's FDTD Simulator, available within EMPro, allows for the simulation of 3D models imported from other CAD tools or created within the software.

5. PIPro specializes in power integrity (PI) analysis of power distribution networks (PDNs), encompassing DC IR drop analysis, AC impedance analysis, and power plane resonance analysis. By leveraging a common setup and analysis environment within ADS, PIPro streamlines the PI analysis workflow. Its EM technology is specifically optimized for PI applications, ensuring faster and more efficient simulations compared to general-purpose EM tools.

2.4.2 Guidance of ADS Simulations for HFEMFs

The main steps involved in high-frequency electromagnetic simulation using ADS are as follows:

1. Create a new electromagnetic simulation project: Open ADS and select "File" > "New" > "Project" from the main interface to create a workspace.
2. Select the simulation tool: In the new project window, choose the appropriate electromagnetic simulation tool, such as Momentum, FEM Element, Planar EM Simulator, etc. The system will automatically load the selected simulation tool.
3. Define the simulation area: Add simulation objects, such as microstrip lines, antennas, filters, etc., to the simulation area. You can use the Layout tool for modeling and editing, or import CAD files for simulation. Pay attention to the geometric shape, position, material, and other parameters of the components during modeling.
4. Set simulation parameters: Configure the simulation settings, including the frequency range, resolution, grid size, boundary conditions, and other parameters. This can be done by right-clicking on the simulation area or using the settings tool in the menu bar.
5. Define ports and measurement points: Define ports and add measurement points based on the characteristics of the simulation objects. In the Layout tool, you can use the Lumped Port tool to define a port and add an S-parameter measurement at the port to analyze the transmission performance of the simulation object. Select appropriate port locations and numbers based on the layout and geometry of the simulation object.
6. Run the simulation: Click the Simulations button to initiate the simulation. Monitor the simulation progress and calculation time, and visualize the S-parameter graph and field distribution of the simulation results during the simulation.
7. Analyze the simulation results: Analyze the simulation results, including S-parameters, power transmission, field distribution, and other relevant information. Use tools like Graph and 3D Viewer for analysis and visualization. Optimize the design performance based on the analysis of the simulation results.
8. Refine the design: Make adjustments to the design based on the simulation results and re-run the simulation for verification. Modify parameters such as size, position, material, etc., to optimize the design performance. Repeat steps 4–7 until the design requirements are met.

It is important to note that different simulation tools and objects may require specific settings and procedures. To achieve the best simulation results, select the appropriate tools and configure the simulation parameters according to the specific design requirements and characteristics of the simulation object. Continuously adjust and verify the design based on the simulation results during the analysis and optimization process to ensure accuracy and reliability. When setting the simulation parameters, keep in mind that the grid size directly affects the simulation accuracy and calculation speed. A finer grid increases the calculation time, while a coarser grid affects the simulation accuracy. As a general recommendation, the mesh size should not be smaller than 1/10 of the smallest feature size.

2.5 Application of HFSS to HFEMFs

In practical applications, the finite element analysis (FEA) method stands out as one of the most effective methods in frequency-domain calculations for electromagnetic simulation. It offers the capability to simulate diverse geometries and material compositions. The ANSYS software provides support for time-harmonic and modal high-frequency analysis. High-frequency electromagnetic field analysis is employed to replicate electromagnetic phenomena in a model where the signal wavelength is of a comparable magnitude to the model size, or in a structure where the wavelength magnitude is smaller than the model size. The ANSYS high-frequency electromagnetic field simulator comprises a pre-processor, a solver, and a post-processor.

The pre-processor is responsible for describing the high-frequency structure of the simulation and applying the necessary excitation boundary conditions and constraints. It offers a wide range of options and flexible settings to accommodate various requirements. The solver generates an element description, incorporates the element matrix into a global finite element matrix, and solves the equations using appropriate boundary conditions, constraints, and excitation sources. In the post-processor, users can conveniently visualize the analysis results and utilize features such as drawing, listing, and other methods of displaying and storing data. Intermediate result data can be exported for seamless integration with other applications. This section will provide an introduction to the electromagnetic simulation tools of HFSS, along with a step-by-step guide to high-frequency electromagnetic simulation in HFSS.

2.5.1 Simulation Steps in HFSS

HFSS is a highly utilized simulation tool for high-frequency electromagnetic fields, renowned for its robust simulation capabilities and flexible modeling. Within HFSS, electromagnetic simulation primarily encompasses two essential steps: geometric modeling and electromagnetic field analysis.

- Geometric Modeling
  In HFSS, geometric modeling serves as the initial step in electromagnetic simulation. HFSS offers a wide range of tools for geometric modeling, including the importation of existing CAD models, manual creation of geometric shapes, and parametric modeling. By supporting the import of popular CAD model formats like STL, IGES, STEP, SAT, and SolidWorks, HFSS simplifies the modeling process while ensuring precise and accurate representations. Once a CAD model is imported, it can be easily modified and edited to align with simulation needs.
  When it comes to manually creating geometric shapes, HFSS provides an array of drawing tools, including line segments, circles, rectangles, and polygons. These versatile drawing tools empower users to manually construct intricate geometric shapes, allowing for subsequent editing and optimization. Furthermore, HFSS supports parametric modeling, which enables the rapid generation of diverse geo-

metric shapes based on varying parameter values, thereby catering to different simulation requirements.

- Electromagnetic Field Analysis
  After completing the geometric modeling, an electromagnetic field analysis is required. HFSS offers a wide range of tools for electromagnetic field analysis, including mesh generation, boundary condition and excitation definition, solver selection, and result analysis. Regarding mesh generation, HFSS supports three types of meshes: Tetra mesh, Hexa mesh, and Hybrid mesh. By generating a grid, the geometric model can be divided into smaller units, facilitating the analysis of electromagnetic fields. For boundary conditions and excitation definition, HFSS provides various options that can be set according to the specific application requirements. Common choices for boundary conditions include Perfect E (Electric) Boundary, Perfect H (Magnetic) Boundary, Radiation Boundary, and Symmetry Boundary. Excitation definition options include Point Source, Waveguide Port, Lumped Port, and Feed Line. HFSS supports both finite element method (FEM) and finite-difference time-domain (FDTD) solvers. When selecting a solver, factors such as the complexity of the geometry, simulation accuracy, and available computing resources need to be considered. In terms of result analysis, HFSS offers a range of tools such as field distribution analysis, power transmission analysis, parameter scanning, and optimization analysis. These tools enable comprehensive analysis and evaluation of simulation results, leading to more accurate results and a deeper understanding of the electromagnetic field behavior.
- Advanced Modeling
  HFSS, apart from its basic geometric modeling and electromagnetic field analysis methods, offers a wide range of advanced modeling capabilities. These include multi-physics coupling simulation, parametric modeling, optimizer, and co-simulations. One notable area of interest is the interaction between the electromagnetic field and the thermal field, which can be effectively simulated using HFSS's multi-physics field coupling simulation. This enables more accurate predictions of performance and behavior in practical applications. Parametric modeling, another powerful feature of HFSS, involves the use of parametric methods to create geometric models. By manipulating different parameter values, various geometric shapes can be generated to fulfill different simulation requirements. HFSS provides diverse parametric modeling, such as Scripting, DesignXplorer, and OptiSLang. Furthermore, HFSS incorporates a range of optimizers to fine-tune simulation models. These optimizers, including genetic algorithm, particle swarm optimization, and adaptive tabu search, enable users to achieve optimal simulation results. Collaborative simulation is a crucial aspect that involves analyzing simulation results from different domains to comprehensively evaluate the performance and behavior of the simulated objects. HFSS supports this collaborative approach, allowing for a holistic analysis of simulation outcomes. Prior to conducting simulations using HFSS, it is essential to thoroughly analyze and evaluate the simulation object. This assessment ensures the selection of the most suitable modeling and simulation method for the given scenario.

2.5.2 Guidance of HFSS Simulations for HFEMFs

The process of high-frequency electromagnetic simulation in HFSS consists of five crucial steps: geometric modeling, mesh generation, boundary conditions, and excitation definition, solver selection, and analysis of simulation results. Firstly, geometric modeling serves as the initial stage of the simulation, which can be accomplished through the import of CAD models or manual creation. During this process, it is essential to ensure the model is accurate and three-dimensional. Additionally, breaking down the model into smaller units facilitates mesh generation. Next, the mesh generation process divides the geometric model into smaller cells. HFSS supports three types of mesh: Tetra, Hexa, and Hybrid. A finer grid yields more precise simulation results, but it also increases computation time. For intricate geometries, the Tetra mesh is preferred, while the Hexa mesh is suitable for simpler geometries. Setting the appropriate boundary conditions and excitation definitions based on the application requirements is crucial. It is essential to ensure consistency with the real-world application and select the appropriate mode for different boundary conditions. The excitation signal should also align with the actual application. In the solver selection stage, HFSS offers two solvers: the finite element method (FEM) solver and the finite-difference time-domain method (FDTD) solver. The FEM solver is well-suited for complex geometries due to its higher accuracy, while the FDTD solver is faster and more suitable for simple geometries. Selecting the appropriate solver depends on the specific application requirements. Lastly, analyzing the simulation results is a pivotal step in evaluating their accuracy. By comparing the simulation results with real-world observations, the accuracy of the simulation can be assessed. Visualizing the simulation results aids in analysis and comprehension. Furthermore, the simulation results can be optimized and refined to achieve the best possible outcomes.

2.6 Co-simulations for HFEMFs

2.6.1 Definition of Co-simulations for HFEMFs

In applications involving high-frequency electromagnetic fields, simulating the electromagnetic field is often necessary to accurately predict its behavior and performance. Co-simulations are a simulation method that combines the simulation models of multiple physical processes or systems to simulate their interaction. In the case of high-frequency electromagnetic fields, the propagation of the electromagnetic field is influenced by the interaction between the electromagnetic wave and an object. Therefore, it is crucial to simulate the interaction between the electromagnetic field and the object. This interaction encompasses factors such as the object's geometry, electrical properties, material properties, and electromagnetic radiation. By utilizing co-simulations, it becomes possible to predict the behavior and performance of high-frequency electromagnetic fields within objects with greater accuracy. This, in turn, facilitates the optimization of electromagnetic field designs and applications. An illustrative example of co-simulations in high-frequency electromagnetic fields is the design of radar antennas. A radar antenna typically consists of a complex metallic structure and an excitation source, such as an antenna feed line. During radar operation, the antenna receives electromagnetic waves and converts them into electrical signals. Consequently, the antenna design must consider numerous factors, including the propagation characteristics of electromagnetic waves, the object's geometric shape and material properties, and the antenna's radiation characteristics. In such cases, co-simulations can be performed using electromagnetic simulation software like CST Studio Suite, in conjunction with structural simulation software such as ANSYS Mechanical. The process begins by establishing the electromagnetic field model of the antenna in the electromagnetic simulation software, incorporating the geometric structure, material properties, and boundary conditions. Similarly, the object model of the antenna is created in the structural simulation software, encompassing the geometric structure, material properties, and boundary conditions. Subsequently, the electromagnetic field model and the object model are coupled to form a complete co-simulation model. Finally, simulation calculations are conducted, and the results are analyzed and optimized. In this example, the principle of co-simulations lies in the coupling of the electromagnetic field model and the object model, taking into account the interaction between the electromagnetic field and the object. Specifically, the electromagnetic field model provides information regarding the electromagnetic waves received by the antenna, while the object model offers the geometry and material characteristics of the antenna. Through co-simulations, the radiation characteristics and performance of the antenna can be predicted with higher accuracy. This co-simulation method can be applied to various high-frequency electromagnetic field applications, including microwave communication, radar systems, antenna design, and more.

Co-simulations within high-frequency electromagnetic fields are typically conducted using one of three primary methods:

1. Pre-post Simulations: This method involves first establishing the electromagnetic field model using specialized simulation software, followed by creating the object model with structural simulation software. Throughout these two separate simulation processes, geometric parameters and boundary conditions of the object are transferred via file exchange. Upon completion of both simulations, their results can be combined for comprehensive co-simulations.

2. Parallel Simulations: Utilizing both electromagnetic and structural simulation software, this approach harnesses shared memory and message passing, among other methods, for parallel computing. This improves simulation efficiency. Throughout the parallel computation process, the models of the electromagnetic field and the object are interconnected using shared memory and message passing.

3. Unified Simulations Software: Certain simulation programs, such as the CST Studio Suite, offer a comprehensive co-simulation function. These can establish both the electromagnetic field and object models simultaneously, streamlining the co-simulation process. As co-simulations can be completed directly within a type

of single software, this method alleviates the complexity of data processing and file transfer.

The principal benefit of co-simulations within high-frequency electromagnetic fields is the precise prediction of electromagnetic field behavior and performance, thereby optimizing electromagnetic field design and application. More specifically, co-simulation aids designers in understanding interactions between electromagnetic fields and objects, foreseeing how the fields will behave within these objects, and fine-tuning designs to satisfy practical needs. Moreover, co-simulations can enhance simulation accuracy and efficiency, while reducing development time and cost. However, when performing co-simulations of high-frequency electromagnetic fields, it's critical to recognize that different simulation software and methods are appropriate for varying application scenarios and simulation needs. Selection of suitable software and method should be based on the specific context. For improved simulation accuracy and reliability, meticulous modeling of the electromagnetic field and the object—including geometric structure, material properties, and boundary conditions—is crucial. This involves considering multiple physical process interactions, such as the propagation of electromagnetic waves and the geometric and material properties of objects. Furthermore, to boost simulation efficiency and precision, it is necessary to optimize simulation parameters and solvers, for instance, mesh accuracy, time step, and iteration count. co-simulations in high-frequency electromagnetic fields are a valuable simulation method. It enables designers to gain a deeper comprehension of the interplay between electromagnetic fields and objects, predict the behavior and performance of these fields within objects, and refine designs to fulfill practical requirements.

The following are the specific implementation method and operation steps of co-simulations in high-frequency electromagnetic field. Taking the design of radar antenna as an example, the target of co-simulations and the physical factors to be considered, such as the geometric structure, electrical characteristics, and radiation characteristics of the antenna, are determined according to the actual needs and application scenarios. The electromagnetic simulation software (such as CST Studio Suite) is used to establish the electromagnetic field model of the antenna, including geometric structure, boundary conditions, material properties, and excitation sources. Set the simulation parameters and solver, such as mesh accuracy, time step, number of iterations, and so on. Structure simulation software (such as ANSYS Mechanical) is used to establish the object model of the antenna, including geometric structure, material properties, and boundary conditions. Set the simulation parameters and solver, such as mesh accuracy, time step, number of iterations, and so on. The electromagnetic field model and the object model are coupled to establish a complete co-simulation model. In this example, the method of file exchange can be used to import the electromagnetic field data from the electromagnetic simulation software into the structure simulation software and import the parameters such as the geometric structure and boundary conditions of the object into the electromagnetic simulation software. Then, the simulation calculation is carried out, and the simulation results are analyzed and optimized.

2.6.2 *Co-simulations of HFEMFs with ADS*

The aforementioned software solutions boast extensive applications and profound impacts within the realm of microwave circuit design and high-frequency electromagnetic simulation, serving a substantial user base alongside a wealth of application scenarios. Prominently, Keysight ADS, CST Studio Suite, and ANSYS HFSS are preeminent players in the market, offering comprehensive simulation features with user-friendly interfaces. These tools are ubiquitously utilized in fields such as microwave circuit and antenna design. COMSOL Multiphysics exemplifies top-tier multi-physics simulation software, characterized by its powerful simulation prowess and an array of practical applications. Meanwhile, Sonnet stands out as a type of dedicated software for high-frequency electromagnetic simulation, catering to a diverse range of application settings and user communities.

2.6.2.1 Co-simulations of CST Studio Suite with ADS

The CST simulation software is a comprehensive three-dimensional electromagnetic simulation tool that operates on the basis of time-domain integral equation. It employs either the finite-difference time-domain (FDTD) method or the time-domain integral equation (TLM) method to tackle electromagnetic field problems [59]. The fundamental principle involves dividing the object into smaller blocks and numerically solving the differential equations to obtain the variations in physical quantities like electric field, magnetic field, charge, and current within each block. Ultimately, it calculates the electromagnetic field distribution for the entire object.

The CST simulation software encompasses several key functionalities. Firstly, it facilitates the simulation of diverse electromagnetic field problems, ranging from antennas and microwave devices to optical devices. It allows for the calculation of physical quantity distributions, including electric field, magnetic field, charge, and current. Secondly, it supports multi-physical field simulation, enabling the simultaneous simulation of various physical fields such as electromagnetic, thermal, and mechanical fields. Moreover, the software offers an optimization design feature, which utilizes automatic optimization algorithms to optimize design parameters and achieve the best possible design outcomes. It also includes a rapid modeling capability, allowing users to swiftly construct models using 3D CAD, 2D CAD, or direct input of coordinates. Additionally, it provides post-processing functionalities, such as generating field distribution maps, calculating field intensity, and assessing radiation power.

Co-simulations of CST and ADS involve combining CST simulation software with ADS circuit simulation software to achieve the simultaneous simulation of circuits and electromagnetic fields. This process can be outlined as follows:

1. Establishing the CST simulation model: Begin by using CST simulation software to create an electromagnetic simulation model, which includes components such as conductors, media, and excitations.

2. Performing CST simulation: Utilize CST simulation software to conduct electromagnetic simulations and obtain data on the distribution of electromagnetic fields.

3. Exporting CST simulation data: Export the electromagnetic field distribution data from CST simulation software into an S-parameter file.

4. Establishing the ADS circuit model: Employ ADS circuit simulation software to construct a circuit model comprising various components and connection modes.

5. Importing CST simulation data: Import the exported S-parameter file into ADS circuit simulation software to facilitate the co-simulations of the circuit and electromagnetic field.

6. Performing ADS circuit simulation: Execute circuit simulations using ADS circuit simulation software to obtain simulation results regarding circuit response.

Throughout the CST and ADS co-simulations, it is crucial to prioritize simulation accuracy. The duration of CST simulation and ADS circuit simulation may differ, necessitating consideration of the time discrepancy to ensure the consistency of simulation data. Additionally, during CST simulation and ADS circuit simulation, it is essential to establish appropriate simulation parameters and calculation methods to guarantee the accuracy of the simulation outcomes.

In this section, we will introduce a case of co-simulations between CST and ADS, and provide detailed steps for setting up the simulation. The case involves a microstrip antenna used in a Ku-band communication system, consisting of an antenna substrate, microstrip line, feed point, and ground plane. The electromagnetic field is simulated using CST simulation software, and the simulation results are imported into ADS circuit simulation software to enable co-simulations of the circuit and electromagnetic field. The purpose of the simulation is to optimize the operating frequency and impedance-matching performance of the microstrip antenna.

The simulation steps are as follows:

1. Establish the CST simulation model: Begin by creating the electromagnetic simulation model of the microstrip antenna in CST simulation software. This model includes components such as the microstrip line, antenna substrate, feed point, and ground plane. The microstrip line is a rectangular shape with dimensions of 25 mm in length, 2 mm in width, and 0.1 mm in thickness. The antenna substrate is a square dielectric substrate measuring 40 mm on each side, with a thickness of 0.8 mm. It has a dielectric constant of 4.4 and a dielectric loss angle tangent value of 0.02. The feed point is a microstrip line with a width of 0.2 mm, located at the center of the antenna substrate. The ground plane is a square metal plate measuring 50 mm on each side.

2. CST simulation: Utilize CST simulation software to simulate the electromagnetic field and calculate the electromagnetic field distribution of the microstrip antenna within the software. The simulation is performed over a frequency range of 10–15 GHz, with a calculation step of 0.05 GHz.

3. Export CST simulation data: Export the S-parameter file from CST simulation software, which will be imported into ADS circuit simulation software.

4. Establish the circuit model in ADS: Set up the circuit model of the microstrip antenna in ADS circuit simulation software, which includes components such as the microstrip line, antenna substrate, feed point, and ground plane. The dimensions and

properties of these components are the same as those defined in the CST simulation model.

5. Import CST simulation data: Import the previously exported S-parameter file from the CST simulation software into ADS circuit simulation software. This step allows for the co-simulation of the circuit and electromagnetic field.

6. Perform ADS circuit simulation: Conduct the circuit simulation within ADS circuit simulation software to optimize the operating frequency and impedance matching performance of the microstrip antenna. Start by opening a new circuit window in ADS and selecting the Layout View to establish the circuit model of the microstrip antenna. Import the S-parameter file derived from CST simulation software into ADS for port matching. Add the port matching circuit, consisting of a chip capacitor and a chip inductor, to the circuit model. These components are used to match the impedance and adjust the operating frequency. By varying the parameters of the chip capacitor and chip inductor, the impedance matching and operating frequency of the microstrip antenna are optimized. Obtain the circuit model and simulation results of the optimized microstrip antenna. Following the co-simulations of CST and ADS, the impedance matching performance of the microstrip antenna is improved, and the bandwidth of the S_{11} parameter with a magnitude less than –20 dB is increased. Overall, the simulation results are excellent.

The co-simulations of CST and ADS allow for the combination of electromagnetic simulation with circuit simulation, enabling a more comprehensive analysis and optimization of electronic device performance. During the co-simulations, attention should be given to simulation accuracy, simulation time, and simulation settings to ensure the accuracy of the results. Furthermore, co-simulations should follow a specific set of steps, including establishing the CST simulation model, performing CST simulation, exporting CST simulation data, establishing the ADS circuit model, importing CST simulation data, and conducting ADS circuit simulation. In practical engineering, the co-simulations of CST and ADS are widely applied in various fields such as antenna design, microwave circuit design, and signal transmission and reception. It serves as an effective simulation and analysis tool for engineers to design and optimize electronic devices more effectively. By combining the strengths of CST's electromagnetic simulation capabilities and ADS's circuit simulation capabilities, engineers can gain deeper insights into the performance of electronic devices. However, it is important to ensure simulation accuracy by validating the simulation results with experimental measurements whenever possible. Simulation time should also be considered, as complex models may require significant computational resources. The co-simulations of CST and ADS provide engineers with a powerful tool to analyze, optimize, and refine electronic devices. By following the specified steps and considering simulation accuracy and time, engineers can make informed design decisions and improve the overall performance of their designs.

2.6.2.2 Co-simulations of COMSOL Multiphysics with ADS

COMSOL Multiphysics is a kind of powerful multi-physics simulation software that is based on finite element analysis theory. It enables the simulation of various physical processes, including heat conduction, structural mechanics, fluid mechanics, and electromagnetic fields [60]. By defining geometric and physical models, users can solve finite element equations to obtain simulation results. COMSOL Multiphysics supports the simulation of multiple physical fields simultaneously, such as heat conduction, structural mechanics, electromagnetic fields, and fluid mechanics. The software offers an intuitive user interface and easy-to-use tools, allowing users to effortlessly model, solve, and visualize tasks. It is compatible with multiple operating systems like Windows, macOS, and Linux.

ADS (Advanced Design System) is a kind of circuit design software, while COMSOL Multiphysics is a kind of multi-physics simulation software. The combination of these two tools enables the integration of circuit design and simulation. With COMSOL Multiphysics, users can simulate various physical phenomena within a circuit, including electromagnetic interference and heat distribution. Compared to traditional circuit simulation software, COMSOL Multiphysics provides higher accuracy and reliability.

Co-simulations of COMSOL Multiphysics and ADS involve the following steps:

1. Design the circuit in ADS and export the model: Design the circuit within ADS and export the circuit model as an S-parameter file or a network parameter file.

2. Import the circuit model into COMSOL Multiphysics: Import the S-parameter file or network parameter file exported from ADS into COMSOL Multiphysics to establish the simulation model.

3. Co-simulations: Define boundary conditions, material properties, and physical models within COMSOL Multiphysics. Run the simulation calculation to obtain the simulation results for the physical phenomena within the circuit.

4. Post-processing: Visualize and analyze the simulation results in COMSOL Multiphysics to evaluate and optimize the circuit design.

During the co-simulations of COMSOL Multiphysics and ADS, it is crucial to ensure the correct format of the network parameter file or S-parameter file. In COMSOL Multiphysics, constructing the geometric model of the simulation model based on the circuit model is necessary to perform the simulation calculations accurately. Additionally, selecting the appropriate physical model is essential for simulating the desired physical phenomena within the circuit.

The objective of co-simulations using COMSOL Multiphysics and ADS is to develop an industrial motor driver resilient to high currents, elevated temperatures, and electromagnetic interference. This design and simulation strategy employs both COMSOL Multiphysics and ADS to create and simulate the circuit design.

The motor driver circuit comprises several components, such as MOSFET, inductors, and diodes, among others. Notably, the MOSFET and inductor must withstand significant current loads and heat. Key performance indices for the circuit encompass current and heat distribution, circuit response time, and resistance to electromagnetic interference. Here are the detailed steps for conducting the COMSOL Multiphysics and ADS co-simulations:

1. Circuit Design and Model Export: Design the motor driver circuit in ADS and export the model as an S-parameter file.

2. Model Import and Construction: Import the circuit model into COMSOL Multiphysics to construct the simulation model, setting model geometry, material properties, and boundary conditions.

3. Physical Model Selection: Select the physical fields for simulation within the circuit, such as heat conduction, structural mechanics, electromagnetic field, etc.

4. Meshing: Undertake meshing within COMSOL Multiphysics for finite element analysis, balancing simulation accuracy with computational efficiency.

5. Solver Configuration: Choose the appropriate solver and set the simulation calculation parameters, such as time step, maximum number of iterations, etc.

6. Simulations: Run the simulation to obtain results representing the circuit's physical phenomena.

7. Post-processing: Visualize and analyze the simulation results within COMSOL Multiphysics to evaluate and optimize circuit design.

This co-simulations of COMSOL Multiphysics and ADS facilitate the acquisition of simulation results for various circuit physical phenomena. The design team can evaluate parameters like response time, current and heat distribution, and susceptibility to electromagnetic interference. Based on these results, they can adjust MOSFET and inductor parameters to enhance circuit performance and stability, optimize PCB layout and design to mitigate electromagnetic interference and heat conduction issues, and determine hotspots and heat paths for optimized heat dissipation schemes. Simultaneously, circuit stability and reliability can be evaluated, while electromagnetic field and interference distribution can be determined for effective electromagnetic compatibility design. The co-simulation approach enables more accurate performance and stability evaluations, averting trial-and-error and testing processes, thereby saving time and cost. It presents promising prospects for circuit design and development, notwithstanding certain challenges, such as data transfer and format conversion, model accuracy, and a balance between simulation precision and computational resources. Interaction between COMSOL Multiphysics and ADS software for co-simulations mainly occurs via the circuit's S-parameters. Once the circuit is designed in ADS, the circuit model can be exported as an S-parameter or network parameter file. These can then be imported into COMSOL Multiphysics to build a simulation model. Through these parameters, COMSOL Multiphysics can simulate a multitude of physical phenomena, including electromagnetic fields, heat conduction, etc., to obtain the respective simulation results. Consequently, the S-parameter file serves as the central parameter for co-simulation interaction between COMSOL Multiphysics and ADS software.

2.6.2.3 Co-simulations Sonnet with ADS

Sonnet is a kind of professional electromagnetic simulation software, which is mainly used for the simulation of microwave circuits, antennas, high-speed digital signal integrity (SI), electromagnetic compatibility (EMC), radio frequency (RF), and millimeter wave (mmWave). The basic principle of Sonnet simulation software is based on the Method of Moments (MoM), which discretizes the calculation area by establishing a grid on the circuit board, converts the electromagnetic characteristics of microwave devices into an equivalent circuit model, and then uses the numerical calculation method to solve the electromagnetic characteristics of the circuit. It can be used to build 2D and 3D electromagnetic models, which enables users to accurately simulate and optimize the electromagnetic characteristics of microwave devices and circuit boards. In addition, Sonnet software can also analyze the transmission and radiation of electromagnetic waves, and calculate the transmission, reflection, and scattering parameters of RF power. The software also provides powerful EMC analysis functions, including the calculation of electromagnetic compatibility and anti-interference capability. It can also be used to simulate high-speed digital signal integrity (SI), including time-domain and frequency-domain simulation. And simulate millimeter-wave (mmWave) devices and circuit boards. The features of Sonnet software are also worth mentioning. First, it has high accuracy and is suitable for complex electromagnetic simulation tasks.In addition, Sonnet software can also carry out fast simulation and support multi-core processors and distributed simulation, thus improving the efficiency of simulation and has strong scalability. It supports script programming and custom electromagnetic models so that users can customize and expand according to their own needs.

Sonnet software can be co-simulated with ADS (Advanced Design System) software, and the Sonnet simulation results can be imported into ADS for analysis and optimization. Co-simulations can analyze the performance of the circuit and optimize the design more comprehensively. The main principle of Sonnet and ADS co-simulations is to import the electromagnetic simulation results of Sonnet into ADS and use them as sub-circuits of the circuit.

The specific steps are as follows:

1. Establish an electromagnetic model in the Sonnet and perform simulation;
2. Export the Sonnet simulation result to generate a Touchstone file;
3. Create a circuit model in ADS and import the Touchstone file of Sonnet into ADS as a sub-circuit;
4. Simulate and optimize in ADS, including frequency response, S-parameter, power, and other analyses.

In the co-simulations, the circuit models of Sonnet and ADS need to be consistent, including the size, layout, and material of the device, and the simulation parameters need to be consistent, such as frequency range, step, simulation accuracy, etc., and the appropriate port and data format need to be selected to specify the correct port and matching network so as to ensure the accuracy of the simulation result.

A specific Sonnet and ADS co-simulations case involves the establishment of an electromagnetic model of a microstrip line filter and the co-simulations. The specific steps are as follows:

1. Build an electromagnetic model in Sonnet: Create an electromagnetic model of the microstrip line filter, including components such as microstrip lines and matching networks. Set the size, material, and port of the model, and simulate to get the S-parameter results. This is done through Sonnet's graphical user interface (GUI).

2. Export Sonnet simulation results: Export the S-parameter Touchstone file in Sonnet and select the correct port and data format. In this case, we chose the S-parameter format and specified the correct ports and port order.

3. Create a circuit model in ADS: Create a circuit model of a microstrip line filter, including components such as microstrip line, capacitor, and inductor, and import Sonnet's Touchstone file into ADS as a sub-circuit. You need to specify the correct port and matching network when importing.

4. Conduct co-simulation: Conduct co-simulation in ADS, including frequency response and S-parameter analysis. We can optimize the circuit model, such as adjusting the parameters or positions of components, to get better performance. In this case, we adjusted the width of the microstrip line, simulated and optimized it, and got better frequency response and insertion loss.

It should be noted that in this case, we need to keep the circuit model and simulation parameters of Sonnet and ADS consistent, such as the size and material of the microstrip line, the frequency range and step of the simulation, etc. When exporting Sonnet's Touchstone file, you need to select the correct port and data format. When importing the Touchstone file, you need to specify the correct port and matching network to ensure the accuracy of the simulation results.

2.6.3 Co-simulations with HFSS for HFEMFs

2.6.3.1 Co-simulations of EMCoS Studio with HFSS

EMCoS Studio is a kind of robust software for electromagnetic compatibility (EMC) and electromagnetic interference (EMI) simulation. It offers engineers an accurate and efficient solution based on computational electromagnetics and electromagnetic radiation theory. This software enables the analysis and optimization of EMC problems in complex electronic systems, helping designers predict and reduce electromagnetic interference during product development. Consequently, it enhances the reliability and performance of products.

EMCoS Studio boasts several unique features and benefits that make it an ideal choice for electromagnetic simulation. Firstly, it provides a comprehensive pre-processing function that supports the import of 3D models from various CAD formats. It simplifies, optimizes, and refines these models, thereby improving simulation accuracy and efficiency. Secondly, EMCoS Studio offers a wide range of electromagnetic field solvers, including the finite element method (FEM), finite-difference time-domain method (FDTD), and integral equation method (IEF). This versatility caters to the diverse needs of solving different types of electromagnetic problems. Additionally, the software offers powerful post-processing capabilities and a rich set of tools to visualize, quantitatively analyze, and optimize simulation results. These features assist engineers in solving practical problems. Moreover, EMCoS Studio supports integration with other professional software, such as HFSS, to efficiently address electromagnetic problems and further enhance simulation performance.

EMCoS Studio enables co-simulations of high-frequency electromagnetic fields with HFSS software. HFSS (High-Frequency Structure Simulator) is a widely used three-dimensional full-wave finite element electromagnetic simulation software for electronic design and electromagnetic compatibility analysis. It excels in solving electromagnetic problems in high-frequency, radio frequency, microwave, and related fields. Through co-simulations, users can utilize the electromagnetic field solution results from HFSS in EMCoS Studio to enhance simulation accuracy and efficiency. The co-simulation process involves establishing and simulating the electromagnetic field model in HFSS. Subsequently, the electromagnetic field distribution solved by HFSS is imported into EMCoS Studio for further analysis and optimization. During this process, EMCoS Studio converts the electromagnetic field data from HFSS into a suitable format for its own solver, facilitating efficient data interaction.

The specific steps for co-simulations of EMCoS Studio and HFSS software for high-frequency electromagnetic fields are as follows:

1. Establish the high-frequency electromagnetic field model in HFSS software and configure boundary conditions, material properties, excitation sources, and other parameters. 2. Utilize the HFSS solver to simulate the electromagnetic field of the model and obtain the electromagnetic field distribution data. 3. Export the simulation results from HFSS to standard data formats such as S-parameters and field distribution files. 4. Create a new project in EMCoS Studio and import the simulation result file from HFSS. 5. Pre-process the imported data, including data conversion and model simplification, as necessary. 6. Select the appropriate solver in EMCoS Studio based on the electromagnetic field data from HFSS and perform further simulation analysis. 7. Utilize the post-processing function in EMCoS Studio to visualize, quantitatively analyze, and optimize the simulation results.

When co-simulating with EMCoS Studio and HFSS, it is crucial to ensure compatibility between the software versions to facilitate smooth data interaction. When importing the simulation results from HFSS, carefully select the correct data format and file type. During data pre-processing, ensure that key information is preserved to maintain the accuracy of the simulation results. When choosing the EMCoS Studio solver, thoroughly consider the characteristics of the HFSS data and the simulation requirements to select the appropriate solution method.

The following presents a specific case of co-simulations of EMCoS Studio and HFSS software for high-frequency electromagnetic fields:

Case Description: In order to design a microwave filter, it is essential to analyze the distribution of its electromagnetic field and assess its electromagnetic compatibility at the operating frequency. The following steps outline the simulation setup:

1. Establish a three-dimensional model of the filter using HFSS, specifying parameters such as the operating frequency, material properties, and boundary conditions.

2. Perform an HFSS simulation to obtain the S-parameters and electromagnetic field distribution data of the filter.

3. Export the simulation results from HFSS as S-parameter and field distribution files.

4. Create a new project in EMCoS Studio and import the HFSS simulation result files.

5. Pre-process the imported data in order to format the field distribution data appropriately for the EMCoS Studio solver.

6. Select the suitable electromagnetic compatibility solver, such as the Finite-Difference Time-Domain (FDTD) method, in EMCoS Studio to simulate the electromagnetic compatibility of the filter.

7. Utilize the post-processing functionality of EMCoS Studio to visualize, quantitatively analyze, and optimize the simulation results. For instance, analyze the electromagnetic field's distribution characteristics, identify areas prone to electromagnetic interference, and optimize the filter structure to enhance electromagnetic compatibility.

8. Adjust the filter design based on the analysis results, such as modifying device dimensions or changing material properties.

9. Re-import the optimized filter model back into HFSS, conduct electromagnetic simulations, and verify the effectiveness of the optimization.

10. Repeat the above steps until satisfactory resolution of the filter's electromagnetic compatibility issue is achieved.

The aforementioned cases demonstrate the powerful capabilities and versatile applications of EMCoS Studio and HFSS software in simulating high-frequency electromagnetic fields. By integrating these two types of software tools closely, engineers can accurately and efficiently analyze and resolve electromagnetic problems in practical engineering, ultimately improving product performance and reliability.

EMCoS Studio supports multiple data formats to accommodate various solvers' requirements during co-simulations. When conducting co-simulations, it is often necessary to convert the imported electromagnetic field data into a format recognized and processed by the solver. Below are some commonly used data formats for the EMCoS Studio solver:

1. S-parameters file: The S-parameters file contains the transmission and reflection characteristics of circuit elements in the frequency domain. S-parameters data is typically stored and transmitted using the Touchstone format (.snp file). In EMCoS Studio, S-parameters files can be utilized by a network analysis solver for electromagnetic compatibility and signal integrity analysis.

2. Field distribution file: The field distribution file contains information about the distribution of the electromagnetic field in three-dimensional space, which includes electric field, magnetic field, or electromagnetic field intensity data. Common formats for field distribution files include CSV and TXT. In EMCoS Studio, field distribution data can be employed by solvers such as the finite element method (FEM) and finite-

difference time-domain method (FDTD) to analyze field distribution characteristics and optimize designs.

3. CAD model files: EMCoS Studio supports 3D model files in various CAD formats, such as STEP (.stp), IGES (.igs), and Parasolid (.x_t), among others. Imported CAD models can be used to construct electromagnetic field models and provide geometric information to solvers like the finite element method (FEM) and finite-difference time-domain method (FDTD).

In the co-simulations of EMCoS Studio and HFSS, it is crucial to select the appropriate data format based on the specific problem and solver characteristics. Additionally, EMCoS Studio offers built-in data conversion capabilities to transform imported electromagnetic field data into formats suitable for specific solvers. This enables efficient interaction and processing of data.

2.6.3.2 Co-simulations of EMPIRE XPU with HFSS

EMPIRE XPU is a kind of advanced three-dimensional high-frequency electromagnetic simulation software based on the finite-difference time-domain (FDTD) method. It is widely utilized in various applications such as antenna design, microwave devices, radio frequency integrated circuits (RFIC), and photonic devices [61]. This powerful software offers an efficient FDTD solver, an intuitive graphical interface, robust post-processing capabilities, and high scalability.

One of the key features of EMPIRE XPU is its ability to simulate complex electromagnetic field structures accurately. It provides user-friendly two-dimensional and three-dimensional structural modeling functions, enabling users to rapidly construct electromagnetic field models. With EMPIRE XPU, users can calculate essential parameters like S-parameters, radiation patterns, and electromagnetic field distributions. This facilitates analysis and optimization design, while also meeting the diverse needs of users in different fields.

EMPIRE XPU also supports co-simulations with HFSS software, enabling the integration of their complementary advantages in calculation methods and optimization algorithms. This co-simulation approach enhances the accuracy and efficiency of electromagnetic field simulations. To perform successful co-simulations of EMPIRE XPU and HFSS, the following steps need to be followed:

1. Model preparation: Create the electromagnetic field structures to be simulated in both EMPIRE XPU and HFSS, ensuring consistency in geometric dimensions and material parameters between the two types of software.

2. Export/import data: Export simulation results from EMPIRE XPU, such as S-parameters and field distributions, to standard formats like Touchstone or CSV. Then, import these results into HFSS.

3. Set co-simulation parameters: Configure the co-simulation parameters in HFSS, including frequency range, grid division, boundary conditions, and other relevant settings. It is crucial to ensure these parameters align with the settings in EMPIRE XPU.

4. Run co-simulations: Initiate the simulation in HFSS while simultaneously reading the real-time exported data from EMPIRE XPU. To enhance simulation accuracy and convergence speed, optimization algorithms and iteration numbers can be adjusted in HFSS.

5. Analyze co-simulation results: Once the simulation is completed, analyze the co-simulation results obtained from both EMPIRE XPU and HFSS. Evaluate the benefits of co-simulations compared to simulations performed solely in EMPIRE XPU or HFSS.

When performing a co-simulation using EMPIRE XPU and HFSS, it is essential to ensure consistency in model parameters to prevent deviations in simulation results due to different settings. Additionally, setting co-simulation parameters such as mesh generation and boundary conditions appropriately can further enhance simulation accuracy and efficiency.

Now, let's consider a specific case of a co-simulation involving EMPIRE XPU and HFSS software for the design of a microstrip antenna:

1. Build the antenna model in EMPIRE XPU: Utilize the graphical interface of EMPIRE XPU to draw the antenna structure and set material parameters and boundary conditions according to the design requirements.

2. Create the antenna model in HFSS: Export the antenna model from EMPIRE XPU in STEP or IGES format and import it into HFSS, ensuring consistency between the models in both software.

3. Perform EMPIRE XPU simulation: Simulate the antenna in EMPIRE XPU, calculating important parameters such as S-parameters and radiation patterns. Export the simulation results to a standard format.

4. Import EMPIRE XPU data into HFSS: Import the simulation results (e.g., S-parameters) exported from EMPIRE XPU into HFSS.

5. Configure HFSS co-simulation parameters: Set the co-simulation parameters in HFSS, including frequency range, grid division, boundary conditions, and other relevant parameters.

6. Run HFSS co-simulations: Initiate the co-simulations in HFSS, simultaneously reading the simulation results from EMPIRE XPU in real-time during the simulation.

7. Analyze co-simulation results: After the co-simulations are completed, analyze the results obtained from both EMPIRE XPU and HFSS. Evaluate the advantages brought by the co-simulation approach. Based on the simulation results, optimize the antenna structure as necessary.

In practical applications, users can choose to utilize EMPIRE XPU, HFSS, or their co-simulations based on specific requirements. These tools provide strong support for electromagnetic field design, analysis, and optimization. It is important to note that the co-simulation case described in this context is just one typical application, and the co-simulations of EMPIRE XPU and HFSS can be applied to various other fields as well.

2.6.3.3 Co-simulations of Sonnet with HFSS

Sonnet is a kind of advanced electromagnetic simulation software that utilizes the transmission line matrix method (TLM) and finite-difference time-domain method (FDTD). Its primary purpose is the design and analysis of radio frequency (RF) and microwave circuits, with a particular focus on planar circuits and transmission line structures. The main functionalities of Sonnet encompass electromagnetic field analysis, transmission parameter analysis (S-parameter), impedance matching, filter design, antenna design, and more. Sonnet, in conjunction with HFSS, enables co-simulations and facilitates highly accurate high-frequency electromagnetic field analysis. Co-simulations involve converting the simulation results between Sonnet and HFSS, establishing an effective information transmission channel between the two types of software. In Sonnet, one designs the circuit and sets simulation parameters such as frequency range and grid division. The Sonnet simulation generates S-parameter results, which can then be exported as Touchstone format files. These Sonnet simulation results are imported into HFSS and utilized as boundary conditions. Subsequently, the corresponding simulation parameters, such as frequency range and grid division, are set in HFSS. The HFSS simulation produces the electromagnetic field distribution and other performance parameters. Finally, the HFSS post-processing tool is employed for data analysis and visualization. During the co-simulations, it is crucial to maintain consistent frequency range and grid division settings in both software packages, ensure that the Sonnet simulation results align with HFSS boundary conditions, and carefully verify the format and unit of import/export data.

As an example, let's consider the co-simulation of high-frequency electromagnetic fields using Sonnet and HFSS for microstrip antenna design:

1. Employ Sonnet software to design the microstrip antenna structure, encompassing components such as microstrip lines, feed points, and antenna radiation elements.

2. Configure Sonnet simulation parameters, specifying details like frequency range (e.g., 2.4–2.5 GHz) and grid division (e.g., 0.5 mm).

3. Run the Sonnet simulation to obtain the S-parameter result (specifically S_{11}) for the antenna.

4. Export the Sonnet simulation result as a Touchstone format file.

5. Create a three-dimensional model of the antenna in HFSS, incorporating microstrip lines, feed points, and antenna radiation elements.

6. Import the Sonnet simulation result into HFSS, utilizing it as the boundary condition for the port.

7. Set HFSS simulation parameters, specifying details like the frequency range (2.4–2.5 GHz) and grid division (e.g., 0.5 mm).

8. Establish the radiation boundary conditions and waveguide ports of the antenna within HFSS.

9. Run the HFSS simulation to obtain the performance parameters of the antenna, including the electromagnetic field distribution, gain, and radiation efficiency.

10. Utilize HFSS post-processing tools for data analysis and visualization, such as generating gain-frequency curves and electromagnetic field distribution maps.

Filters are key components in RF and microwave systems for selectively transmitting signals. In the filter design, the electromagnetic field distribution has an important influence on the performance. Co-simulations are an effective method to improve the design performance. Through co-simulations, we can more accurately predict the electromagnetic field distribution and optimize the filter design. Similarly, antenna arrays are widely used in modern communication and radar systems, and co-simulations can also help us to more accurately predict the performance of antenna arrays, such as radiation patterns and gains, and analyze the effects of mutual coupling for optimal design. In radio frequency integrated circuit (RFIC) design, we can evaluate the overall performance of FEM by analyzing parasitic effects, including gain and noise figure. Co-simulations provide us with a powerful tool to predict and optimize performance in the design of key components in RF and microwave systems.

2.6.3.4 Co-simulations of QuickWave with HFSS

QuickWave is a kind of electromagnetic simulation software that utilizes the finite-difference time-domain (FDTD) algorithm. It enables the simulation of materials and equipment performance in high-frequency electromagnetic field environments by solving Maxwell's equations [62]. This powerful software offers various functionalities, including antenna design, microwave device simulation, RF circuit analysis, and electromagnetic compatibility analysis. QuickWave boasts several key features, such as providing highly accurate and stable electromagnetic field calculation results, supporting multiple boundary conditions, material models, and vector signal sources, and utilizing multi-core, multi-CPU, and GPU-accelerated calculations to enhance simulation efficiency. Additionally, it facilitates simulation modeling and result visualization through a user-friendly graphical user interface (GUI) and supports simulation task automation through scripting languages.

To achieve multi-physical field calculations, QuickWave enables the co-simulation of high-frequency electromagnetic fields with HFSS software. The process involves establishing a simulation model in QuickWave and configuring relevant parameters, such as materials and boundary conditions. Similarly, a corresponding simulation model is created in HFSS, and the necessary parameters are set. The interface tool between QuickWave and HFSS facilitates the import of QuickWave's simulation results into HFSS. Subsequently, the imported data is post-processed in HFSS to complete the co-simulation calculation. Finally, the simulation results from both software are analyzed and compared to assess the accuracy and reliability of the co-simulations. It's crucial to maintain consistency in the simulation models of both software to ensure accurate simulation data. When importing data, attention should be paid to proper data format conversion to avoid errors.

To illustrate the specific simulation setup steps for analyzing the performance of a microwave filter in a high-frequency electromagnetic field environment, consider the following:

1. Create a three-dimensional model of the filter in QuickWave, specifying material properties and electromagnetic boundary conditions. 2. Set the signal source and define the operating frequency range and input power of the filter. 3. Perform simulation calculations in QuickWave to obtain electromagnetic field results such as S-parameters and transmission characteristics of the filter. 4. Save the simulation results from QuickWave in standard file formats (e.g., CSV, S2P). 5. Establish an identical three-dimensional model of the filter in HFSS and configure corresponding material properties and boundary conditions. 6. Utilize HFSS's import tool to import the simulation result file from QuickWave. 7. Set the signal source in HFSS, referencing the imported QuickWave data, and simulate the high-frequency electromagnetic field of the filter. 8. Analyze and compare the simulation results from HFSS with the calculation data from QuickWave to evaluate the filter's performance in the high-frequency electromagnetic field environment. 9. Based on the simulation results, optimize the filter's design to improve its performance.

The combination of QuickWave and HFSS software provides enhanced flexibility and accuracy for simulating high-frequency electromagnetic fields. This co-simulation approach empowers engineers and researchers to more effectively evaluate equipment performance, optimize design parameters, shorten product development cycles, and reduce costs. Furthermore, employing scripting languages to automate simulation tasks can further improve simulation efficiency and convenience.

2.7 Future Prospects of Simulations for HFEMFs

With the rapid advancements in science and technology, high-frequency electromagnetic simulation is assuming an increasingly significant role in various domains such as communication, radar, navigation, and electronic warfare. Going forward, it is poised to achieve crucial advancements, offering more precise, efficient, and dependable simulation methods for related fields. The key areas of focus for future developments in high-frequency electromagnetic simulation include the following.

2.7.1 Model Refinement and Simulation Accuracy

The advancement and refinement of high-frequency electromagnetic simulation require attention to several key aspects. Firstly, to improve the relevance and practicality of simulations, it is crucial to customize specialized simulation models and methods according to specific objectives and scenarios. In wireless communication, for instance, high-frequency electromagnetic simulation models can be tailored to adapt to diverse application scenarios, considering factors such as spectrum resources, propagation environment, and channel models. Similarly, in radar systems, simulation models for radar beamforming, signal processing, and target recognition can be optimized based on variables like targets, detection range, and weather conditions. Also, material properties and the electromagnetic environment significantly influence the accuracy of high-frequency electromagnetic simulations. To align the simulation model with real-world conditions, it is necessary to incorporate complex material parameters such as permittivity and permeability while also considering electromagnetic environment factors like terrain, buildings, and atmosphere. Additionally, parameter data for actual materials and electromagnetic environments can be obtained through experiments and measurements, providing more reliable input information for simulation models.

The accuracy and reliability of high-frequency electromagnetic simulation can be ensured through effective verification and evaluation methods. Comparative studies between simulations, experiments, and measurements are essential to assess the accuracy and applicability of the simulation model. Statistical analysis and error analysis can also be employed to evaluate the stability and reliability of simulation results. To comprehensively evaluate the performance of simulation, appropriate evaluation indices and standards, such as absolute error, relative error, and signal-to-noise ratio, can be formulated for different simulation tasks and application scenarios. Refining model parameters and optimization algorithms can enhance the accuracy of high-frequency electromagnetic simulation. Intelligent optimization methods, such as genetic algorithms and particle swarm optimization algorithms, can automatically adjust simulation model parameters to achieve better results. Moreover, optimization algorithms can balance accuracy, computational efficiency, and stability according to specific application requirements, facilitating comprehensive optimization of simulation performance. Uncertainty analysis and sensitivity analysis are crucial approaches for comprehensive evaluation of the accuracy of high-frequency electromagnetic simulation. Uncertainty analysis focuses on assessing the impact of various uncertainties in the simulation model on the simulation results, including parameter errors, model assumptions, and measurement errors, to determine the credibility of the simulation results. Sensitivity analysis, on the other hand, explores the sensitivity of simulation results to model parameters and input data, guiding the optimization and adjustment of simulation models. Through continuous and in-depth research, high-frequency electromagnetic field simulation will better meet the practical needs of communication, radar, navigation, electronic warfare, and other fields, providing robust support for scientific research and engineering applications.

2.7.2 Computational Efficiency and Simulation Speed

With the advancements in high-performance computing platforms, we can significantly enhance the processing capacity of simulation tasks and reduce simulation time through the utilization of supercomputers, cluster computing, and cloud computing resources. By optimizing the allocation and scheduling strategies of computing resources, we can improve the utilization of these resources and enhance system performance. Furthermore, the utilization of parallel computing and accelerators is crucial. Accelerators like multi-core processors, graphics processing units

(GPUs), and programmable gate arrays (FPGAs) can greatly expedite the processing speed of simulation tasks. By designing parallel algorithms and data structures that align with the characteristics of different simulation tasks, we can achieve efficient task decomposition, data transmission, and result integration. Additionally, utilizing accelerators to expedite specific computing tasks in hardware can further enhance simulation speed.

Algorithm optimization and adaptive algorithms also play a key role in improving computational efficiency. By carefully selecting or designing appropriate computational methods, such as fast Fourier transform (FFT) and multiple boundary element method (MBE), we can reduce the computational complexity for various simulation tasks and scenarios. Simultaneously, adaptive algorithms dynamically adjust grid settings, time steps, and other parameters based on actual computing requirements, enabling dynamic resource allocation and load balancing. Data compression and storage optimization can reduce the costs associated with data transmission and storage during simulations. Employing lossless or lossy compression allows for the compression of simulation data, reducing data volume and improving data transmission speed. Additionally, optimizing storage strategies, such as implementing distributed storage and cache technologies, helps to decrease data access latency and enhance computing efficiency. Efficient pre-processing and post-processing also play a vital role in improving simulation speed. Pre-processing, including mesh generation and parameter extraction, provides accurate and efficient input data for the simulation process. During the post-processing stage, data mining and visualization facilitate rapid analysis and interpretation of simulation results, providing decision-making support and optimization capabilities. Through the use of automatic and intelligent pre-processing and post-processing methods, manual intervention and errors are minimized, ultimately enhancing simulation speed.

Model simplification and multi-scale simulation are important approaches for improving computational efficiency while maintaining simulation accuracy. Model simplification methods, such as geometric simplification and parameter simplification, reduce computational complexity during the simulation process. Multi-scale simulation employs models with varying levels of accuracy at different scales, tailored to the specific requirements of simulation tasks, enabling efficient utilization of computing resources.

2.7.3 *Artificial Intelligence and Machine Learning for Simulations of HFEMFs*

The application of artificial intelligence (AI) and machine learning (ML) in high-frequency electromagnetic simulations has gained a significant attention due to their rapid development. This integration holds great potential for enhancing simulation efficiency, accuracy, and adaptability. Several important aspects can benefit from the integration of AI and ML.

Intelligent optimization algorithms based on AI and ML, such as genetic algorithms, particle swarm optimization, and simulated annealing, can automatically optimize model parameters and allocate computational resources. This optimization improves the effectiveness of simulations. By learning from historical simulation data and results, intelligent optimization algorithms can accelerate convergence and enhance optimization accuracy. The utilization of artificial neural networks (ANN) and deep learning (DL) enables more efficient data processing and information extraction. For instance, convolutional neural networks (CNN) can automatically extract electromagnetic field distribution characteristics, thereby enhancing the accuracy and real-time performance of radar target recognition. Additionally, recurrent neural networks (RNN) and long short-term memory networks (LSTM) can analyze and predict time-series electromagnetic field data, improving channel modeling accuracy and propagation environment simulation. The introduction of reinforcement learning (RL) provides intelligent decision support for high-frequency electromagnetic simulation. By designing appropriate state space, action space, and reward functions, RL algorithms continuously learn and optimize during the simulation process, enabling adaptive control and strategy optimization of the electromagnetic environment. In the field of electronic warfare, RL can optimize intelligent jamming and countermeasure strategies.

Furthermore, AI and ML facilitate the development of data-driven high-frequency electromagnetic simulation models and methods. By learning from a vast amount of experimental and measured data, data-based electromagnetic environment modeling, material characterization, and channel modeling can be achieved, leading to improved simulation accuracy and fidelity. The data-driven approach allows adaptive adjustment of model parameters during the simulation process, enhancing simulation efficiency. Similarly, model fusion and ensemble learning can integrate multiple AI and ML-based high-frequency electromagnetic simulation models, comprehensively improving simulation performance. Combining different models, such as neural networks, support vector machines (SVMs), and decision trees, help construct more powerful and stable integrated models, reducing the risks of uncertainty and overfitting associated with individual models. The incorporation of adaptive and online learning empowers the high-frequency electromagnetic simulation system with real-time learning and adaptive capabilities. By continuously collecting real-time simulation data and updating model parameters and structures, online learning and real-time optimization are achieved. Adaptive and online learning enables the simulation system to maintain high accuracy and flexibility in the face of complex and evolving electromagnetic environments.

2.7.4 *Standardization and Interoperability*

Standardization and interoperability have become increasingly vital in the field of high-frequency electromagnetic simulation. Standardization reduces compatibility issues between different simulation systems, improving simulation efficiency and reliability. Interoperability enables cooperative work among various simulation systems and expands the application scope of simulation. Standardization establishes a unified specification for high-frequency electromagnetic simulation models and algorithms. By creating common standards, it ensures consistency across different simulation systems, reduces redundant development efforts, and lowers research and development costs. Additionally, standardization promotes optimization and innovation of models and algorithms, leading to enhanced performance and accuracy of simulation.

Standardization and interoperability facilitate data sharing and exchange. Data is a crucial element in high-frequency electromagnetic field simulation. By establishing unified data formats and interface standards, researchers can conveniently utilize various simulation data, accelerating the research process and improving efficiency. Interoperability is key to realizing the integration of simulation systems. By promoting interoperability in high-frequency electromagnetic simulation, different simulation subsystems can collaborate effectively, enhancing the integration capabilities of simulation systems. Interoperability also simplifies system expansion, upgrades, and maintenance, reducing associated costs. Moreover, standardization and interoperability support multi-domain co-simulations, enabling collaboration between high-frequency electromagnetic field simulation and other related fields. This broadens the application scope of simulation in communication, radar, navigation, and electronic warfare and fosters innovation and development. Standardization and interoperability contribute significantly to international cooperation and exchange. By promoting international standardization of high-frequency electromagnetic simulation, global technical exchanges and collaboration are facilitated. This raises the research and application capabilities of countries in related fields while establishing an international consensus and improving global electromagnetic environmental protection and management.

Lastly, standardization and interoperability play a critical role in information security and privacy. In the application of high-frequency electromagnetic simulation, concerns about information security and privacy are paramount. Standardization and interoperability address these concerns by implementing unified security standards and protection measures. This reduces the risk of information leakage and abuse during the transmission and storage of simulation data. Furthermore, they contribute to building a secure and reliable simulation environment, ensuring the normal operation and application of simulation.

2.8 Co-simulations of Designer with HFSS for HFEMFs

When dealing with large-scale design problems that require accurate simulations, there is often a trade-off between speed and accuracy for the simulations of HFEMFs. While traditional simulation methods are typically based on equivalent circuits, they may not provide the desired level of accuracy. Therefore, there is a need for a practical solution that combines speed, electromagnetic accuracy, and the ability to handle problems of any scale. In addition to ensuring solution accuracy, an kind of effective engineering design software should also offer an efficient and user-friendly design process. The concept of "field-circuit combination and collaborative simulation" addresses this requirement in real-world engineering scenarios. The figure below illustrates the solution offered by Ansys, which involves the co-simulation design of high-frequency electromagnetic fields for complex passive devices using HFSS and Designer:

Firstly, the design process begins by dividing a complex passive device into several basic units. Each basic unit is then modeled in three dimensions using HFSS for electromagnetic simulation and parametric scanning. The purpose of parametric scanning is to gather essential data for the subsequent automatic optimization design. By appropriately dividing the basic units, their structures become simpler and their electrical sizes smaller. Consequently, achieving convergent simulation results in HFSS becomes easier. This step maximizes the accuracy of field simulation to provide precise foundational data for later stages. Next, the basic units in HFSS are connected to the circuit design schematic in Designer using field-to-path

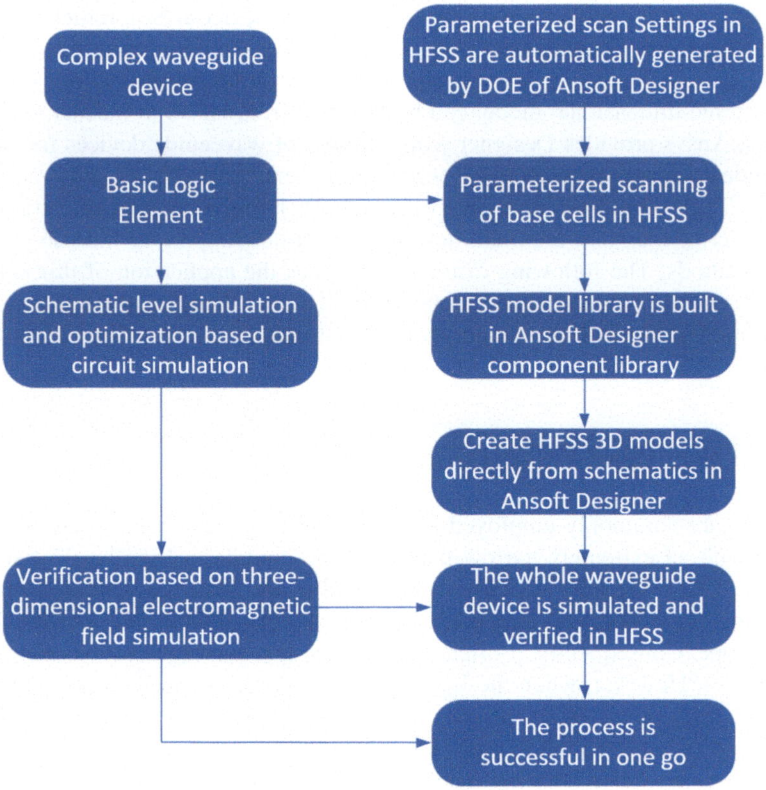

Fig. 2.3 Solution and process for co-simulations of Designer with HFSS

"co-simulations." This approach enables the completion of the guided wave characteristics of the entire complex device through circuit simulation, with the basic units in HFSS representing the elements in the circuit schematic diagram (Fig. 2.3).

However, if "co-design" only relies on circuit simulation based on pseudo-S-parameter files, it can only serve as a means of design verification rather than actual design. When the simulation results fail to meet the design targets, the model cannot be optimized. At the circuit simulation level, only the S-parameters of the basic components are retained, while all the three-dimensional structural information is lost. Consequently, it becomes impossible to optimize the entire structure at the circuit level, especially when different electromagnetic field simulators and circuit simulators from various manufacturers are employed. The greatest advantage of the Ansys co-design software package lies in its inclusion of both the powerful 3D electromagnetic simulation tool HFSS and the circuit simulation tool Designer. When the basic units in HFSS are inserted into the circuit design schematic of Designer as components, all variables (such as dimensions and material properties) and parametric scan results can be dynamically linked, providing fundamental data for optimization design based on circuit simulation. In Designer, even the points without parametric scan in HFSS can be obtained through interpolation algorithms during the optimization process, enabling direct optimization of the entire device at the circuit level. Once the optimization of the entire device is completed at the circuit level, the schematic diagram can be directly exported to HFSS for verification using a script. Additionally, the three-dimensional model can be exported from HFSS to mechanical CAD software. Ansys provides Designer with a library of waveguide devices for several waveguide device types, such as Iris waveguide filters, cavity filters, and branch line couplers. This component library not only includes HFSS models of all components but also script files that facilitate design automation, supporting the use of DOE design methods. The following example illustrates the application of this solution in the design of complex passive components for high-frequency electromagnetic fields using HFSS and Designer co-simulation design.

2.8.1 Iris Waveguide Filter

Iris filters are commonly employed in the design of narrow-band filters [63, 64]. A filter with an extremely narrow bandwidth necessitates a high level of solving accuracy. This is because the operational band of the filter tends to shift slightly towards higher frequencies after each adaptive solution, and the S_{12} curve exhibits a steep slope. Consequently, it often requires multiple solving iterations to achieve convergence. However, when the Iris waveguide filter is segmented into units, the frequency response of each unit does not exhibit band-pass characteristics. Hence, convergence is not a concern in this case. Employing a field-road solution for this design offers cost-effectiveness. The detailed design steps can be summarized as follows:

2.8.2 *Modeling and Simulations in HFSS*

The following figure, as shown in Fig. 2.4, depicts a typical Iris waveguide filter [65]. Based on this configuration, we can easily decompose it into two fundamental units. The fundamental unit represents the IRIS component of the filter, which should only include the IRIS diaphragm section in terms of electrical characteristics. However, due to the issue of higher-order modes, it is necessary to include two consecutive waveguides during the modeling process. The de-embed function of the waveport in HFSS can be used to shift the reference plane of the S-parameters to the root of the IRIS diaphragm.

The waveguide adopts the WR-90 standard waveguide, and the fillet on the side is created using the Fillet function in HFSS: select the edges that need to be filleted (multiple edges can be selected at once), go to 3D Modeler > Fillet. Next, we will model and solve for these two basic units. First, the waveguide structure can be solved using the "Driven Model" solving type. The fillet rounding process is used at the waveguide insertion diaphragm. At these fillet locations, corresponding manual mesh divisions can be used. Select the waveguide structure, right-click and choose "Assign Mesh Operation" and then "Surface Approximation," and specify the "Normal deviation" as 5° (18 surfaces in total, with each angle being 90°). The solution settings for the waveguide component can be done according to the following steps: 1. Insert a new solution setup. Taking this waveguide filter as an example, if the desired frequency band for sweeping is 8–12 GHz, select 12 GHz as the solution frequency.

2. In order to avoid cumulative errors with simulations in Designer, a higher solution accuracy should be set in HFSS:

–Maximum Number of Passes: 20
–Maximum Delta S: 0.01
–Maximum Refinement Per Pass: 30% of the previous mesh
–Minimum Number of Passes: 3
–Port Field Accuracy: 0.1%

3. Sweep settings: Considering the subsequent collaborative simulation with Designer, each solution setup must contain only one sweep setting (when using HFSS alone, there is no limitation on the number of sweep settings). The sweep setting should use the default name "Sweep1". For waveguide structures, a fast sweep

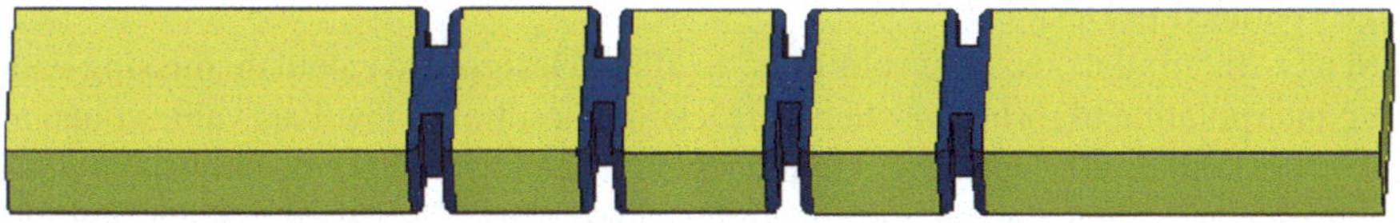

Fig. 2.4 Iris waveguide bandpass filter

is recommended. Only when the simulation frequency range reaches the cut-off frequency of the waveguide, an interpolated sweep is recommended. Deselecting "Save Fields" reduces the disk space requirement.

4. Port settings: Use wave ports and solve only for the fundamental mode (main mode). Define integral lines on both ports of each basic unit to ensure that the calculated electric field phases do not differ by 180°. By doing this, only the main mode is solved at each port. We know that the presence of IRIS in the waveguide will generate higher-order modes (evanescent modes). However, these modes decay rapidly. When the wave port is sufficiently far from the iris, these higher-order modes have attenuated to a small magnitude before being reflected back to the port (usually below –20 dB), making them negligible in engineering terms. In the basic unit of the IRIS, the waveguide length can be set as 2a, where a is the width of the waveguide cross-section. In this example, we simplify the waveguide walls as ideal conductors, so there is no need to depict the waveguide walls. HFSS will automatically assume the simulated object's interface with the background as a Perfect Electric Conductor (PEC), neglecting metal losses. If there are two IRIs placed closely together in the structure, in such cases, except for the main mode, a few of the higher-order modes may not have attenuated sufficiently before reaching the wave port. In that case, you need to simulate both IRIS together as one basic unit in HFSS.

2.8.3 Dynamic Link Between HFSS and Designer

After the solution and parameterized sweep of basic units in HFSS are completed, you can proceed to establish a dynamic link between HFSS and Designer. In Designer, insert a circuit design by selecting "Project > Insert Circuit Design" and choose "None" for the layout. In the project manager, right-click on the circuit and insert an HFSS subcircuit. This way, the HFSS design will be inserted into Designer's circuit design as an N-port element.

Next, you will see a window for dynamic connection settings called "Dynamic NPort import." In the "File" field, browse and select the HFSS basic unit project that needs to be linked. In the "Design" field, choose the specific design because an HFSS project can manage multiple designs simultaneously. (Tip: You can store the designs of all HFSS basic units in a single HFSS project and give each basic unit's design a different name to facilitate searching.) In the "Solution" field, when a design contains multiple solution setups (e.g., solving multiple frequency ranges), you can select a specific solution and frequency sweep settings based on the frequency scan range you need in Designer.

Select "Interpolate existing solutions" to allow Designer to calculate missing data using interpolation algorithms when the Designer's solution involves values outside the HFSS simulation results (e.g., frequency points in linear network analysis or optimized variable values during optimization in Designer). This feature is particularly useful for frequency sweeps and optimizations. Alternatively, you can select "Simulate missing solutions," which automatically triggers HFSS's solver when encountering unsolved data. The interpolation algorithm also includes advanced settings, such as selecting the corresponding interpolation algorithm based on different curve characteristics, but the details of this part are beyond the scope here. Interested designers can refer to the relevant content in the online help. Otherwise, the default settings should be sufficient.

In the "Information" section, information about the number of pins is provided, usually matching the number of ports in HFSS. In the "Parameter" section, you can see all the variables defined in HFSS, which can be passed to Designer for optimization.

Another type of basic unit is an open-ended waveguide, which can be used to adjust the spacing between adjacent IRIS. There are two methods to implement this open-ended waveguide. The first method, which everyone can think of, is to create a two-port waveguide and parameterize its length for a sweep. This serves as the basic data for optimization in Designer, but it will not be further elaborated here.

We will focus on introducing a more convenient and efficient method: utilizing the "transmission line model" to easily connect a uniform transmission line to one end. When an HFSS design is linked to Designer, not only the characteristic impedance of the port but also the propagation coefficients of the port mode (including attenuation and phase shift factors) are transferred to Designer. For a uniform transmission line, as shown in Fig. 2.5, in HFSS, only one port needs to be set, and Designer will calculate the S-parameters of the transmission line for different lengths based on the characteristic impedance and wave propagation coefficients of this port.

During this process, the post-processing engine of HFSS is continuously called without the need for re-solving, resulting in a very fast speed. Once the dynamic linking is set up, the basic unit can be easily copied multiple times using simple Ctrl+C and Ctrl+V commands. For example, if we want to design a filter containing four IRSs, after placing the components, we add the ports as shown in Fig. 2.6.

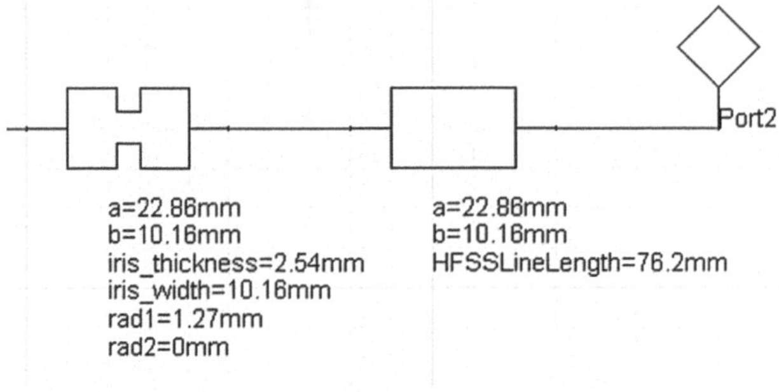

Fig. 2.5 Port setup for calculation in HFSS and Designer

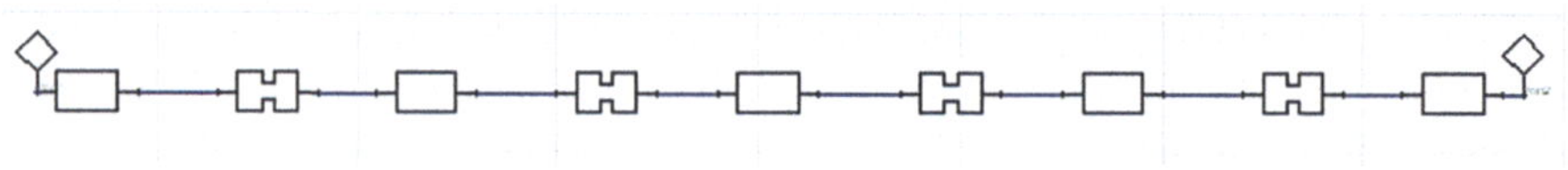

Fig. 2.6 Four-units IRS filters

2.8.4 *Solution by Designer*

The setup process in Designer is much simpler compared to HFSS. To begin, include a "Linear Network Analysis" and specify the scanning range from 8.2 GHz to 12 GHz. The simulation results for S_{11} and S_{21} obtained from the initial design are displayed below. It is evident that the desired band-pass characteristics of the filter have not been achieved, highlighting the need for further optimization (Fig. 2.7).

2.8.5 *Optimization in Designer*

After transferring the variables from HFSS to Designer, they are displayed as properties of the basic unit. Before tuning, it is necessary to reset these variables. Let's consider the optimization of the distance between IRIs as an example. In Fig. 2.8, the iris band-pass filter structure exhibits symmetrical characteristics. We can define the lengths of the five hollow waveguides as variables: L_0, L_1, and L_2, as shown below. It is important to note that the length of L_0 is independent of the bandpass characteristics, so our optimization efforts will focus solely on L_1 and L_2.

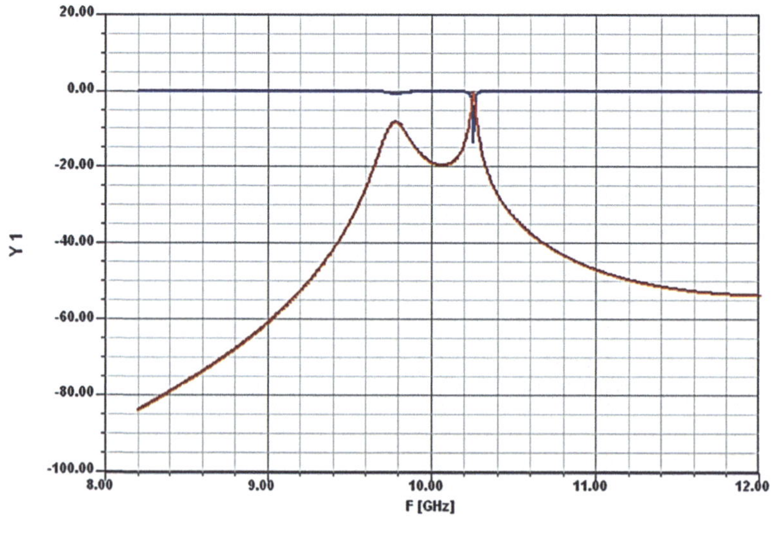

Fig. 2.7 Simulation results of S parameters for the filter

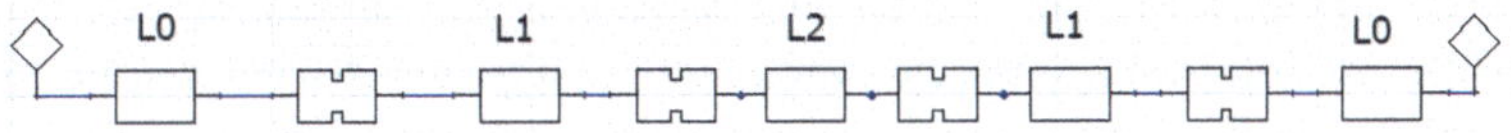

Fig. 2.8 Definition of variables based on symmetric characteristics of filter structure

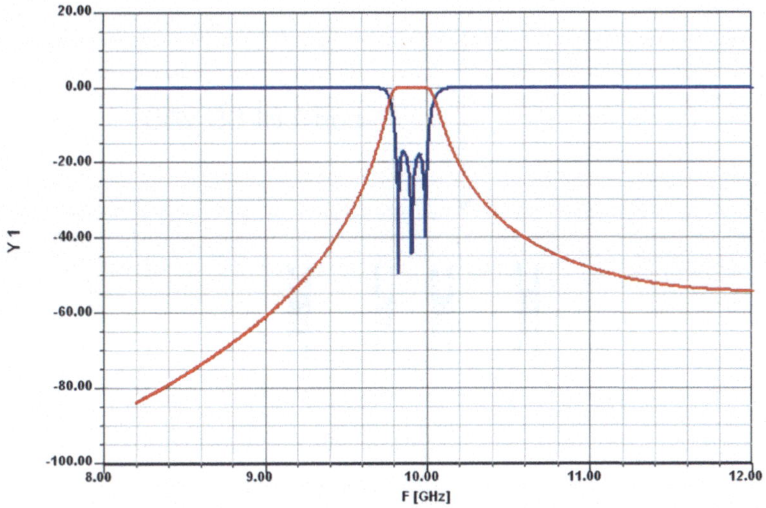

Fig. 2.9 Simulation results of optimized S parameters for the filter

Before establishing the optimization objective, it is necessary to choose the variables that will be involved in the optimization process. Referring to the diagram below, we observe that L_1 and L_2 have been selected as "local variables" under the "design properties" category. Furthermore, it is essential to assign suitable initial values and value ranges to L_1 and L_2 in order to enhance the optimization process in terms of efficiency and effectiveness (Fig. 2.9).

2.8.6 *Export to HFSS for Verification*

The schematic diagrams generated in Designer through dynamic linking with HFSS can be directly output to HFSS as a "ready to solve" design using a script program. Designer's waveguide library includes such script programs, creating a seamless loop for engineering design. In Designer's menu bar, go to Tools>Run Script and execute the ExportToHFSS.vbs script (which needs to be used in conjunction with the waveguide library). This will directly lay out the Iris filter schematic from Designer to HFSS. Next, we can compare the simulation results of the entire filter between HFSS and Designer (Fig. 2.10).

As shown in Fig. 2.11, the dashed line represents the simulation results from HFSS, while the solid line represents the simulation results from Designer. It can be observed that both simulation results are highly consistent, whether in terms of passband range, out-of-band suppression, or ripple magnitude.

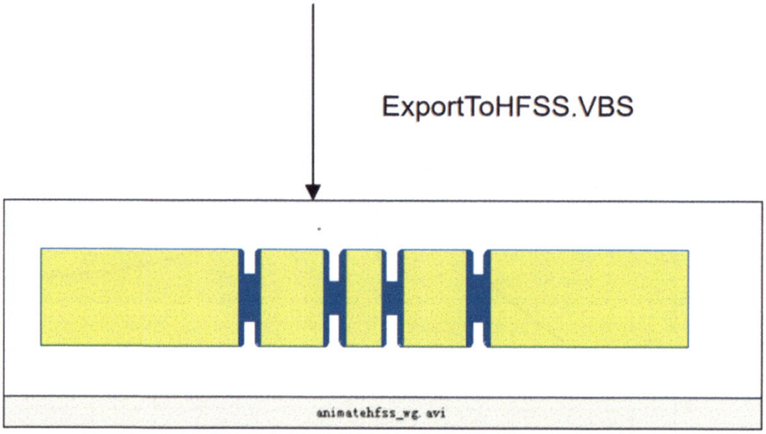

Fig. 2.10 Schematic diagram output to HFSS via script program

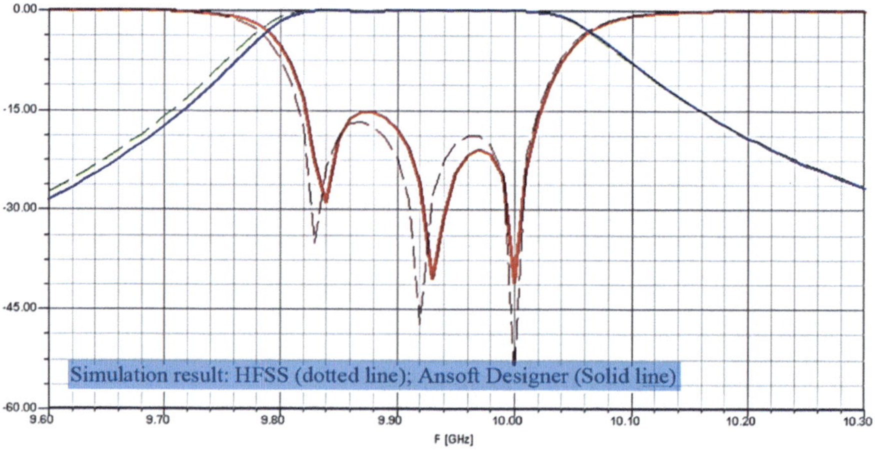

Fig. 2.11 Comparative analysis of co-simulation results in HFSS and Designer

Problems

P2.1 ADS co-simulation with Sonnet for 2–18 GHz power divider

1. Design Steps
   Confirmation of impedance transformer stages and their respective impedances, confirmation of isolation resistor values, electromagnetic simulation of individual impedance transformers, electromagnetic simulation of T-sections, and co-simulation of the entire power divider.
2. Impedance Transformer Stages and Parameter Extraction
   The corresponding parameters are obtained through modeling and simulation in ADS. The effective dielectric constant can be approximately obtained using transmission line calculation tools, with an initial length of 1/4 wavelength of the center frequency. The other impedance values are arranged from high to low, and simple optimization is performed to obtain the impedance values for each stage.
3. Isolation Resistor Simulation
   Based on the results from the second step, a complete theoretical model of the power divider is established in ADS. The transmission line parameters are fixed, and an isolation resistor is added. By providing an initial value and optimizing the standing wave at the 2-port once, the numerical value of the isolation resistor can be conveniently obtained.
4. Electromagnetic Simulation of Impedance Transformers and T-sections
   Using the calculated parameters from the previous steps, electromagnetic simulation is performed in Sonnet. Since the model is large and the simulation time is long, it can be divided into two parts for separate simulation and optimization to reduce resource consumption. The two divided circuits are the impedance transformer from 100 to 50 Ω and the T-section from 50 Ω to two 100 Ω terminations. Based on the impedance transformer dimensions calculated in step (2), an impedance transformer model is constructed (bending is applied to reduce the length, and the bending angles can be calculated independently).
5. Overall Simulation of the Power Divider Model
   The impedance transformers and T-sections are combined to form the complete power divider for simulation, using the netlist circuit simulation in Sonnet. The sub-circuits from step (4) are interconnected for simulation. This method can be applied to other large circuits by dividing them into smaller circuits and then combining them using netlist, which accelerates the simulation and optimization process. The results of the netlist circuit simulation are compared with the results of the full circuit simulation to verify the correctness of circuit segmentation and the accuracy of the entire power divider simulation process. Finally, with the addition of resistors, the overall simulation of the power divider is essentially completed.

P2.2 Conduct a co-simulation using Ansys HFSS and EMPIRE XPU
Import an existing Ansys HFSS simulation model of your choice into EMPIRE XPU, ensuring that the geometries and materials are retained. Perform simulations in both

Ansys HFSS and EMPIRE XPU, and then conduct a comparative analysis of the results obtained from both simulation environments. Discuss any discrepancies and provide insights into the potential advantages and limitations of each tool for the specific model you have chosen.

References

1. Krathwohl DR (1993) Methods of educational and social science research: an integrated approach. Longman/Addison Wesley Longman
2. Wilson EB (1990) An introduction to scientific research. Courier Corporation
3. Bauer I et al (2000) Numerical methods for optimum experimental design in DAE systems. J Comput Appl Math 120(1–2):1–25
4. Villegas FR (2007) Experimental number theory, No 13. Oxford University Press
5. Christensen LB et al (2011) Research methods, design, and analysis
6. Carrie W (2007) Research methods. J Bus Econ Res (JBER) 5(3)
7. Bilbao S (2004) Wave and scattering methods for numerical simulation. Wiley
8. Beer G (2015) Advanced numerical simulation methods: from CAD data directly to simulation results. CRC Press
9. Kennard RW, Stone LA (1969) Computer-aided design of experiments. Technometrics 11(1):137–148
10. Yu W (2009) Electromagnetic simulation techniques based on the FDTD method. Wiley
11. Paul CR, Scully RC, Steffka MA (2022) Introduction to electromagnetic compatibility. Wiley
12. Hannan P, Balfour M (1965) Simulations of a phased-array antenna in waveguide. IEEE Trans Antennas Propag 13(3):342–353
13. Balaguru S, Selvam R, Desai MS, Negra R (2012) Investigation of the spinal cord as a natural receptor antenna for incident electromagnetic waves and possible impact on the central nervous system. Electromagn Biol Med 31(2):101–111
14. Ansys (2019) High frequency electromagnetic field simulation. Accessed from (2019)
15. Sahan N, Inal ME, Demir S, Toker C (2008) High-power 20–100-MHz linear and efficient power-amplifier design. IEEE Trans Microw Theory Tech 56(9):2032–2039. https://doi.org/10.1109/TMTT.2008.2002238
16. Zheng Z, Li D, Tan X, Chen Q (2023) Single-layer Dual-/Tri-band SIW Filtenna based on multifunctional cavity-backed slots. IEEE Trans Antennas Propag 71(5):4498–4503. https://doi.org/10.1109/TAP.2023.3242110
17. Hamed A, Saeed M, Negra R (2020) Graphene-based frequency-conversion mixers for high-frequency applications. IEEE Trans Microw Theory Tech 68(6):2090–2096. https://doi.org/10.1109/TMTT.2020.2978821
18. Zhou Y, Zhao G, Li XY, Tong MS (2022) A liquid-metal-based crossed-slot antenna with polarization and continuous-frequency reconfiguration. IEEE Open J Antennas Propag 3:1102–1108. https://doi.org/10.1109/OJAP.2022.3206899
19. Li XY, Jiang D, Liu J, Tong MS (2022) A Ka-band multilayer beaming-scanning antenna using liquid crystals. IEEE Antennas Wirel Propag Lett 21(1):44–48. https://doi.org/10.1109/LAWP.2021.3116661
20. Grosdidier S, Baussard A, Khenchaf A (2010) HFSW radar model: simulations and measurement. IEEE Trans Geosci Remote Sens 48(9):3539–3549. https://doi.org/10.1109/TGRS.2010.2047022
21. Frangi AF, Tsaftaris SA, Prince JL (2018) Simulations and synthesis in medical imaging. IEEE Trans Med Imaging 37(3):673–679
22. Josephs O, Henson RNA (1999) Event-related functional magnetic resonance imaging: modelling, inference and optimization. Philosoph Trans R Soc Lond Ser B Biol Sci 354(1387):1215–1228

23. Cangellaris AC (1996) Frequency-domain finite element methods for electromagnetic field simulations: fundamentals, state of the art, and applications to EMI/EMC analysis. In: Proceedings of symposium on electromagnetic compatibility
24. Hirano T, Itoh K, Tsunekawa K (2012) Accuracy investigation of de-embedding techniques based on electromagnetic simulations for on-wafer RF measurements. In: Numerical simulations-from theory to industry, pp 233–258
25. Cheng X, Yan H, Hu Y (2015) Electromagnetic interference analysis of integrated radar antennas on stealth mast. In: 2015 7th Asia-Pacific Conference on Environmental Electromagnetics (CEEM)
26. Zhang HH, Yang HF, Liu XF, Wang XF, Chen ZW (2020) Electromagnetic-circuital-thermal multiphysics simulations method: a review. Prog Electromagn Res 169:87–101
27. Yam CY, Tam PHT, Mok PKT, Wong SP (2011) Multiscale quantum mechanics/electromagnetics simulations for electronic devices. Phys Chem Chem Phys 13(32):14365–14369
28. Pena B, Huang L (2021) A review on the turbulence modelling strategy for ship hydrodynamic simulations. Ocean Eng 241:110082
29. Nikolova NK, Bandler JW, Bakr MH (2004) Adjoint techniques for sensitivity analysis in high-frequency structure CAD. IEEE Trans Microw Theory Tech 52(1):403–419
30. Zhang N, Wu YM, Yang Y (2017) Recent advances on the high frequency electromagnetic modeling and simulation techniques. In: 2017 International applied computational electromagnetics society symposium (ACES)
31. Zhao K, Petersson LR (2018) Overview of hybrid solver in HFSS. In: 2018 IEEE international symposium on antennas and propagation & USNC/URSI national radio science meeting
32. Sullivan DM (2013) Electromagnetic simulations using the FDTD Method. Wiley
33. Hajimiri A (2009) The future of high frequency circuit design. In: 2009 Proceedings of ESSCIRC
34. Andersh D et al (2000) Xpatch 4: the next generation in high frequency electromagnetic modeling and simulation software. In: Record of the IEEE 2000 international radar conference [Cat. No. 00CH37037]
35. White JF (2004) High frequency techniques: an introduction to rf and microwave design and computer simulation. Wiley
36. LeVeque R (1998) Finite difference methods for differential equations
37. Lee J-F, Lee R, Cangellaris A (1997) Time-domain finite-element methods. IEEE Trans Antennas Propag 45(3):430–442
38. Reddy JN (2019) Introduction to the finite element method. McGraw-Hill Education
39. Dyczij-Edlinger R, Peng G, Lee J-F (1998) A fast vector-potential method using tangentially continuous vector finite elements. IEEE Trans Microw Theory Tech 46(6):863–868
40. Jin J-M (2015) Theory and computation of electromagnetic fields. Wiley
41. Weile DS et al (2004) A novel scheme for the solution of the time-domain integral equations of electromagnetics. IEEE Trans Antennas Propag 52(1):283–295
42. Lynch DR, Paulsen KD (1990) Time-domain integration of the Maxwell equations on finite elements. IEEE Trans Antennas Propag 38(12):1933–1942
43. Chadebec O, Coulomb J-L, Janet F (2006) A review of magnetostatic moment method. IEEE Trans Magn 42(4):515–520
44. Monorchio A et al (2004) A hybrid time-domain technique that combines the finite element, finite difference and method of moment techniques to solve complex electromagnetic problems. IEEE Trans Antennas
45. Huray PG (2011) Maxwell's equations. Wiley
46. Sumithra P, Thiripurasundari D (2017) Review on computational electromagnetics. Adv Electromagn 6(1):42–55
47. Holland R (1983) Finite-difference solution of Maxwell's equations in generalized nonorthogonal coordinates. IEEE Trans Nucl Sci 30(6):4589–4591
48. Grigoriev AD (2017) Computer simulations of high-frequency electromagnetic fields. Comput Simul 193

49. Mittra R (2016) Computational electromagnetics. Springer
50. Monk P (2003) Finite element methods for Maxwell's equations. Oxford University Press
51. Yee KS, Chen JS (1997) The finite-difference time-domain (FDTD) and the finite-volume time-domain (FVTD) methods in solving Maxwell's equations. IEEE Trans Antennas Propag 45(3):354–363
52. Bernardi C, Maday Y (1997) Spectral methods. In: Handbook of numerical analysis, vol 5, pp 209–485
53. Guo B (1998) Spectral methods and their applications. World Scientific
54. Johnson SG, Joannopoulos JD (2001) Block-iterative frequency-domain methods for Maxwell's equations in a planewave basis. Opt Express 8(3):173–190
55. Buffa A, Costabel M, Schwab C (2002) Boundary element methods for Maxwell's equations on non-smooth domains. Numer Math 92(4):679–710
56. Zhu JZ, Zienkiewicz OC (1988) Adaptive techniques in the finite element method. Commun Appl Numeric Methods 4(2):197–204
57. Hatamzadeh-Varmazyar S (2015) The error analysis and convergence evaluation of a computational technique for solving electromagnetic scattering problems. Adv Comput Techn Electromagn 66–69
58. Smith JT (1996) Conservative modeling of 3-D electromagnetic fields, Part I: properties and error analysis. Geophysics 61(5):1308–1318
59. CST (2002) CST microwave studio. Darmstadt, Germany
60. COMSOL Multiphysics (1998) Introduction to COMSOL multiphysics. Burlington, MA. (9 Feb 2018)
61. Simon W et al (2018) CAD data selection and simplification techniques for EM automotive simulations using EMPIRE XPU. In: 2018 IEEE international symposium on antennas and propagation & USNC/URSI national radio science meeting
62. Morrison DG (1995) Quickwave: an impulse algorithm to calculate preliminary design shear and moment envelopes for compliant towers, pp 85–90
63. Rezaee M, Zaman AU, Kildal P-S (2015) A groove gap waveguide iris filter for V-band application. In: 2015 23rd Iranian conference on electrical engineering
64. Papziner U, Arndt F (1993) Field theoretical computer-aided design of rectangular and circular iris coupled rectangular or circular waveguide cavity filters. IEEE Trans Microw Theory Tech 41(3):462–471
65. http://www.mweda.com/HFSS/917.html

Chapter 3
Co-simulations of Microwave Circuits

3.1 Introduction

Microwave circuits, typically operating within the frequency range from 300 MHz to 300 GHz [1, 2], are the fundamental electronic circuits in radio frequency (RF) and microwave devices and systems [3]. Due to their high-frequency nature, the microwave circuits exhibit several unique characteristics and requirements in terms of materials, structures, circuit configurations, components, and design methodologies when compared to conventional low-frequency digital circuits. As an independent and specialized field, the technologies of microwave circuits have been significantly improved in theories, materials, processes, components, and design methods in recent years, and their applications have also been greatly expanded.

Moreover, with the rapid progress of technologies on large-scale integrated circuits, the operating speeds of contemporary chips have surpassed 1.0 GHz. Consequently, the design techniques for microwave circuits have been mostly updated and extensively employed in the designs of chips and packaging structures with high-speed circuits. The boundaries between traditional low-frequency circuits, digital circuits, and microwave circuits have become increasingly blurred, leading to a greater interdependence and mixture of these circuits and their related technologies. The microwave technologies have served as fundamental components for other technologies like wireless communication technologies and found broad applications in the fields such as wireless communications, radar navigation, remote sensing, geophysical exploration, medical imaging, and wireless power transfer. In these domains, the microwave circuits and related technologies play a pivotal role in designing some key electronic devices and systems, encompassing amplifiers, mixers, filters, antennas, couplers, and impedance matching networks. This chapter intends to present up-to-date simulation technologies of microwave circuits for readers so that they can well grasp the basic knowledge of this area which may be needed in practical applications.

M. S. Tong and X. Y. Li, *Co-simulations of Microwave Circuits and High-Frequency Electromagnetic Fields*, https://doi.org/10.1007/978-981-99-8307-0_3

3.1.1 Basic Concepts of Microwave Circuits

The design, analysis, and optimization of microwave circuits rely on the good understanding of some fundamental concepts as referenced in previous works [1–6] and they are listed as follows.

1. Frequency: Microwave circuits typically operate within the frequency range from 300 MHz to 300 GHz. The selection of the operating frequency depends on various factors, including application requirements, signal transfer characteristics, and device performance.

2. Wavelength: Wavelength represents the distance traveled by an electromagnetic wave during one cycle and is inversely proportional to the operating frequency. The wavelength significantly influences the size, shape, and material choices in the designs of microwave circuits.

3. Bandwidth: In microwave circuits, bandwidth refers to the frequency range in which a signal can be transmitted through the circuit. It is typically expressed in hertz (Hz). The bandwidth determines the frequency response range of the circuit, i.e., the range of frequencies over which the circuit can effectively transmit signals. In microwave circuits, bandwidth can refer to different concepts, such as the -3 dB bandwidth, which is the most common bandwidth definition. It represents the frequency range at which the circuit's gain (or power) decreases to 70.7% (approximately half) of the input signal's original value. The -3 dB bandwidth defines the frequency response range of the circuit. Transmission bandwidth refers to the frequency range over which the circuit can effectively transmit signals. In this definition, bandwidth represents the frequency range in which the signal experiences minimal distortion or attenuation within the circuit. Coupling bandwidth, in microwave communication systems, refers to the maximum effective frequency range between antennas or different circuits. It involves the transmission and reception of signals, as well as the connection and coupling between circuits.

4. Input impedance: The input impedance of a microwave circuit refers to the impedance requirements or characteristics of the input signal at the input port. It represents the resistance and reflection characteristics of the microwave circuit to the input signal. The input impedance is a complex number and is typically represented by impedance value and phase angle. The impedance value indicates the resistance magnitude of the circuit to the input signal, while the phase angle represents the phase change of the circuit with respect to the input signal. In microwave circuits, the matching of input impedance is crucial. If the input impedance does not match the output impedance of the signal source, reflection and signal loss will occur. By adjusting the parameters of circuit components, such as resistors, inductors, and capacitors, the matching of input impedance can be achieved, thereby maximizing the transmission of signals into the circuit, and minimizing signal loss and reflection.

5. Transmission line: A transmission line serves as the medium for propagating electromagnetic waves in microwave circuits and encompasses technologies like microstrip lines, coplanar waveguides, and waveguides. Designing a transmission line involves considerations of characteristic impedance, transmission speed, and signal loss.

6. Characteristic impedance: The characteristic impedance of a microwave circuit refers to the impedance encountered when transmitting electromagnetic signals in the circuit. It is a circuit parameter used to describe the behavior of electromagnetic waves in transmission lines, such as microwave cables. The characteristic impedance represents the equivalent resistance encountered by electromagnetic waves as they propagate along the transmission line. It is influenced by factors such as the geometric shape of the transmission line, material properties, and operating frequency. The numerical value of the characteristic impedance depends on the structure and characteristics of the transmission line, such as the diameter of the conductor, dielectric constant, and magnetic permeability of the conductor material. Common types of transmission lines, such as coaxial cables and microstrip lines, have specific characteristic impedance values.

7. Impedance matching: Impedance matching aims to adjust the input and output impedances of a circuit to maximize the power transfer. This critical step in microwave circuit design can be achieved through impedance transformers, tuning elements, and other techniques.

8. Telegraph equation: The telegraph equation is a partial differential equation that describes the propagation of voltage and current on a transmission line. It can describe the spatial and temporal variations of voltage and current along the line. The equation takes into account the inductance, capacitance, and resistance of the transmission line, and describes the propagation speed and attenuation characteristics of signals on the line. In general, a transmission line can be seen as a linear element composed of inductance, capacitance, and resistance. By performing differentiation and derivative operations on the voltage and current on the transmission line, the Telegraph equation can be obtained. The analytical solution of the Telegraph equation can be used to analyze the voltage and current waveforms on the transmission line, understand the propagation characteristics, reflection, and interference phenomena of signals on the transmission line.

9. Attenuation and gain: Attenuation indicates the degree of signal loss during transmission, while gain represents the amplification of the signal. Both attenuation and gain serve as essential metrics for evaluating the performance of microwave circuits.

10. Propagation constant: The propagation constant is an important parameter in microwave circuits, used to describe the propagation characteristics of electromagnetic waves in microwave circuits. The propagation constant is typically represented by the symbol γ and is a complex number consisting of a real part and an imaginary part. The real part represents the phase change rate of the electromagnetic wave during propagation, while the imaginary part represents the extent of attenuation or amplification of the electromagnetic wave during propagation.

3.1.2 Main Components of Microwave Circuits

The understanding of microwave circuits encompasses several aspects, including circuit classification, commonly used microwave transmission line circuit elements and discontinuities, and frequently employed microwave components. When considering circuit classification, microwave circuits can be categorized from three perspectives based on the nature of transmission lines, technology, and whether they are active or passive circuits. According to the nature of transmission lines, microwave circuits can be classified as microstrip, stripline, coaxial line, waveguide, and coplanar waveguide. From a process standpoint, microwave circuits can be divided into microwave hybrid integrated circuits, microwave integrated circuits (MIC), and microwave monolithic integrated circuits (MMIC). Additionally, microwave circuits can be classified as either active circuits (e.g., amplifiers, oscillators) or passive circuits (e.g., splitters, couplers, phase shifters, switches, mixers, filters).

Commonly used microwave transmission line circuit elements and discontinuities include transmission line segments, coupled lines, open lines, short lines, right-angle turn lines, step lines, tapered lines, slots, T-junctions, and cross junctions. Other components, such as fan lines, Lange couplers, interdigital capacitors, and spiral inductors, also play crucial roles in microwave circuits, constituting their fundamental components. Microwave components commonly employed in microwave circuits include mounted passive devices and microwave semiconductor devices. These devices encompass microwave transmission lines, and components, microwave resonators, microwave mixers, up-converters and frequency multipliers, microwave transistor amplifiers, microwave negative resistance oscillators, PIN diodes, and microwave control circuits.

Microwave circuits typically comprise the following main components [1]:

1. Source: The source generates electromagnetic wave signals in a microwave circuit and includes components such as oscillators and signal generators.

2. Load: The load consumes electromagnetic wave energy in the microwave circuit and can consist of components such as antennas and resistors.

3. Transmission line: As mentioned earlier, transmission lines serve as the medium for transmitting electromagnetic waves in microwave circuits and include microstrip lines, striplines, coaxial lines, waveguides, and coplanar waveguides.

4. Tuning elements: Tuning elements adjust the working state and performance of microwave circuits and can include capacitors, inductors, and adjustable impedances.

5. Active devices: Active devices amplify, modulate, or detect signals and can include amplifiers, mixers, and detectors.

6. Passive devices: Passive devices primarily adjust the transmission characteristics of signals and can include filters, couplers, isolators, circulators, bias networks, and more.

Microwave circuits play a vital role in modern communication and radar systems. Researchers need to design and select appropriate microwave circuits based on specific application requirements. For instance, wireless communication systems often require microwave circuits capable of processing high-frequency signals with good bandwidth performance to meet the demands of high-speed data transmission. Similarly, radar systems rely on microwave circuits with high power and sensitivity characteristics to achieve long-range target detection and localization. During the design process of microwave circuits, researchers must consider various factors such as the circuit's operating frequency, transmission characteristics, and electromagnetic wave losses and radiation. To optimize microwave circuit performance, researchers commonly employ computer-aided design (CAD) tools for circuit simulation and optimization. These tools assist researchers in accurately predicting circuit performance, thereby guiding actual circuit production and testing.

3.1.3 Design Method of Microwave Circuits

There are several methods for microwave circuit design, as referenced in previous studies [2, 6, 7]:

1. Traditional network analysis method: This approach relies on network parameters, such as scattering parameters and impedance parameters, along with equivalent circuit models. It is commonly employed in low-frequency and intermediate-frequency microwave circuit design.

2. Transmission line theory method: This method involves decomposing the microwave circuit into multiple transmission line segments and passive devices. The circuit design is achieved by solving the transmission line equations and meeting matching conditions. It is particularly suitable for high-frequency microwave circuit design, especially for larger-sized circuits.

3. Electromagnetic field method: This method directly solves electromagnetic field equations, such as Maxwell's equations, to obtain the electromagnetic field distribution, impedance, and scattering parameters of the circuit. It is well-suited for designing complex structures and broadband microwave circuits.

The scattering parameter, denoted as S-parameter, is often used to describe microwave networks. For instance, consider a two-port network as an example.

In the two-port microwave network depicted in Fig. 3.1, a_1 and b_1 respectively represent the normalized incident voltage wave and the reflected voltage wave at port 1, while a_2 and b_2 correspond to the normalized incident voltage wave and the reflected voltage wave at port 2. The relationship between the input and output of a two-port microwave network can be expressed as

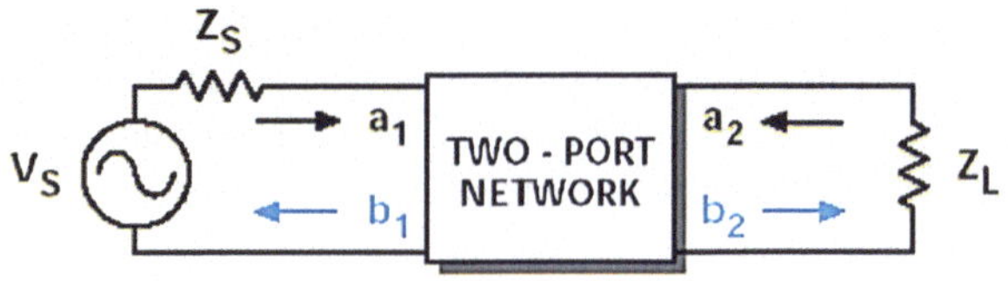

Fig. 3.1 Two-port network

$$b_1 = s_{11}a_1 + s_{12}a_2 \tag{3.1}$$

$$b_2 = s_{21}a_1 + s_{22}a_2 \tag{3.2}$$

where

$$[b_1] = \begin{bmatrix} b_1 & b_2 \end{bmatrix} \tag{3.3}$$

$$[S] = \begin{bmatrix} s_{11} & s_{12} \\ s_{21} & s_{22} \end{bmatrix} \tag{3.4}$$

$$[a_1] = \begin{bmatrix} a_1 & a_2 \end{bmatrix} \tag{3.5}$$

Equation 3.1 represents the scattering equation, with $[S]$ denoting the scattering matrix or scattering parameter. The S-parameters of the two-port network can be defined as follows:

$$S_{11} = \left.\frac{b_1}{a_1}\right|_{a_2=0} \tag{3.6}$$

This equation represents the reflection coefficient of port 1 when port 2 is matched ($Z_L = Z_0$).

$$S_{22} = \left.\frac{b_2}{a_2}\right|_{a_1=0} \tag{3.7}$$

where we have the reflection coefficient of port 2 when port 1 is matched ($Z_S = Z_0$).

$$S_{12} = \left.\frac{b_1}{a_2}\right|_{a_1=0} \tag{3.8}$$

This equation represents the transmission coefficient from port 2 to port 1 when port 1 is matched.

$$S_{21} = \left.\frac{b_2}{a_1}\right|_{a_2=0} \tag{3.9}$$

Lastly, we have the transmission coefficient from port 1 to port 2 when port 2 is matched. Through the above analysis, it becomes evident that the S-parameters of the microwave network hold a definite physical interpretation. In fact, the Z-parameter, Y-parameter, and H-parameter that were commonly used in the past can be calculated to obtain the S-parameters. However, at microwave frequencies, only the S-parameters can be directly measured, thus resolving the issue of measuring microwave network parameters.

3.1.4 Development of Microwave Circuits

With the advancement of science, technology, and increasing application demands, microwave circuits are experiencing several development trends [8–14]:

1. Interconnection and manufacturing of microwave circuits

The technology for interconnecting and manufacturing microwave circuits with frequencies above 1 GHz has witnessed rapid growth and widespread usage. In modern information systems and military electronic equipment like radar, navigation systems, and communication devices, microwave circuits serve as the "lifeline" for high-speed information transmission. Therefore, microwave circuit interconnection and manufacturing technology play a pivotal role in the development and production of information systems and military electronic equipment.

Microwave circuit interconnection and manufacturing technology encompass various aspects, including microwave circuit substrate materials and manufacturing techniques, microwave circuit design and manufacturing techniques, packaging and assembly technology for microwave devices or components, as well as interconnection and debugging technology for microwave components or systems. These areas involve multiple disciplines such as microelectronics, material science, computer-aided design, electronic engineering, and mechanical engineering. It is a multidisciplinary and comprehensive field characterized by high scientific and technological content, significant technical challenges, rapid development, wide applications, and its crucial role in information systems and military electronic equipment.

With the rapid progress in microelectronics, component technology, material science, computer-aided design, and manufacturing science and technology, new processes and technologies for microwave circuit interconnection and manufacturing are continuously emerging. Examples include multilayer microwave integrated circuits, three-dimensional microwave integrated circuits, low-loss transmission lines, shielding film microstrip circuits, multi-chip microwave modules, interconnection and manufacturing technology of micro-electromechanical systems for microwave circuits, novel resin microwave PCB technology, novel microwave circuit protective coating technology, three-dimensional circuit simulation technology applied to microwave circuit design, as well as microwave circuit CAD and optimization technology based on intelligent methods.

2. Photonic bandgap structure of microwave circuits

In recent years, photonic bandgap (PBG) structure has been extended to the microwave frequency range, garnering significant attention. When an electromagnetic wave propagates through a material with a periodic structure, it experiences modulation that gives rise to a photonic bandgap. Within the bandgap, there are no available transmission states for the operating frequency of the electromagnetic wave. Photonic bandgap structures can be achieved by incorporating metal, dielectric, ferromagnetic, or ferroelectric materials into the substrate, or by directly arranging various materials in a periodic manner.

Numerous types of microwave photonic bandgap structures have been proposed both domestically and internationally. Currently, research on photonic bandgap structures spans from three-dimensional to one-dimensional and two-dimensional structures. This research has expanded into the fields of electronics and communications due to the ease of implementation and integration. The design of photonic bandgap structures requires attention to the unit shape, periodicity conditions, combinations of different periodic structures, and advancements in materials. The emergence of planar photonic bandgap (PBG) structures has revolutionized traditional design methods, offering new possibilities for designing high-performance and highly integrated circuits. One-dimensional (1D) and two-dimensional (2D) planar bandgap structures have found widespread use in microwave circuits due to their flexibility, ease of implementation, and integration, thus accelerating the development of microwave integrated circuits.

3. Micro-Electromechanical-System (MEMS) switch for microwave circuits
According to the latest definition of MEMS, it refers to miniaturized devices or arrays that integrate electrical and mechanical components and can be mass-produced using IC processes. While there are similarities between conventional IC fabrication processes and MEMS fabrication processes, the latter involves three-dimensional technology instead of planar technology. Currently, widely used MEMS fabrication technologies include bulk micromachining, surface micromachining, bonding micromachining, and LIGA (photolithography and electroforming).

Switches play a crucial role in microwave signal conversion. Modern RF MEMS switches, in comparison to traditional PIN diode switches and FET switches, offer superior microwave characteristics and inherent advantages such as light weight, small size, and low power consumption. With advancements in MEMS fabrication technology and process theory, RF MEMS switches are expected to find extensive use in microwave systems, overcoming limitations related to short working life and low switching rates. Currently, RF MEMS switches have been employed in front-end circuits of some microwave systems, digital capacitor banks, and phase-shifting networks.

4. Lumped components and two-dimensional planarization of microwave circuits
Another trend in microstrip circuits is the utilization of lumped elements. In addition to lumped elements and one-dimensional transmission line elements, two-dimensional planar elements have been proposed for microwave circuits. These components are compatible with stripline and microstrip, providing a valuable alternative for microwave circuit design. Presently, there are three main approaches to achieve two-dimensional planar circuits: three-element structures, open structures, and cavity structures. Compared to stripline circuits, two-dimensional planar circuits offer advantages such as increased design flexibility and lower input resistance. Furthermore, compared to waveguide circuits, they are easier to analyze and design. With the aid of powerful computing capabilities, these circuits can be analyzed according to specific requirements, significantly improving work efficiency. It is anticipated that their application will become increasingly widespread in the near future.

Likewise, a new generation of MICs (monolithic microwave integrated circuits) may be developed on semiconductor substrates such as high-resistivity silicon, high-resistivity gallium arsenide, and low-resistivity silicon with a silicon dioxide layer. Gallium arsenide metal-semiconductor field-effect transistors (MESFETs) will dominate in analog amplifiers with gigahertz bandwidths and in digital integrated circuits with gigabit rates. The advantages of both hybrid and monolithic microwave integrated circuits are similar to those of low-frequency integrated circuits, including high system reliability, reduced volume and weight, and ultimately cost reduction in scenarios requiring a large number of standardized components. Similar to low-frequency integrated circuits, MICs possess significant potential for expanding existing markets and developing numerous new applications, including civilian projects. With the increasing prevalence of various integrated circuits, the future of microwave circuits is undoubtedly promising.

In microwave circuit design, simulation technology plays a crucial role in verifying and optimizing circuit performance, reducing experiment and test costs, improving design speed and flexibility, fostering innovative abilities and professional qualities, supporting the development of new technologies and applications, guiding experiments and tests, and aiding in education and training [3, 15–17]. By utilizing simulation technology, engineers can predict and analyze circuit performance during the design stage, including gain, bandwidth, loss, and isolation. This enables more accurate determination of design parameters, avoiding repetitive trial and error and rework, and ultimately improving design efficiency and success rates.

3.1.5 Characteristics of Low-Frequency Simulations for Microwave Circuits

Low-frequency microwave circuits exhibit specific characteristics and present unique challenges in terms of simulation technology [18]. This section provides an overview of the distinctive features of low-frequency microwave circuit simulations [6, 18–23], including:

1. Influence of circuit parameters: Low-frequency microwave circuits are sensitive to variations in circuit parameters. Simulations should consider the impact of parameter uncertainties, tolerances, and variations to ensure accurate performance prediction.

2. Consideration of coupling and radiation effects: Mutual coupling between circuit elements and radiation effects become more significant in low-frequency microwave circuits. Simulations' techniques should account for these effects to accurately analyze circuit behavior and performance.

3. Selection of simulation methods: Various simulation methods, such as circuit-level simulation, electromagnetic simulation, and mixed-level simulation, can be employed to model and analyze low-frequency microwave circuits. The appropriate choice of simulation method depends on the circuit complexity, accuracy requirements, and computational resources available.

4. Optimization of computing resources and time: Low-frequency microwave circuits often involve large-scale simulations that require significant computational resources and time. Optimizing simulation algorithms, utilizing parallel computing

techniques, and employing model order reduction methods can enhance simulation efficiency.

5. Analysis of error and uncertainty: Simulation results may contain errors and uncertainties due to various factors, including modeling assumptions, numerical approximations, and parameter variations. Understanding and analyzing these errors and uncertainties are crucial for reliable circuit performance evaluation.

6. Experimental verification and testing: Experimental validation is essential to verify simulation results and ensure the accuracy of the design. Comparing simulation results with measurement data helps identify discrepancies and refine the simulation models.

Considering these characteristics and challenges, engineers can effectively guide their simulation and design work for low-frequency microwave circuits.

- Influence of Circuit Parameters
  The parameters of low-frequency microwave circuits, such as resistance, capacitance, inductance, and transmission line characteristics, significantly impact circuit performance and reliability. To enhance simulation accuracy and reliability, it is essential to utilize high-precision component libraries, develop detailed equivalent circuit models, and perform sensitivity analysis. Additionally, factors such as temperature characteristics, nonlinear effects, and component lifespan should be considered to ensure circuit stability and reliability in real-world operating environments.
- Consideration of Coupling and Radiation Effects
  Due to the large size of low-frequency microwave circuits, the effects of coupling and radiation become more pronounced. Electromagnetic field simulation methods, such as the finite-difference time-domain method (FDTD) and finite element method (FEM), are necessary for accurately modeling and analyzing the distribution and interaction of electromagnetic fields. It is also crucial to employ shielding and grounding techniques to optimize circuit layout and structure, minimizing coupling and radiation losses while improving overall circuit performance.
- Selection of Simulation Methods
  Low-frequency microwave circuit simulation requires the integration of both circuit-level and electromagnetic field methods. At the circuit level, techniques like network analysis and transmission line theory can be employed to quickly evaluate circuit performance and parameters. At the electromagnetic field level, methods such as FDTD and FEM enable detailed simulation of electromagnetic field distribution and interactions. By combining various simulation methods, it is possible to achieve more accurate, efficient, and flexible simulations.
- Optimization of Computing Resources and Times
  Simulating low-frequency microwave circuits often demands significant computational resources and time due to their large size. To improve simulation efficiency and reduce computational costs, efficient algorithms, hierarchical modeling, and parallel computing techniques can be utilized. Efficient algorithms can reduce calculation complexity and iteration counts, thereby enhancing convergence speed. Hierarchical modeling allows the division of complex circuits into subsystems

for more manageable analysis and optimization. Parallel computing harnesses the power of multi-core processors and high-performance computers to expedite computation time.

- Analysis of Errors and Uncertainties
  Low-frequency microwave circuit simulations may introduce errors and uncertainties stemming from component parameters, algorithms, numerical formats, boundary conditions, and other factors. To ensure accuracy and reliability, error analysis and uncertainty evaluation are essential. Error analysis encompasses absolute, relative, and maximum errors, providing insights into the accuracy and bias of simulation results. Uncertainty evaluation involves statistical analysis, Monte Carlo simulations, and other techniques to predict fluctuations and variations in simulation results within real-world operating environments.
- Experimental Verification and Testing
  Combining low-frequency microwave circuit simulation with experimental verification enhances accuracy and reliability. During the experimental phase, it is crucial to design reliable test plans, select appropriate testing instruments and conditions, and accurately measure circuit performance and parameters. Analyzing the disparities and consistencies between experimental and simulation results enables optimization of simulation models and methods, improvement of circuit design and manufacturing processes, and reduction of risks and losses in practical engineering.
  Low-frequency microwave circuit simulation possesses distinct characteristics and challenges. Continued development and innovation are necessary to address the demands of emerging technologies and applications, contributing to the advancement and progress of wireless communications and related fields.

3.1.6 Development History of Simulation Technologies for Low-Frequency Microwave Circuits

Microwave circuit computer-aided-design (CAD) technology has made significant advancements since the 1970s. CAD software vendors have introduced a variety of general and specialized microwave circuit CAD software products, offering features such as circuit schematic diagram input, microwave circuit graphic input, circuit simulation and optimization, tolerance analysis, layout generation and output, and interface integration with test instruments. These software packages are equipped with extensive circuit model libraries, component libraries, linear model libraries, and nonlinear model libraries for semiconductor devices, providing powerful, user-friendly, and readily available resources [6, 8, 23–27]. Consequently, microwave circuit CAD software has been widely adopted in the design of various microwave circuits, becoming an essential tool for microwave engineers.

Microwave circuit CAD software can be categorized as follows:

1. Linear/nonlinear microwave circuit simulation software: This category includes software specifically designed for simulating the behavior of microwave circuits, both in linear and nonlinear operating regimes.
2. 2.5D planar circuit electromagnetic field simulation software: Software in this category focuses on simulating the electromagnetic fields and interactions within planar microwave circuits, typically using 2.5D simulation techniques.
3. 3D electromagnetic field simulation software: This type of software enables comprehensive simulation and analysis of the electromagnetic fields and wave propagation in three-dimensional microwave circuits.
4. System simulation software: System-level simulation software allows for the simulation and analysis of microwave systems, taking into account the interactions and behavior of various interconnected microwave circuits.
5. Design software for special circuits: This category includes software tailored for the design and analysis of specific types of microwave circuits, such as filters, amplifiers, mixers, and antennas.
6. Typesetting software: This software focuses on facilitating the layout and arrangement of circuit elements, enabling efficient and visually appealing presentation of circuit designs.

Early simulation of low-frequency microwave circuits primarily relied on analytical methods such as transmission line theory, network analysis, and electromagnetic field analysis. These methods, based on mathematical formulas and theories, provided engineers with intuitive predictions of circuit performance and parameters. However, the accuracy and applicability of analytical methods were limited, making it challenging to handle complex circuit structures and practical working conditions. The CAD software began to be applied to the simulations of low-frequency microwave circuits. The CAD technology facilitated the transfer of circuit analysis and optimization from manual calculations to computers, significantly enhancing the speed and flexibility of simulations. Additionally, the CAD technology introduced new numerical methods such as the finite-difference method, finite element method, and finite integration method, enabling more accurate and comprehensive electromagnetic field simulations. Commercial simulation software gained widespread use in low-frequency microwave circuit simulation, driven by the rapid development of the electronics industry and communication technology. These types of software packages provided extensive component libraries, user-friendly interfaces, and powerful computing capabilities, offering convenient simulation platforms for engineers and researchers. Prominent examples of commercial simulation software include ADS (Advanced Design System), HFSS (High-Frequency Structure Simulator), CST (Computer Simulation Technology), and others.

Since the 21st century, low-frequency microwave circuit simulation technology has entered a new stage of development characterized by continuous integration and innovation to meet the demands of complex systems and new applications. Noteworthy technological advancements include circuit electromagnetic field coupling simulation, multiphysics field simulation, and artificial intelligence (AI)-assisted simulation. Circuit electromagnetic field coupling simulation combines circuit analysis and electromagnetic field simulation to achieve more accurate and comprehensive performance predictions, enabling engineers to consider the electromagnetic field characteristics and interactions of circuit components and analyze their impact on circuit performance. Moreover, circuit electromagnetic field coupling simulation enhances design optimization and verification efficiency, reducing the number of iterations and associated costs.

Simultaneously, multiphysics field simulation technology integrates electromagnetic field, thermal field, structural field, and other physical fields to consider the complex factors present in real-world working environments. This approach helps engineers evaluate circuit stability and reliability under extreme conditions such as high temperature, high voltage, and high frequency, providing crucial guidance for design and manufacturing. Additionally, multiphysics simulation supports research and application of new materials and processes, fostering innovation and development in low-frequency microwave circuit technology. With the rapid advancement of artificial intelligence (AI) technology, an increasing number of AI methods have been incorporated into low-frequency microwave circuit simulation. These methods, including neural networks, genetic algorithms, and deep learning, improve simulation speed and accuracy, assisting engineers in solving complex design problems. AI-assisted simulation also enables automated and intelligent design optimization, reducing labor and time costs while enhancing circuit performance. Microwave circuit CAD technology, a branch of electronic design automation (EDA), specializes in computer simulation and optimization design of radio frequency and microwave circuits. In comparison to other EDA techniques, microwave circuit CAD software exhibits several notable features. It requires accurate transmission line models and various device models, sometimes employing numerical simulation tools such as electromagnetic field simulation. Additionally, microwave circuit CAD software often incorporates S-parameter analysis functionality.

Microwave circuit CAD encompasses a range of analysis methods. Linear circuits are typically computed using equivalent circuit models and cascade S-parameter matrices. Nonlinear circuits may employ Spice, harmonic balance methods, envelope simulation methods, and others. Numerical methods like the moment method and finite element method are frequently used for electromagnetic field simulation. In terms of optimization, CAD software automatically adjusts the values of circuit elements to meet specified requirements, considering the network topology, initial values, and target parameters of the design index.

3.2 Principle of Microwave Circuit Simulations

3.2.1 Electromagnetic Field Equations

The electromagnetic field equation is the fundamental theoretical basis in microwave circuit simulations [2, 28–31, 33]. It relies on Maxwell's equations, which describe the behavior of the electromagnetic field, and incorporates the wave equation and boundary conditions. Maxwell's equations consist of four fundamental equations that govern the electromagnetic field:

1. Gauss's law (electric field):

$$\nabla \cdot \mathbf{D} = \rho_v \tag{3.10}$$

2. Gauss's law (magnetic field):

$$\nabla \cdot \mathbf{B} = 0 \tag{3.11}$$

3. Faraday's law:

$$\nabla \times \mathbf{E} = -\frac{\partial \mathbf{B}}{\partial t} \tag{3.12}$$

4. Ampere's loop law (with displacement current):

$$\nabla \times \mathbf{H} = \mathbf{J} + \frac{\partial \mathbf{D}}{\partial t} \tag{3.13}$$

where **D** represents the electric flux density, ρ_v represents the charge density , **B** represents the magnetic flux density, **E** represents the electric field strength, **H** represents the magnetic field strength, and **J** represents the current density. Additionally, the interaction between matter and the electromagnetic field can be described by the following relations:

1. Electric polarizability:

$$\mathbf{P} = \chi_e \mathbf{E} \tag{3.14}$$

2. Susceptibility:

$$\mathbf{M} = \chi_m \mathbf{H} \tag{3.15}$$

Using these relations, we can derive the following expressions: 1. Electric displacement vector:

$$\mathbf{D} = \epsilon_0 \mathbf{E} + \mathbf{P} = \epsilon_0 (1 + \chi_e) \mathbf{E} = \epsilon_0 \epsilon_r \mathbf{E} \tag{3.16}$$

2. Magnetic induction:

$$\mathbf{B} = \mu_0 \mathbf{H} + \mathbf{M} = \mu_0 (1 + \chi_m) \mathbf{H} = \mu_0 \mu_r \mathbf{H} \tag{3.17}$$

where ϵ_0 represents the vacuum permittivity, μ_0 represents the vacuum permeability, and ϵ_r and μ_r represent the relative permittivity and permeability, respectively. Through Maxwell's equations, we can derive the wave equation for electromagnetic waves in microwave circuits. First, by taking the curl of Faraday's law equation, we obtain

$$\nabla \times (\nabla \times \mathbf{E}) = -\frac{\partial}{\partial t}(\nabla \times \mathbf{B}) \tag{3.18}$$

Table 3.1 Analysis Methods through Electromagnetic Fields

Maxwell's equations	Wave equation and boundary conditions	Solution of electric and magnetic field distribution
$\begin{cases} \nabla \times \mathbf{H} = j\omega\varepsilon\mathbf{E} \\ \nabla \times \mathbf{E} = -j\omega\mu\mathbf{H} \\ \nabla \cdot \mathbf{D} = 0 \\ \nabla \cdot \mathbf{B} = 0 \end{cases}$	$\begin{cases} \nabla^2\mathbf{E} + k^2\mathbf{E} = 0 \\ \nabla^2\mathbf{H} + k^2\mathbf{H} = 0 \\ \text{Electricwall } \mathbf{n} \times \mathbf{E}\vert_\Omega = 0,\ \mathbf{n} \cdot \mathbf{H}\vert_\Omega = 0 \\ \text{Magneticwall } -\mathbf{n} \times \mathbf{H}\vert_\Omega = 0,\ \mathbf{n} \cdot \mathbf{E}\vert_\Omega = 0 \end{cases}$	$\begin{cases} \mathbf{E}(\mathbf{r}, t) = \mathbf{E}(\mathbf{r})e^{j\omega t} \\ \mathbf{H}(\mathbf{r}, t) = \mathbf{H}(\mathbf{r})e^{j\omega t} \end{cases}$

Table 3.2 Analysis Methods through Circuits

Kerseyhoff's theorem	Wave equation and boundary conditions	Solution of voltage and current distribution
$\begin{cases} \Sigma I = 0 \\ \Sigma U = 0 \end{cases}$	$\begin{cases} \frac{d^2U}{dz^2} - \gamma^2 U = 0 \\ \frac{d^2I}{dz^2} - \gamma^2 I = 0 \end{cases}$	$\begin{cases} U(z)e^{-\gamma z}e^{j\omega t} \\ I(z)e^{-\gamma z}e^{j\omega t} \end{cases}$

Using the vector identity $\nabla \times (\nabla \times \mathbf{E}) = \nabla(\nabla \cdot \mathbf{E}) - \nabla^2\mathbf{E}$ and substituting Gauss's law and Faraday's law, we can simplify the equation as follows:

$$\nabla^2\mathbf{E} - \epsilon_0\mu_0\frac{\partial^2\mathbf{E}}{\partial t^2} = 0 \tag{3.19}$$

Similarly, by taking the curl of Ampere's loop law equation, we obtain the magnetic wave equation:

$$\nabla^2\mathbf{H} - \epsilon_0\mu_0\frac{\partial^2\mathbf{H}}{\partial t^2} = 0 \tag{3.20}$$

In the simulation of low-frequency microwave circuits, boundary conditions need to be considered. These boundary conditions can be categorized into electric field boundary conditions and magnetic field boundary conditions.

In the specific simulation process, the structure, materials, and process characteristics of the actual microwave circuit also need to be taken into account. By solving the wave equation and applying the boundary conditions, we can determine the distribution and propagation characteristics of the electromagnetic field in the microwave circuit. This information is crucial for circuit design, optimization, and performance evaluation.

3.2.2 *Transmission Line Theory*

Transmission line theory serves as the foundation for the analysis and design of microwave circuits, providing a simplified one-dimensional model to describe the propagation of electromagnetic waves on transmission lines [2, 34]. In the simulation of low-frequency microwave circuits, the application of transmission line theory

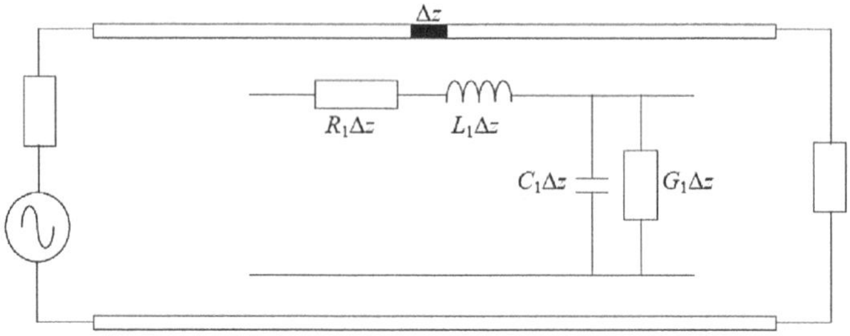

Fig. 3.2 Equivalent circuit of transmission line

enables us to comprehend fundamental propagation characteristics of electromagnetic waves, as well as phenomena like reflection and impedance matching.

3.2.2.1 Transmission Line Parameters

Transmission lines are of significant importance in low-frequency microwave circuits, as they directly impact the transmission quality of signals. The fundamental parameters of transmission lines, including resistance R, inductance L, capacitance C, and conductance G, are crucial in characterizing various aspects of signal transmission. These parameters reflect properties such as resistive loss, electromagnetic coupling, electric field coupling, and dielectric loss within the transmission line. To describe these characteristics effectively, it is common practice to express the transmission line parameters as parameters per unit length, denoted as R', L', C', and G', respectively, considering a transmission line segment of length ΔZ that is sufficiently small (Fig. 3.2).

The resistance R' characterizes the resistance encountered by the current flow in the transmission line due to the conductor material. The resistance of a conductor is influenced by factors such as the material's resistivity, cross-sectional area, and length. In the microwave frequency range, the skin effect causes the current to concentrate near the conductor's surface, resulting in an increase in resistance with frequency.

The inductance L' reflects the coupling between the conductor's magnetic field and the current flowing through it. When an electric current passes through a conductor, it generates a magnetic field around the conductor. Changes in this magnetic field induce an electromotive force within the conductor, affecting current propagation. The inductance is influenced by factors such as the conductor's geometry, relative position, and permeability.

The capacitance C' describes the electric field coupling between adjacent conductors in the transmission line. Capacitance and charge have a proportional relationship, meaning that the charge between two conductors is proportional to the potential difference between them. Capacitance depends on parameters such as the distance between conductors, their geometry, and the dielectric constant of the insulating material.

The conductance G' represents the dielectric loss in the transmission line and primarily arises from charge movement and polarization phenomena within the dielectric material. The conductance depends on the conductivity and geometry of the dielectric. In practical applications, minimizing dielectric loss is desirable to reduce signal loss during transmission.

There is a fundamental relationship between the transmission line parameters R', L', C', and G'. For instance, in the case of a lossless transmission line, we have the following expression:

$$Z_0 = \sqrt{\frac{R' + j\omega L'}{G' + j\omega C'}} \tag{3.21}$$

where Z_0 represents the characteristic impedance of the transmission line, and ω denotes the angular frequency. The characteristic impedance of the transmission line is closely associated with these four parameters. By comprehending and controlling these parameters, we can optimize the performance of the transmission line.

In the simulation of low-frequency microwave circuit, the calculation and model of transmission line parameters are the key links. By understanding and controlling the transmission line parameters, we can predict and optimize the propagation characteristics of electromagnetic waves on the transmission line, thus improving the quality of signal transmission and system performance.

In order to obtain accurate transmission line parameters, we need to consider the following:

1. Material properties: The material properties of conductors and dielectrics, such as resistivity, permeability, and permittivity, affect the transmission line parameters. In the process of simulation, it is very important to choose the right material model and parameters.
2. Geometric shape and size: The geometric shape and size of the transmission line directly affect the inductance and capacitance. In the simulation process, it is the key to ensure the accurate establishment of the geometric model of the transmission line.
3. Frequency dependence: In the microwave band, the transmission line parameters may vary with frequency. For example, conductor resistance increases due to skin effects, and dielectric losses may also vary with frequency. Therefore, it is necessary to consider the frequency dependence during the simulation.
4. Boundary conditions: In practical microwave circuits, transmission lines often interact with other components and structures. These interactions may lead to changes in the transmission line parameters. In the simulation process, it is very important to consider the boundary conditions and coupling effects.

3.2.2.2 Transmission Line Equation

The transmission line equations describe the distribution of voltage V and current I along the transmission line. For a lossless transmission line, we can express the following equations:

$$\frac{\partial V(z,t)}{\partial z} = -L' \frac{\partial I(z,t)}{\partial t} \tag{3.22}$$

$$\frac{\partial I(z,t)}{\partial z} = -C' \frac{\partial V(z,t)}{\partial t} \tag{3.23}$$

where Z represents the distance along the transmission line, t denotes time, and L' and C' are the inductance and capacitance per unit length, respectively.

To solve Eqs. 3.22 and 3.23, we can employ the method of separation of variables. Let the voltage $V(Z,t) = V(Z)T(t)$ and the current $I(Z,t) = I(Z)T(t)$, where $V(Z)$ and $I(Z)$ represent the spatial distribution of voltage and current along the transmission line, respectively, and $T(t)$ represents the time-dependent term.

Substituting $V(Z,t)$ and $I(Z,t)$ into Eqs. 3.22 and 3.23, we obtain

$$\frac{\partial V(z)T(t)}{\partial z} = -L' \frac{\partial I(z)T(t)}{\partial t} \tag{3.24}$$

$$\frac{\partial I(z)T(t)}{\partial z} = -C' \frac{\partial V(z)T(t)}{\partial t} \tag{3.25}$$

Separating the variables in Eqs. 3.24 and 3.25, we have

$$\frac{1}{V(z)} \frac{\partial V(z)}{\partial z} = -L' \frac{1}{I(z)} \frac{\partial I(z)}{\partial t} \tag{3.26}$$

$$\frac{1}{I(z)} \frac{\partial I(z)}{\partial z} = -C' \frac{1}{V(z)} \frac{\partial V(z)}{\partial t} \tag{3.27}$$

Since the left-hand side of Equations 3.26 and 3.27 only depend on Z and the right-hand side only depends on t, they must be equal to a constant, which we set as $-k^2$. This yields the following system of ordinary differential equations:

$$\frac{1}{V(z)} \frac{dV(z)}{dz} = -k^2 L' I(z) \tag{3.28}$$

$$\frac{1}{I(z)} \frac{dI(z)}{dz} = -k^2 C' V(z) \tag{3.29}$$

Combining Eqs. 3.28 and 3.29, we can obtain the second-order ordinary differential equations for $V(Z)$and $I(Z)$:

$$\frac{d^2V(z)}{dz^2} + k^2LC'V(z) = 0 \tag{3.30}$$

$$\frac{d^2I(z)}{dz^2} + k^2LC'I(z) = 0 \tag{3.31}$$

Equations 3.21 and 3.31 represent homogeneous linear ordinary differential equations with complex characteristic roots K. By solving these equations, we can derive the general solutions as follows:

$$V(z) = V_0^+e^{-jkz} + V_0^-e^{jkz} \tag{3.32}$$

$$I(z) = I_0^+e^{-jkz} + I_0^-e^{jkz} \tag{3.33}$$

where V_0^+ and V_0^- denote the amplitudes of the right-propagating and left-propagating voltage waves, respectively. Similarly, I_0^+ and I_0^- represent the amplitudes of the right-propagating and left-propagating current waves, respectively. The term $K = \Omega\sqrt{L'C'}$ corresponds to the wavenumber, where Ω represents the angular frequency.

By solving Eqs. 3.32 and 3.33, we can determine the waveform, phase, and amplitude distributions of voltage and current along the transmission line. This information is crucial for analyzing transmission characteristics in microwave circuits, including insertion loss, reflection, and impedance matching. Furthermore, the transmission line equation can be applied to study more complex microwave circuits, such as coupled lines, power dividers, and filters. In practical applications, it becomes necessary to modify and solve the transmission line equation due to factors like transmission line losses, variations in distribution parameters, and boundary conditions. For instance, for a lossy transmission line, we introduce resistance R' and admittance G' to account for line losses, resulting in a more intricate transmission line equation. Additionally, during the simulation process, several factors need to be considered, including electromagnetic field-circuit coupling, nonlinear characteristics of microwave devices, and temperature effects. Accounting for these factors allows us to obtain more accurate electromagnetic field distribution and transmission characteristics. The transmission line equation is a fundamental theory for investigating the transmission characteristics of microwave circuits, providing significant guidance for simulating and designing low-frequency microwave circuits. By solving the transmission line equation, we can comprehend the fundamental propagation characteristics of electromagnetic waves on transmission lines, enabling us to optimize circuit design and performance.

3.2.2.3 Microstrip Line and Stripline

Microstrip and stripline are two commonly used types of transmission lines in low-frequency microwave circuits. The geometric dimensions and medium properties

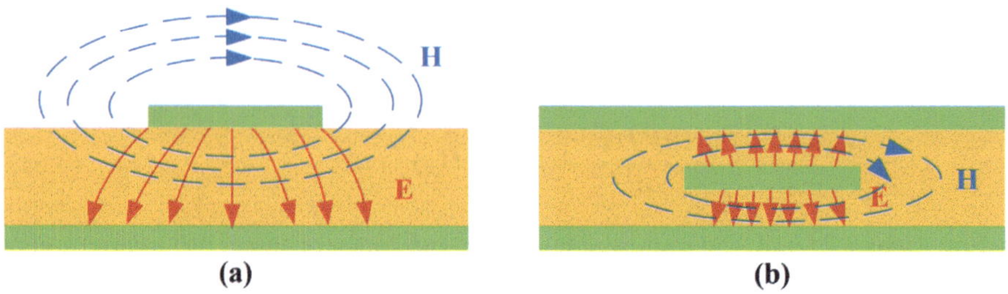

Fig. 3.3 Microstrip line and stripline. (a) Microstrip (b) Stripline

have a significant impact on their basic characteristics and performance parameters (Fig. 3.3).

1. Microstrip line

A microstrip line is a planar transmission line consisting of a narrow conductor strip (signal line) and a large ground plane. It is characterized by a simple structure, easy integration, and manufacturing. The main parameters of a microstrip line include the characteristic impedance (Z_0), the phase velocity (V_p), and the propagation loss (α). The characteristic impedance is an important parameter of the transmission line, which is related to the width of the signal line (w), the dielectric thickness (h), and the dielectric permittivity (ε_r). The characteristic impedance can be approximated as follows:

When $w/h \leq 1$:

$$Z_0 = \frac{60}{\sqrt{\varepsilon_{\text{eff}}}} \ln\left(\frac{8h}{w} + \frac{w}{4h}\right) \tag{3.34}$$

When $w/h > 1$:

$$Z_0 = \frac{120\pi}{\sqrt{\varepsilon_{\text{eff}}}[w/h + 1.393 + 0.667\ln(w/h + 1.444)]} \tag{3.35}$$

where ε_{eff} is the effective dielectric constant, which can be calculated as

$$\varepsilon_{\text{eff}} = \frac{\varepsilon_r + 1}{2} + \frac{\varepsilon_r - 1}{2}\left[1 + \frac{12h}{w}\right]^{-\frac{1}{2}} \tag{3.36}$$

The phase velocity is closely related to the characteristic impedance and capacitance and can be calculated using the following equation:

$$v_p = \frac{c}{\sqrt{\varepsilon_{\text{eff}}}} \tag{3.37}$$

where c is the speed of light. Propagation loss mainly includes conductor loss and dielectric loss. Conductor loss is mainly caused by the resistance of the signal line and is related to the material, size, and frequency. Dielectric loss is mainly caused by the loss factor of the dielectric material. The total propagation loss can be calculated as the sum of the conductor loss and dielectric loss:

$$\alpha = \alpha_c + \alpha_d \tag{3.38}$$

where α_c is the conductor loss and α_d is the dielectric loss.
2. Stripline
A stripline consists of two parallel conductor strips with a medium between them. Stripline exhibits low radiation loss, high-power capacity, and a good shielding effect. The main parameters of stripline include the characteristic impedance (Z_0), the phase velocity (v_p), and the propagation loss (α). The characteristic impedance is related to the distance between the conductor strips (s), the width of the conductor strips (w), and the dielectric permittivity (ϵ_r). The characteristic impedance can be approximated as

$$Z_0 = \frac{60}{\sqrt{\varepsilon_r}} \ln\left[\frac{4h}{w} + \sqrt{\left(\frac{4h}{w}\right)^2 + 2}\right] \tag{3.39}$$

The transmission speed is closely related to the dielectric constant and can be calculated using the following formula:

$$v_p = \frac{c}{\sqrt{\varepsilon_r}} \tag{3.40}$$

where c is the speed of light. Propagation loss mainly includes conductor loss and dielectric loss. Conductor loss is primarily caused by the resistance of the signal line and is influenced by the material, size, and frequency. Dielectric loss is mainly caused by the loss factor of the dielectric material. The total propagation loss can be calculated as the sum of the conductor loss and dielectric loss:

$$\alpha = \alpha_c + \alpha_d \tag{3.41}$$

where α_c represents the conductor loss, and α_d represents the dielectric loss. Both microstrip and stripline are widely used in low-frequency microwave circuits. Their characteristic impedance, transmission speed, and propagation loss are closely related to their geometric dimensions and dielectric properties. By designing the transmission line structure and parameters appropriately, the requirements of different circuits and systems can be met.

3.2.2.4 Reflection and Impedance Matching

In microwave circuits, reflection and impedance matching are crucial factors that affect performance. When the characteristic impedance of a transmission line does not match the impedance of a circuit element, reflected waves are generated, leading to power loss and signal distortion. The objective of impedance matching is to make the input impedance equal to the characteristic impedance of the transmission line, thus eliminating reflections.

1. Reflection coefficient and VSWR

To assess the level of reflection and matching, we introduce two parameters: the reflection coefficient (Γ) and the voltage standing wave ratio (VSWR). The reflection coefficient is defined as

$$\Gamma = \frac{Z_L - Z_0}{Z_L + Z_0} \tag{3.42}$$

where Z_L represents the load impedance and Z_0 is the characteristic impedance of the transmission line. The reflection coefficient takes values between -1 and 1. A perfect match is achieved when $\Gamma = 0$, while $\Gamma = 1$ or $\Gamma = -1$ indicates a perfect reflection. The voltage standing wave ratio (VSWR) is defined as the ratio of the maximum voltage to the minimum voltage:

$$VSWR = \frac{V_{max}}{V_{min}} = \frac{1 + |\Gamma|}{1 - |\Gamma|} \tag{3.43}$$

VSWR values range from 1 to infinity, with VSWR = 1 representing a perfect match and VSWR approaching infinity indicating a perfect reflection.

2. Impedance matching methods

In low-frequency microwave circuit simulation, impedance matching is typically achieved using the following methods:

a. Adjusting the impedance of a circuit element: By altering the size, shape, or material of a circuit element, its impedance can be modified to achieve matching.

b. Using an impedance transformer: Impedance transformers, such as quarter-wave transformers or double-reactance transformers, can be employed to transform between different impedances and achieve matching.

c. Adding matching networks: Matching networks, such as L-type, π-type, and T-type matching networks, can transform the input impedance to match the characteristic impedance of the transmission line.

3. Designing a matching network

The design of a matching network needs to consider various factors, including frequency range, impedance range, insertion loss, and complexity. Here are some common types of matching networks and their design considerations:

a. L-type matching network: The L-type matching network consists of two series and parallel passive components (inductors or capacitors) that facilitate impedance transformation. It can be configured as either high-resistance low-reactance or low-resistance high-reactance. During the design process, the values of the matching elements are calculated based on the desired impedance transformation and the load impedance.

b. π-type matching network: The π-type matching network comprises two parallel inductors (or capacitors) and one series capacitor (or inductor). This network enables a wider range of impedance transformation but increases the network's complexity.

During the design process, the values of the matching elements are iteratively calculated to meet the impedance transformation and frequency response requirements.

c. T-type matching network: The T-type matching network is similar to the π-type matching network but consists of two series inductors (or capacitors) and one shunt capacitor (or inductor). In certain scenarios, the T-type matching network can offer better matching performance compared to the π-type matching network. The design process also involves iteratively calculating the values of the matching elements.

d. Quarter-wave transformer: A quarter-wave transformer is a special transmission line segment that is one-quarter wavelength long at the operating frequency. When the input impedance is the complex conjugate of the characteristic impedance of the transmission line, the quarter-wave transformer achieves perfect matching. During the design process, it is necessary to calculate the appropriate transmission line length and characteristic impedance to achieve the desired impedance transformation.

4. Simulations and optimizations

Microwave circuit simulation typically involves the use of electromagnetic simulation software such as ADS, HFSS, or CST to analyze and optimize impedance matching. The simulation process encompasses the following steps:

a. Build a circuit model: Construct an electromagnetic simulation model based on the actual circuit structure and material parameters.

b. Set boundary conditions and ports: Define appropriate boundary conditions and ports based on the circuit's working conditions and connectivity.

c. Run the simulation: Solve the electromagnetic field equations and transmission line equations to obtain circuit performance indicators such as S-parameters and impedance.

d. Analyze the results: Evaluate the matching performance and identify optimization directions by examining simulation results, including S-parameters, impedance curves, and electric field distributions.

e. Optimization design: Adjust circuit parameters or modify the matching network structure to enhance the matching performance based on simulation results. Multiple iterations may be necessary until satisfactory performance is achieved.

f. Perform tolerance analysis: Evaluate the stability of circuit performance under component parameter variations.

g. Design the layout: Generate a layout design and output detailed design drawings for production and manufacturing purposes.

h. Review performance indicators and check the layout to ensure that design requirements are met. Address structural design considerations to ensure the circuit meets expected technical performance in practical applications.

Microwave circuit CAD technology provides a powerful tool for electronic design automation, allowing designers to simulate and optimize RF and microwave circuit designs on computers. These technologies reduce design complexity, shorten design cycles, and enhance circuit performance, leading to a significant revolution in RF and microwave circuit design. In summary, microwave circuit CAD technology offers an efficient platform for engineers to achieve more accurate and efficient circuit designs.

3.2.3 Equivalent Circuit Model of Microwave Circuits

In the simulations of low-frequency microwave circuits, the equivalent circuit model is a common method [35–38]. It simplifies complex electromagnetic field problems into circuit problems, making them easier to comprehend and calculate. This section presents equivalent models of circuit elements, the construction, and simplification of circuit models, as well as the utilization of SPICE models and S-parameters.

- Equivalent Model of Circuit Elements
  In low-frequency microwave circuits, common components such as inductors, capacitors, resistors, and transformers play vital roles. To facilitate analysis, these components are often represented by equivalent circuit models. Here are the equivalent models for some common components:
  (1) Resistor R: The ideal resistor can be represented by a simple equivalent circuit model consisting of a single resistance element R.
  (2) Capacitor C: The ideal capacitor can be modeled by including a capacitive element C, in its equivalent circuit. In practical circuits, capacitors also exhibit parasitic resistance and inductance. Therefore, the equivalent circuit model of a real capacitor typically includes a capacitive element $C_E Q$ in series with a resistance element $R_E SR$, and an inductance element $L_E SL$.
  (3) Inductor L: The equivalent circuit model of an ideal inductor consists of an inductive element L. In practical circuits, an inductor also exhibits parasitic resistance and capacitance. Therefore, the equivalent circuit model of a real inductor typically includes an inductive element L, in series with a resistance element R, and a shunt capacitance element C.
- Construction and Simplification of Circuit Model
  In low-frequency microwave circuits, the construction and simplification of circuit models are crucial steps in the simulation process. The following are the general steps involved in building and simplifying a circuit model:
  (1) Divide the circuit into several sub-circuits based on the actual circuit structure, such as amplifiers, filters, and mixers.
  (2) Establish the equivalent circuit model for each sub-circuit, including the equivalent models of components and their interconnections.
  (3) Combine the equivalent circuit models of the sub-circuits to form an overall equivalent circuit model of the entire circuit.
  (4) Simplify the equivalent circuit model by techniques such as merging series and parallel components, and eliminating redundant components. The aim of simplification is to reduce computational complexity while preserving critical information.
  (5) Perform circuit simulation analysis using the simplified equivalent circuit model, such as calculating parameters related to voltage, current, and power.
- SPICE Models and S-parameters
  In the simulation of low-frequency microwave circuits, SPICE models and S-parameters are commonly used description methods. The following is a brief description of these two methods:
  (1) SPICE model

SPICE (Simulations Program with Integrated Circuit Emphasis) is a widely used circuit simulation tool. The SPICE model provides a mathematical description based on circuit components and their interconnections. It includes information such as DC characteristics, AC characteristics, temperature characteristics, and more for the components. Through the SPICE model, various circuit simulation analyses such as DC analysis, AC analysis, and transient analysis can be easily performed. In low-frequency microwave circuits, SPICE models are used to describe components and interconnect structures, such as resistors, capacitors, inductors, transmission lines, switches, and more. When using the SPICE model for simulation, it is important to select appropriate component models and parameters to ensure accurate results.

(2) S-parameters

S-parameters are parameters that describe the characteristics of linear microwave networks. They represent the energy transmission and reflection relationships between different ports of the network. For an n-port network, the S-parameters can be represented by an $n \times n$ matrix as follows:

$$[S] = \begin{bmatrix} S_{11} & S_{12} & \cdots & S_{1n} \\ S_{21} & S_{22} & \cdots & S_{2n} \\ \vdots & \vdots & \vdots & \vdots \\ S_{n1} & S_{n2} & \cdots & S_{nn} \end{bmatrix} \tag{3.44}$$

$$S_{ij} = \left. \frac{b_i}{a_j} \right|_{a_j \neq 0,\ a_k = 0 \text{ for } k \neq j,\ k=1,2,\cdots,n} \tag{3.45}$$

where a_j is the input voltage wave at the jth port and b_i is the output voltage wave at the ith port in the n-port network. When $i = j$, S_{ii} represents the reflection coefficient at the input port; when $i \neq j$, S_{ij} represents the transmission coefficient from the jth port to the ith port. In the simulation of low-frequency microwave circuits, S-parameters are commonly used as a description method. They provide a convenient way to analyze the frequency response, impedance matching, isolation, and other performance indicators of the circuit. Additionally, S-parameters can be used to cascade and interconnect multiple networks, enabling the realization of more complex circuit functions. By employing these methods, the complex electromagnetic field problems can be simplified into circuit problems, making them easier to understand and calculate. In practical applications, it is essential to select the appropriate model and description method based on specific requirements to ensure the accuracy and effectiveness of the simulation results (Figs. 3.4 and 3.5).

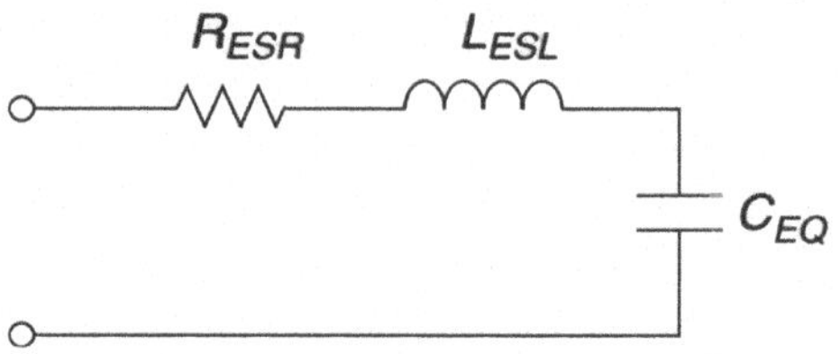

Fig. 3.4 Equivalent circuit of capacitor

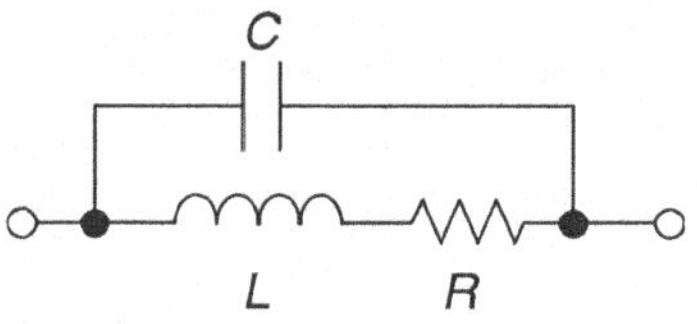

Fig. 3.5 Equivalent circuit of inductor

3.2.4 Finite Element Method and Finite-Difference Time-Domain Method

3.2.4.1 Finite Element Method (FEM)

The finite element method (FEM) is a widely used numerical method in the simulation of low-frequency microwave circuits [39–41]. It involves dividing the electromagnetic field region into numerous small finite elements, approximating the field distribution using interpolation functions (such as Lagrange interpolation functions), and establishing algebraic equations to solve the electromagnetic field problem. The FEM offers a significant advantage in handling complex geometries and material distributions.

The first step in FEM is to divide the electromagnetic field region into small finite elements, typically triangles or quadrilaterals. This process is known as meshing or mesh generation. To enhance calculation accuracy, an adaptive mesh technique can be employed to generate a denser mesh in areas requiring higher resolution, such as boundary layers and discontinuities. For each element, an interpolation function is utilized to approximate the field distribution. Commonly used interpolation functions include Lagrange interpolation function, Hermite interpolation function, and NURBS interpolation function. Taking the Lagrange interpolation function as an example, the field distribution, denoted by $E(x, y)$, can be approximated as

$$E(x, y) \approx N_1(x, y)E_1 + N_2(x, y)E_2 + N_3(x, y)E_3 \tag{3.46}$$

where N_1, N_2, and N_3 represent the interpolating functions, and E_1, E_2, and E_3 are the field values at the vertices of the element. The interpolation function satisfies the following property:

$$N_i(x_i, y_j) = \delta_{ij} \tag{3.47}$$

where δ_{ij} is the Kronecker delta function.

Next, the Maxwell equations are transformed into a weak form. The weak form implies that for a set of test functions (or weight functions) $\varphi_i(x, y)$, the following relation is satisfied

$$\int_\Omega \nabla \times (\nabla \times \mathbf{E}) \cdot \varphi_i d\Omega - k^2 \int_\Omega \varepsilon_r \mathbf{E} \cdot \varphi_i d\Omega = \int_\Omega \mathbf{J} \cdot \varphi_i d\Omega \tag{3.48}$$

where Ω denotes the electromagnetic field region, and $\mu(x, y)$ and $\varepsilon(x, y)$ denote the permeability and permittivity, respectively. The interpolation functions are substituted into the weak form equations, and the weighted residual method is used for their discretization. Specifically, the field distribution $E(x, y)$ is approximated by an interpolation function, and the test function $\varphi_i(x, y)$ is also approximated by an interpolation function. This leads to a system of linear algebraic equations:

$$[A]\cdot[E] = [b] \tag{3.49}$$

where $[A]$ is the stiffness matrix whose elements are determined by the interpolation function and the material parameters, $[E]$ is the vector formed by unknown field components (i.e., the field values at the vertexes of elements), and $[b]$ is a source vector whose elements are determined by the interpolation function and the source term.

Finally, by solving the system of linear algebraic equations, the values of the field at the vertices of each element can be obtained. Commonly used solution methods include direct methods (such as Gaussian elimination, LU decomposition) and iterative methods (such as conjugate gradient method, GMRES method). Iterative methods are often more computationally efficient for large-scale problems. Preconditioning techniques can be employed to accelerate convergence during the solution process. These techniques improve the condition number of the new system of equations by left-multiplying the original system of equations with an inverse matrix that approximates the stiffness matrix. Common preconditioners include incomplete LU factorization (ILU) preconditioner, incomplete Cholesky factorization (IC) preconditioner, etc.

After obtaining the field values at the vertices of each element, the field distribution of the entire electromagnetic field region can be obtained through the interpolation function. For example, for a Lagrange interpolating function, the field value at any point (x, y) can be calculated as

$$E(x, y) = N_1(x, y)E_1 + N_2(x, y)E_2 + N_3(x, y)E_3 \tag{3.50}$$

Thus, through the finite element method (FEM), the electromagnetic field problem in low-frequency microwave circuits can be solved. The advantage of FEM lies in its flexibility and ability to handle complex geometries, material distributions, and boundary conditions. However, the computational cost of FEM significantly increases with the size of the problem, necessitating efficient solvers and preconditioning techniques to improve computational efficiency. In practical applications,

FEM is often combined with other numerical methods (such as finite-difference time-domain method and time-domain integral equation method) to achieve efficient and accurate simulation of low-frequency microwave circuits.

3.2.4.2 Finite-Difference Time-Domain (FDTD) Method

The finite-difference time-domain (FDTD) method is a numerical method used to directly solve Maxwell's equations in the time domain [42]. FDTD discretizes the field components and their derivatives in space and time, transforming Maxwell's equations into difference equations. These difference equations are then solved iteratively to obtain the time-domain response of the field.

1. Yee grid

In the FDTD method, the electromagnetic field region is divided into small spatial grids. A commonly used grid is the Yee grid, which is a staggered and orthogonal grid. In the Yee grid, the electric and magnetic field components are located at the edges and centers of the grid, respectively. This staggered layout enhances the accuracy of discretizing the derivatives.

2. Central difference formula

To discretize the derivative terms in Maxwell's equations, the central difference formula is applied. For instance, in one dimension, Faraday's law:

$$\begin{aligned}\frac{E^{n+1}(x)-E^{n}(x)}{\Delta t}&=-\frac{H^{n+\frac{1}{2}}\left(x+\frac{\Delta x}{2}\right)-H^{n+\frac{1}{2}}\left(x-\frac{\Delta x}{2}\right)}{\Delta x}\\ \frac{H^{n+\frac{1}{2}}(x)-H^{n}(x)}{\Delta t}&=-\frac{E^{n+1}\left(x+\frac{\Delta x}{2}\right)-E^{n}\left(x-\frac{\Delta x}{2}\right)}{\Delta x}\end{aligned} \tag{3.51}$$

can be discretized as

$$\frac{E(x,t+\Delta t)-E(x,t)}{\Delta t}=-\frac{H\left(x+\frac{\Delta x}{2},t+\frac{\Delta t}{2}\right)-H\left(x-\frac{\Delta x}{2},t+\frac{\Delta t}{2}\right)}{\Delta x} \tag{3.52}$$

where Δt and Δx represent the discrete steps in time and space, respectively.

3. Time update

The FDTD method solves the difference equations through point-by-point iteration. Starting from the initial conditions, the electric field components at each grid point are updated iteratively. When updating the electric field component, the spatial variation of the magnetic field component, as dictated by Maxwell's equations, is taken into account. For example, in the one-dimensional case, the update equation for the electric field component is

$$E\left(x,t+\Delta t\right)=E\left(x,t\right)-\frac{\Delta t}{\epsilon_0\Delta x}\cdot\left[H\left(x+\frac{\Delta x}{2},t+\frac{\Delta t}{2}\right)-H\left(x-\frac{\Delta x}{2},t+\frac{\Delta t}{2}\right)\right] \tag{3.53}$$

Similarly, the magnetic field components are updated, considering the spatial variation of the electric field component. For example, in the one-dimensional case, the update equation for the magnetic field component is

$$H\left(x+\frac{\Delta x}{2}, t+\Delta t\right)=H\left(x+\frac{\Delta x}{2}, t\right)-\frac{\Delta t}{\mu_0 \Delta x} \cdot\left[E\left(x+\Delta x, t+\frac{\Delta t}{2}\right)-E\left(x, t+\frac{\Delta t}{2}\right)\right] \tag{3.54}$$

4. Boundary conditions
In practical problems, boundary conditions play a crucial role in FDTD simulations. The method must account for various boundary conditions, including conductive boundaries, magnetic boundaries, and absorbing boundaries. For a conductive boundary, the tangential component of the electric field at the surface of the conductor should be zero. For a magnetic boundary, the tangential component of the magnetic field at the surface of the magnetic material should be zero. Absorbing Boundary Conditions (ABC) are used to simulate infinite space. Common types of ABCs include Mur ABCs and Perfectly Matched Layer (PML).
5. Source excitation
In FDTD simulations, it is necessary to introduce source excitation to model practical excitation sources. The source stimulus can be a current source, a voltage source, a magnetic field source, or other types. The FDTD method typically employs hard sources or soft sources to introduce the excitation. A hard source directly replaces the field component, while a soft source adds the excitation source to the field component.
6. Post-processing and analysis
Once the iteration process is completed, the time-domain response at each grid point can be obtained. By Fourier transforming the time-domain response, the frequency-domain response of the field can be obtained. Post-processing tools enable the calculation of electromagnetic parameters such as reflection coefficient, transmission coefficient, and impedance by extracting the field components. These parameters help analyze the performance of low-frequency microwave circuits.

In summary, the finite-difference time-domain (FDTD) method is a powerful and flexible numerical method. It requires handling various boundary conditions, introducing source excitation, and performing post-processing analysis for practical applications involving low-frequency microwave circuits.

3.2.4.3 Time-Domain Integral Equation (TDIE)

Time-domain integral equation (TDIE) Method is a numerical method based on the time-domain integral equations [43]. TDIE directly obtains the time-domain response of the field by solving the integral equation of the electromagnetic field in the time domain. Compared with the FDTD method, the TDIE method offers higher accuracy and flexibility, but it requires more complex numerical algorithms, such as the time-domain method of moments (TDMoM). The basic principle and implementation of the TDIE method are described in detail below.

1. Derivation of time-domain integral equation
The time-domain integral equation is derived from Maxwell's equations and boundary conditions. Let's consider a simplified field problem where the electromagnetic field satisfies the following equations:

$$\begin{aligned} \nabla \times \mathbf{H}(\mathbf{x}, t) &= \mathbf{J}(\mathbf{x}, t) + \varepsilon \frac{\partial \mathbf{E}(\mathbf{x}, t)}{\partial t} \\ \nabla \times \mathbf{E}(\mathbf{x}, t) &= -\mu \frac{\partial \mathbf{H}(\mathbf{x}, t)}{\partial t} \end{aligned} \tag{3.55}$$

where $\mathbf{E}(\mathbf{x}, t)$ and $\mathbf{H}(\mathbf{x}, t)$ denote the electric and magnetic fields, respectively, and $\mathbf{J}(\mathbf{x}, t)$ represents the current source. The parameters ε and μ represent the permittivity and permeability, respectively. In this derivation, we assume a linear, isotropic, and lossless medium. To obtain the time-domain integral equation, we integrate Maxwell's equations in the time domain. For example, integrating the electric field equation yields

$$\int (\nabla \times \mathbf{H}(\mathbf{x}, t))\, dt = \int (\mathbf{J}(\mathbf{x}, t))\, dt + \varepsilon \int (\mathbf{E}(\mathbf{x}, t))\, dt \tag{3.56}$$

Then, using the Green function and boundary conditions, we transform the time-domain integral equation into integral equation form. For the electric field equation, we obtain the following time-domain integral equation:

$$\begin{aligned} \mathbf{E}(\mathbf{x}, t) = \iint G\left(\mathbf{x}, \mathbf{x}', t, t'\right) \cdot \Big[\mathbf{J}\left(\mathbf{x}', t'\right) \\ + \varepsilon \frac{\partial \mathbf{E}\left(\mathbf{x}', t'\right)}{\partial t'}\Big] d\mathbf{x}' dt' \end{aligned} \tag{3.57}$$

where $G(\mathbf{x}, \mathbf{x}', t, t')$ represents the Green's function, which describes the field propagation characteristics from the source point $\mathbf{x}'$ to the observation point $\mathbf{x}$ in the time domain. Similarly, we can obtain the time-domain integral equation for the magnetic field.
2. Time-domain method of moments (TDMoM)
The time-domain method of moments (TDM) is a common numerical algorithm for solving time-domain integral equations (TDIEs). TDMoM discretizes the time-domain integral equation into a system of algebraic equations and solves them using the method of moments [44]. The following describes the basic steps of TDMoM.

First, the electromagnetic field region is divided into many small cells, such as triangular cells or quadrilateral cells, and the discrete values of the field on the vertices or edges of the cells are defined. For each cell, the field distribution can be approximated using basis functions, such as linear or higher-order Lagrangian basis functions. For example, for the electric field $\mathbf{E}(\mathbf{x}, t)$:

$$\mathbf{E}(\mathbf{x}, t) \approx \sum_i \mathbf{E}_i(t)\psi_i(\mathbf{x}) \tag{3.58}$$

where $\mathbf{E}_i(t)$ denotes the value of the electric field at the vertex i, and $\psi_i(\mathbf{x})$ is the basis function.

Next, the time-domain integral equation is substituted into the basis function approximation, and the integral equation is transformed into a discrete equation using the weighted residual method or the Galerkin method. For the electric field equation, we obtain the following discrete equation:

$$\sum_j [Z_{ij}]\,\mathbf{E}_j(t) = \sum_j [V_{ij}]\,\frac{d\mathbf{E}_j(t)}{dt} + [I_i] \tag{3.59}$$

where the integral is performed over the region of the electromagnetic field. Writing the discrete system of equations in matrix form, we obtain the following system of linear algebraic equations:

$$[Z]\{\mathbf{E}\} = [V]\frac{d\{\mathbf{E}\}}{dt} + \{\mathbf{J}\} \tag{3.60}$$

where $[Z]$ and $[Y]$ denote the impedance and admittance matrices, respectively, $\mathbf{E}$ denotes the unknown electric field components, and $\mathbf{J}$ denotes the current source vector.

Finally, by solving the system of linear algebraic equations, we can obtain the time-domain response of the electric field at each vertex. Similarly, we can solve for the time-domain response of the magnetic field. Through the basis functions, we can obtain the field distribution of the entire electromagnetic field region.

3.2.4.4 Frequency-Domain Integral Equation (FDIE)

The frequency-domain integral equation method is based on the Fourier transform of the time-domain integral equation, which converts the impulse response in the time domain into a transfer function in the frequency domain [45]. This method has many advantages, such as high computational efficiency, good stability, and wide application.

1. Basic principle of integral equation in the frequency domain

The fundamental idea of the frequency-domain integral equation method is to transform the time-domain integral equation into the frequency domain using the Fourier transform, solve the field quantity in the frequency domain, and finally obtain the field quantity in the time domain through the inverse Fourier transform.

The Fourier transform is a mathematical method that converts a time domain signal into a frequency-domain signal and is expressed as

$$F(\omega) = \int_{-\infty}^{\infty} f(t)e^{-j\omega t}dt \tag{3.61}$$

where $F(\omega)$ represents the frequency-domain signal, $f(t)$ represents the time-domain signal, j represents the imaginary unit, and ω represents the angular frequency. The inverse Fourier transform is the method of converting a frequency-domain signal into a time-domain signal and is expressed as

$$f(t) = \frac{1}{2\pi}\int_{-\infty}^{\infty} F(\omega)e^{j\omega t}d\omega \tag{3.62}$$

In microwave circuit analysis, the time-domain integral equation is a mathematical model that describes the propagation process of the electromagnetic field and its general form is

$$E(t) = \int_{-\infty}^{\infty} h(t-\tau)I(\tau)d\tau \tag{3.63}$$

where $E(t)$ denotes the electric field strength, $I(\tau)$ denotes the current source, and $h(t-\tau)$ denotes the impulse response function. By converting the time-domain integral equation to the frequency domain using the Fourier transform, we obtain the frequency-domain integral equation:

$$E(\omega) = H(\omega)I(\omega) \tag{3.64}$$

where $E(\omega)$ and $I(\omega)$ represent the frequency-domain representations of the electric field strength and the current source, respectively, and $H(\omega)$ denotes the transfer function.

2. Frequency-domain integral equation solution methods

After obtaining the frequency-domain integral equation, the next step is to solve for the unknown field quantities. Generally, iterative methods, method of moments, and other techniques can be employed to solve the frequency-domain integral equation.

(1) Iterative method

The iterative method is a solution approach that decomposes complex problems into simpler ones. In the frequency-domain integral equation, we can decompose the electric field intensity into the internal source electric field and the scattered electric field:

$$E(\omega) = E_{\text{in}}(\omega) + E_{\text{sc}}(\omega) \tag{3.65}$$

where $E_{\text{in}}(\omega)$ represents the internal electric field, and $E_{\text{sc}}(\omega)$ represents the scattered electric field. Substituting Eq. 3.65 into Eq. 3.64, we obtain

$$E_{\text{in}}(\omega) + E_{\text{sc}}(\omega) = H(\omega)I(\omega) \tag{3.66}$$

The basic idea of the iterative method is to first solve for the internal electric field and then iteratively compute the total electric field using the scattered electric field. The specific process is as follows:

Initialize the number of iterations $K = 0$, set the convergence criterion ϵ, and initialize the electric field intensity $E^{(0)}(\omega) = E_{\text{in}}(\omega)$.

Calculate the scattered electric field $E_{sc}^{(K)}(\omega) = H(\omega)[I(\omega) - I_{in}^{(K)}(\omega)]$. Update the total electric field $E^{(K+1)}(\omega) = E_{in}(\omega) + E_{sc}^{(K)}(\omega)$. Check for convergence: if $||E^{(K+1)}(\omega) - E^{(K)}(\omega)|| < \epsilon$, stop the iteration; otherwise, set $K = K + 1$ and return to Step 2.

(2) Method of Moments

The method of moments (MoM) is a numerical integration-based method for solving integral equations in the frequency domain. In a microwave circuit, we can express the field quantities as linear combinations of basis functions:

$$E(\omega) = \sum_{n=1}^{N} a_n \phi_n(\omega) \tag{3.67}$$

where a_n represents the weight coefficient of the basis function, $\phi_n(\omega)$ represents the basis function, and N represents the number of basis functions.

By substituting Eq. 3.67 into Eq. 3.64, we can obtain

$$\sum_{n=1}^{N} a_n \phi_n(\omega) = H(\omega)I(\omega) \tag{3.68}$$

Through discretization and numerical integration, we can transform Eq. 3.68 into a system of linear equations:

$$\mathbf{A} \cdot \mathbf{x} = \mathbf{b} \tag{3.69}$$

where **A** denotes the matrix, **x** denotes the unknown vector of weight coefficients, and **b** denotes the vector of source terms. We can solve for the unknown field quantities by solving the system of linear equations. By dividing the frequency-domain interval into N subintervals, we can discretize Eq. 3.68 using the Gaussian integral formula as

$$\sum_{n=1}^{N} a_n \phi_n(\omega_i) = H(\omega_i)I(\omega_i), \quad i = 1, 2, \ldots, N \tag{3.70}$$

By substituting the discretized equations into the basis functions, we obtain

$$\sum_{n=1}^{N} a_n \int_{\omega_{i-1}}^{\omega_i} \phi_n(\omega)\phi_m(\omega)d\omega = \int_{\omega_{i-1}}^{\omega_i} H(\omega)I(\omega)\phi_m(\omega)d\omega, \quad m = 1, 2, \ldots, N \tag{3.71}$$

The system of linear Eq. 3.69 can be obtained by expressing Eq. 3.71 in matrix form. By solving the linear system Eq. 3.69, we can obtain the weight coefficient vector **x**. The field quantity $E(\omega)$ in the frequency domain can be obtained by substituting the weight coefficients into the linear combination Eq. 3.67 of the basis functions.

After solving the field in the frequency domain, we need to transform the field back to the time domain by inverse Fourier transform. According to the inverse Fourier transform formula Equation, we can obtain the field quantity in the time domain:

$$E(t) = \frac{1}{2\pi} \int_{-\infty}^{\infty} E(\omega) e^{j\omega t} d\omega \tag{3.72}$$

Through the above process, we can transform the time-domain problem in the microwave circuit into a frequency-domain problem and solve the field quantity in the frequency domain. The frequency-domain integral equation method has been widely used in microwave circuit design due to its high computational efficiency, good stability, and wide application [46–48].

3.2.5 *Electromagnetic Field and Circuit Coupling*

Electromagnetic-circuit coupling is a significant challenge in the design and analysis of microwave circuits as it involves the interaction between electromagnetic field equations and circuit equations. This section provides a detailed introduction to the simulation method of electromagnetic field-circuit coupling, as well as the coupling of electromagnetic field with thermal field and structural field. It covers the coupling principles, equation derivation, and solution methods for each coupling scenario [49–51].

3.2.5.1 Direct Coupling

The direct coupling method combines the electromagnetic field equations with the circuit equations to form a system of simultaneous equations. In microwave circuits operating at low frequencies, the electromagnetic field equations are typically represented by Maxwell's equations, while the circuit equations involve Kirchhoff's voltage law (KVL) and Kirchhoff's current law (KCL).

$$\sum V_k = 0 \quad \text{(KVL)} \tag{3.73}$$

$$\sum I_k = 0 \quad \text{(KCL)} \tag{3.74}$$

where V_K and I_K represent the voltage and current in the circuit, respectively.

The direct coupling method has the advantages of high precision and good stability. However, it needs to solve large-scale simultaneous equations, and the computational complexity is high.

3.2.5.2 Indirect Coupling

The indirect coupling method involves solving the electromagnetic field equation and the circuit equation separately. First, the electromagnetic field equations are solved to obtain the field response. This response is then used as an input to the circuit equation, which is subsequently solved to obtain the circuit's response. The advantage of this method is its relatively simple calculation process. However, it may encounter numerical stability issues. In practical applications, the choice of an appropriate electromagnetic field-circuit coupling simulation method depends on the problem's characteristics and the available computational resources. The direct coupling method is suitable for scenarios with high-precision requirements, while the indirect coupling method is suitable for scenarios with limited computing resources.

3.2.5.3 Coupling of Electromagnetic Field and Thermal Field

In low-frequency microwave circuits, the coupling between the electromagnetic field and the thermal field is a significant problem [52]. The energy loss in the electromagnetic field can cause temperature rise in structural materials, thereby affecting the circuit's performance and reliability. This coupling mainly arises from energy losses in the electromagnetic field, including conductor loss and dielectric loss. Conductor loss occurs due to Joule heating caused by the current flowing through the conductor, while dielectric loss results from the heat generated by the electric field in the medium. These energy losses are converted into heat, leading to an increase in the temperature of the structural material.

To describe the coupling between the electromagnetic field and the thermal field, it is necessary to solve the electromagnetic field equation and the heat conduction equation simultaneously. The heat conduction equation describes the propagation and scattering of heat in the structural material:

$$\frac{\partial T}{\partial t} = \alpha \nabla^2 T + \frac{Q(x,t)}{\rho c} \tag{3.75}$$

where T denotes temperature, α is the thermal conductivity, $Q(x,t)$ is the heat source, ρ is the density, and c is the specific heat capacity.

When solving the coupling problem between the electromagnetic field and the thermal field, both the direct coupling method and the indirect coupling method can be used. In the direct coupling method, the electromagnetic field equation and the heat conduction equation are solved simultaneously. In contrast, the indirect coupling method involves solving the electromagnetic field equation first and then using the response of the field as an input to the heat conduction equation as a heat source. The advantage of the indirect coupling method is its relatively simple calculation process, but it may encounter numerical stability issues. Moreover, the nonlinear and dynamic characteristics of the coupling problem need to be taken into account.

The analysis of the coupling between the electromagnetic field and the thermal field is of great significance in the design and optimization of low-frequency microwave circuits. For instance, in components such as power amplifiers, filters, and antennas, the coupling between the electromagnetic field and the thermal field may result in structural deformation, performance degradation, and reduced lifespan. By analyzing the coupling between the electromagnetic field and the thermal field, the design of these components can be optimized, thereby enhancing system performance and reliability.

3.2.5.4 Coupling of Electromagnetic Field and Structural Field

In low-frequency microwave circuits, the coupling between electromagnetic fields and structural fields can significantly impact circuit performance [53]. The electromagnetic forces can induce deformation in the structural material, thereby altering the electromagnetic response of the microwave device. The coupling between the electromagnetic field and the structural field manifests in two main aspects. First, the electromagnetic forces in the electromagnetic field act on the structural material, causing its deformation. This deformation can affect the electromagnetic response of microwave devices. Second, changes in the structural field result in variations in the geometric parameters of the electromagnetic field, which, in turn, influence the distribution and response of the electromagnetic field.

To describe the coupling between the electromagnetic field and the structural field, it is necessary to solve the electromagnetic field equation and the structural equation simultaneously. The structural equation is commonly expressed as a linear elasticity equation:

$$\rho \frac{\partial^2 \mathbf{u}}{\partial t^2} = \nabla \cdot \overline{\sigma} + \mathbf{F} \tag{3.76}$$

where $\mathbf{u}$ denotes the displacement, ρ is the density, $\overline{\sigma}$ is the stress tensor, and $\mathbf{F}$ represents the electromagnetic force.

When solving the coupling problem, the direct coupling method or the indirect coupling method can be used. The direct coupling method needs to solve the electromagnetic field equation and the structural equation simultaneously. The Maxwell equations are first combined with the linear elasticity equations. The simultaneous equations are then solved by the finite element method, the finite-difference method, or other numerical methods to obtain the response of the electromagnetic field and the structural field. The indirect coupling method needs to solve the electromagnetic field equation and the structural field equation separately. Then the electromagnetic force $\mathbf{F}$ is calculated according to the response of the electromagnetic field as the input, and the response of the structural field is obtained by solving the linear elastic equation. The calculation process of the indirect coupling method is relatively simple, but there may be numerical stability problems.

The coupling of the electromagnetic field and the structural field is widely used in low-frequency microwave circuits. For example, in high-power microwave devices, electromagnetic forces may cause thermal expansion and deformation of structural materials, thus affecting the electromagnetic response of the device. Additionally, in microwave transmission structures such as microstrip lines and waveguides, the deformation of structural materials may lead to impedance mismatch and signal attenuation due to the electromagnetic force.

3.2.6 Error Analysis and Compensation

In low-frequency microwave circuit simulation, errors can significantly impact the simulation results and reduce overall accuracy. Therefore, analyzing errors and implementing appropriate compensation methods are crucial for improving simulation accuracy [54]. This section primarily addresses error sources, error analysis methods, and commonly employed error compensation techniques.

The simulation errors in low-frequency microwave circuit analysis can arise from various sources. Firstly, geometric modeling errors play a significant role. Simplification and discretization of the geometric model during simulation can introduce deviations between the obtained results and the actual devices. Secondly, material parameter errors are also influential factors. Due to challenges in accurately measuring material parameters or mismatches between measurement conditions and actual working conditions, the material parameters employed in the simulation may differ from the real-world scenario. Thirdly, boundary condition errors can impact the simulation results. Describing the actual boundary conditions accurately is often challenging, and discrepancies in boundary condition settings can introduce errors in the electromagnetic field distribution. Lastly, numerical errors can be a potential concern. When employing numerical methods like the finite element method or the finite-difference time-domain method to solve microwave circuits, errors can arise due to discretization and truncation processes. Hence, it is crucial to consider and address these sources of error to enhance the accuracy of simulation results in low-frequency microwave circuit analysis.

To accurately assess simulation errors, the following error analysis methods can be employed. Firstly, error sensitivity analysis can be conducted by systematically varying model parameters, boundary conditions, and other factors to observe the corresponding changes in simulation results. This helps evaluate the impact of these factors on simulation accuracy. Secondly, grid independence analysis can be performed by comparing simulation results obtained using different grid sizes. This enables the evaluation of how grid size influences simulation accuracy. Next, convergence analysis can be conducted by observing the changes in simulation results during the iterative process, thereby assessing the impact of convergence on simulation accuracy. Finally, the simulation results can be compared with experimental data to evaluate the simulation error. By comprehensively applying these methods, a more comprehensive understanding of the sources and effects of simulation errors can be obtained, leading to improvements in simulation accuracy.

For mitigating errors in the simulation of low-frequency microwave circuits, the following compensation methods can be applied:

1. Geometric modeling refinement: By employing more detailed geometric modeling, the disparities between the simulated model and the actual device can be reduced.

2. Material parameter correction: Material parameters can be adjusted based on experimental data, aligning them more closely with the real-world conditions.

3. Optimization of boundary conditions: Optimizing boundary conditions according to the actual operating environment helps minimize errors introduced by boundary conditions.

4. Mesh refinement and adaptive mesh: Enhancing solution accuracy through mesh refinement, and utilizing adaptive mesh techniques to automatically adjust mesh size based on error distribution, enables improved simulation accuracy while reducing computational overhead.

5. High-order numerical methods: Leveraging high-order finite element methods, high-order finite-difference time-domain methods, and other similar approaches can enhance calculation accuracy.

6. Iterative convergence optimization: Optimizing iterative algorithms, improving convergence speed, and setting appropriate convergence criteria ensure accurate simulation results.

3.3 ADS for Low-Frequency Simulations of Microwave Circuits

3.3.1 Microwave Circuit Component Library

The Advanced Design System (ADS) offers a comprehensive library of low-frequency microwave circuit components, including various categories to cater to different design requirements:

1. Passive component library: This library comprises essential components like resistors, capacitors, and inductors. Designers can choose the appropriate models based on their specific needs.

2. Active component library: This library consists of transistors, field-effect transistors (FETs), diodes, and other semiconductor devices. These components are utilized for functions such as amplification, oscillation, and mixing.

3. Coupler and power divider library: This library encompasses a wide range of components used for signal distribution, synthesis, and detection. It provides various couplers and splitters for signal splitting and combining purposes.

4. Filter library: The filter library includes diverse types of filters for noise suppression, signal channel isolation, and signal quality enhancement.

5. Antenna library: This library comprises different types of commonly used antennas for wireless communication, facilitating transmission and reception tasks.

6. Transmission line library: This library contains various transmission line models such as microstrip, stripline, and coplanar waveguide. These transmission lines serve the purpose of signal transmission and impedance matching.

7. Nonlinear component library: This library encompasses mixers, frequency multipliers, limiters, and other components that play a vital role in signal modulation, demodulation, and frequency conversion.

ADS provides users with a wide range of component resources, supporting different models and customizable components. This extensive component library aids in efficient circuit simulation and optimization, ensuring that designers have the necessary resources to meet their evolving design needs.

3.3.2 Main Functions and Settings

- S-Parameter Simulations
  Scattering parameters (S-parameters) are vital for describing the performance of microwave circuits. In ADS, the S-parameter simulation function is extensively employed for the design and analysis of low-frequency microwave circuits. This simulation enables the quick evaluation of transmission characteristics, reflection characteristics, and insertion loss of the circuit. Additionally, S-parameter simulation facilitates impedance matching of the circuit and waveguide structure design. Within ADS, users can simulate the S-parameters of microwave circuits by specifying the simulation frequency range, port characteristics, and other related parameters. The simulation results can be visualized in formats such as Smith Chart, impedance matrix, transfer matrix, and more. These representations provide a foundation for optimizing and adjusting the circuit.
- Steady-State and Transient Analysis
  Steady-state analysis and transient analysis are two commonly used time-domain analysis methods in ADS. Steady-state analysis is primarily employed to solve the output response of microwave circuits in the steady state and is typically used for designing and optimizing frequency-domain circuits. On the other hand, transient analysis is used to solve the response of microwave circuits during transient processes, such as circuit performance during power-on, power-off, or signal switching. Transient analysis allows engineers to comprehensively understand the response characteristics of the circuit, facilitating circuit design optimization. Within ADS, users can select steady-state analysis or transient analysis based on their requirements and configure corresponding parameters, such as source signal type, initial conditions, and simulation time step. The simulation results are displayed in the form of time-domain waveforms, providing users with an intuitive evaluation of circuit performance.
- Harmonic Balance Analysis
  Harmonic Balance Analysis (HBA) is a frequency-domain analysis method used for nonlinear microwave circuits. By employing harmonic balance analysis, engineers can effectively solve the response of nonlinear circuits at multiple harmonic frequencies, enabling assessment of the impact of nonlinear effects on circuit performance. In ADS, harmonic balance analysis is applicable to the design and analysis of nonlinear circuits, including mixers, power amplifiers, and oscillators.

When performing harmonic balance analysis, users need to specify the harmonic order, port characteristics, and other relevant parameters. ADS will simulate the nonlinear circuit in the frequency domain based on the set parameters, generating performance indicators such as harmonic response and intermodulation distortion. These indicators provide a more accurate evaluation of the circuit's nonlinear characteristics, guiding circuit optimization and improvement.

- Electromagnetic Field Simulations
  Electromagnetic field simulation is a crucial capability in ADS that calculates the electromagnetic field distribution, radiation characteristics, and electromagnetic wave propagation properties of microwave circuits. In low-frequency microwave circuits, electromagnetic field simulation is employed for antenna design, electromagnetic shielding, electromagnetic compatibility, and other applications. ADS offers various built-in electromagnetic field solvers, including finite element method (FEM), finite-difference time-domain (FDTD) method, and integral equation method (IEM). Users can choose the appropriate electromagnetic field solver based on the circuit type and solution requirements. During electromagnetic field simulation, users need to configure parameters such as simulation frequency range, boundary conditions, and grid division.
- Analysis of Coupling between Circuit and Electromagnetic Field
  The analysis of coupling between the circuit and electromagnetic field is an advanced simulation method for complex microwave circuits in ADS. In low-frequency microwave circuits, analyzing the coupling between circuits and electromagnetic fields is applicable to the design and analysis of RF front-ends, couplers, filters, and other devices. ADS supports two-way coupling between the circuit and electromagnetic field, considering the effects of both circuit components.
- Optimization and Sensitivity Analysis
  Optimization and sensitivity analysis are essential functions in ADS that help in finding optimal design parameters for microwave circuits. Optimization analysis assists engineers in identifying optimal component parameters while meeting specific performance indicators. Sensitivity analysis evaluates the circuit's performance sensitivity to component parameters, guiding designers in conducting tolerance analysis and reliability assessment. Within ADS, users can set optimization objectives, constraints, optimization algorithms, and other parameters for optimization analysis. ADS supports various optimization algorithms such as gradient descent, genetic algorithm, particle swarm optimization, suitable for different types of microwave circuit optimization problems. When performing sensitivity analysis, users need to specify the circuit parameters and performance indicators for analysis. ADS calculates the impact of parameter changes on performance indicators and provides sensitivity coefficients and other information. The results of optimization and sensitivity analysis aid engineers in enhancing circuit reliability and stability while ensuring optimal performance.
- Parameter Scanning and Variable Setting
  Parameter sweep is a common feature in ADS used to evaluate the performance trend of a circuit with changing parameters. In low-frequency microwave circuit simulation, parameter scanning finds application in impedance matching, filter

design, oscillator stability analysis, and other fields. Users can define scanning parameters, scanning range, scanning step, and other information within ADS. ADS performs scanning simulations based on the specified parameters and generates performance index curves illustrating the changes in the parameters. ADS also allows users to set global or local variables for flexible control of circuit parameters. With variable settings, engineers can easily adjust circuit component parameters to meet different design requirements.

- Post-processing and Visualization of Simulations Results
  ADS provides a wide range of post-processing and visualization tools for analyzing and presenting simulation results. Users can utilize ADS's built-in data processing functions to smooth, differentiate, and integrate simulation results to obtain more valuable information. Furthermore, ADS supports various chart types, such as waveform chart, Smith Chart, vector chart, and contour chart, enabling users to intuitively view simulation results. ADS also facilitates data export, allowing users to export simulation results to text files, images, or other formats for further analysis and reporting. In conclusion, the post-processing and visualization capabilities of ADS in low-frequency microwave circuit simulation provide robust support for engineers to better understand and optimize circuit performance.
- Design Process and Automation
  In low-frequency microwave circuit simulation, ADS offers design flow and automation functions to enhance engineers' efficiency throughout the design cycle. ADS supports user-defined design flows, enabling the organization and execution of complex simulation tasks through an orderly flowchart. Additionally, ADS provides powerful automation scripting capabilities, supporting programming languages such as Python and Visual Basic. This allows users to write custom scripts for specific simulation tasks or data processing. The design process and automation capabilities not only improve engineers' productivity but also ensure consistency and repeatability of simulation results. By effectively utilizing the design process and automation capabilities of ADS, engineers can efficiently complete complex low-frequency microwave circuit design and analysis within a relatively short timeframe.

3.3.3 Practical Functions for Low-Frequency Design of Microwave Circuits

- Component Layout
  Component placement is the initial step in circuit placement and layout design, involving the arrangement of circuit components on the circuit board. When placing components, it is crucial to minimize component interactions, particularly for parts with high-frequency signals, high current, and high heat. Considerations should include signal integrity, electromagnetic compatibility, and heat dissipation. The location and spacing of components should comply with design rules

to prevent signal interference and short circuits. The component layout should also consider the convenience of the production process and assembly, aiming to minimize production costs and maintenance difficulties. Partitioning components according to circuit function facilitates management and debugging. In ADS software, users can drag components from the component library onto the circuit board, optimize the layout using alignment, spacing, and other tools, and enhance layout accuracy by setting grids, reference lines, and other auxiliary tools.

- Drawing of Conductor and Signal Layers
  The signal layer is responsible for carrying signal wires on the circuit board, including the top layer, bottom layer, and inner layers. When drawing wires and designing the signal layer, parameters such as wire width, spacing, and length, as well as considerations for crosstalk, reflections, trends, and layout, impact electromagnetic interference and anti-interference capabilities. In ADS software, users can draw wires using tools like lines and arcs, set parameters such as wire width and spacing, and design the signal layer using wire models and layer structures.
- Circuit Board Layering
  Circuit board layering involves dividing the circuit board into layers, each serving a specific function such as signal transmission, power distribution, and ground planes. Proper coupling between signal layers, ground layers, and power supply layers helps reduce electromagnetic interference and improve signal integrity. The signal layer should have a stable reference plane to ensure reliable signal transmission.
- Rule Check
  Design rule checking (DRC) is the final step in circuit placement and layout design. It is used to verify whether the design complies with specified design rules, such as component spacing, wire width, and line spacing. Ensuring qualified DRC results is essential for successful circuit production, effectively avoiding problems and risks during manufacturing. During the DRC process, it is necessary to establish appropriate design rule sets to diagnose and address any issues identified. This ensures that the circuit design adheres to the required rules. When resolving DRC problems, caution must be exercised to avoid any negative impact on circuit performance.

3.3.4 Circuit Modeling and Parameter Extraction

Circuit modeling and parameter extraction play a crucial role in low-frequency microwave circuit simulation. Achieving accurate circuit modeling and extracting device parameters are essential for ensuring the precision of simulation results. This section provides a comprehensive introduction to the theory and methods of constructing equivalent circuit models, extracting parameters, and modeling circuit elements.

3.3.4.1 Construction of Equivalent Circuit Model

The construction of an equivalent circuit model is a crucial step in simulating low-frequency microwave circuits, offering important functions and settings. The following steps outline the detailed process and common methods involved in building an equivalent circuit model:

1. Analyze the structure and operating principles of the microwave circuit: Thoroughly analyze the structure of the microwave circuit to be modeled, understand its working principles, and identify the main performance indicators. This step helps determine the circuit's key electromagnetic characteristics, such as transmission characteristics, impedance characteristics, scattering parameters, and more.

2. Select appropriate equivalent circuit elements: Based on the microwave circuit's characteristics, select suitable equivalent circuit elements to accurately represent its electromagnetic structure. For instance, an antenna may be represented by a radiation resistor and resonant capacitor, while a filter could be represented by a series or parallel combination of LC resonators. Choosing the appropriate circuit components is critical for establishing an accurate equivalent circuit model.

3. Extract the parameter values of circuit elements: Extract the parameter values of the equivalent circuit elements using electromagnetic field simulation software or experimental measurement data. The accuracy of parameter extraction directly affects the performance of the equivalent circuit model. Parameters such as resistance, capacitance, inductance, and transmission line characteristics can be obtained through simulation software or experimental measurements.

4. Build the equivalent circuit model: Using circuit simulation software, construct the equivalent circuit model based on the extracted parameters. Through simulation, analyze the circuit's performance, such as impedance characteristics, frequency response, phase response, and more. The construction of the equivalent circuit model simplifies the electromagnetic structure into a combination of circuit elements for circuit-level analysis and simulation.

5. Validate and optimize the equivalent circuit model: Compare the simulation results with experimental data to validate the accuracy of the equivalent circuit model. If necessary, optimize the parameters to achieve better fitting results. Validating and optimizing the equivalent circuit model is crucial to ensure its accuracy and reliability. Constructing an equivalent circuit model offers the advantage of significantly simplifying the electromagnetic field problem, reducing computational complexity, and improving simulation efficiency.

However, it is important to acknowledge the limitations of an equivalent circuit model. For microwave circuits with high frequencies or complex structures, the equivalent circuit model may not accurately capture their electromagnetic characteristics. Therefore, in practical applications, it's necessary to choose appropriate modeling methods based on specific problems and requirements, taking into account the coupling effects between the electromagnetic field and the circuit.

3.3.4.2 Parameter Extraction Method

In low-frequency microwave circuit simulation, the parameter extraction method is a crucial step in obtaining the necessary parameters for the equivalent circuit model. The parameter values of circuit elements can be obtained by processing electromagnetic field simulation data or actual measurement data. This section will provide a detailed description of commonly used parameter extraction methods, as well as recent research progress and references.

1. Network parameter method:

The network parameter method is a widely used parameter extraction method that describes the performance parameters of microwave circuits, such as S-parameters, Y-parameters, and Z-parameters. The parameters of the equivalent circuit model can be extracted by analyzing the network parameters in electromagnetic field simulation data or actual measurement data. This method is applicable to various microwave circuit elements, such as filters, couplers, and amplifiers.

2. Time-domain reflectometry (TDR):

Time-domain reflectometry (TDR) is a parameter extraction method based on time-domain signal analysis. By analyzing the TDR waveform from simulation data or actual measurement data, parameters of circuit elements, such as characteristic impedance and transmission line length, can be extracted. This method is suitable for microwave circuit components such as transmission lines and interconnection structures.

3. Optimization algorithm method:

The optimization algorithm method uses global or local optimization algorithms (such as genetic algorithm, particle swarm optimization, and simulated annealing) to search for the best parameter values within a given range. The objective is to minimize the error between the simulation results of the equivalent circuit model and the actual measurement or electromagnetic field simulation data. This method can handle nonlinear and multivariable problems, but it may be computationally expensive and time-consuming.

4. Analytic method:

For microwave circuits with regular structures, the parameter values can be directly calculated using analytic methods. For example, a transmission line filter's circuit parameters can be calculated by resolving its transmission matrix and scattering matrix. This approach may provide high accuracy for specific types of circuits, but it may not be generally applicable.

5. Parameter extraction method based on machine learning:

In recent years, machine learning technology has achieved remarkable results in various fields, including microwave circuit parameter extraction. Machine learning methods, such as support vector machines, neural networks, and deep learning, have made progress in this field. By training the machine learning model with a large amount of input and output data, circuit parameters can be quickly and accurately extracted. The advantage of this method is its ability to handle nonlinear and high-dimensional problems with fast computation. However, it requires a substantial amount of training data and an appropriate model structure.

6. Parameter extraction method based on computational electromagnetics:
The parameter extraction method based on computational electromagnetics calculates the distribution and interaction of electromagnetic fields by solving Maxwell's equations. By processing the electromagnetic field data, the parameters of circuit components can be extracted. This method offers higher accuracy and is suitable for various circuit types. However, it may be computationally expensive and time-consuming.

3.3.4.3 Characteristic Modeling of Circuit Elements

In low-frequency microwave circuit simulation, the characterization of circuit element behavior is a crucial aspect. Different types of components exhibit different characteristics, thus requiring various modeling methods to accurately describe their behavior. Here are some commonly used approaches for modeling component characteristics:
1. Frequency-dependent model:
For elements like resistors, capacitors, and inductors, their impedance typically varies with frequency. Frequency-dependent models can describe the frequency response characteristics of these components. Common approaches include distributed parameter models, equivalent circuit models, and transmission line models. In recent years, researchers have also proposed frequency-dependent models based on fractional differentiation and integration to capture microwave components with non-integer power frequency dependence.
2. Temperature-dependent model:
Many circuit element parameters change with temperature. Temperature-dependent models can capture the temperature characteristics of components. By introducing the temperature coefficient into the element model, the variation of element parameters with temperature can be accounted for. Additionally, some researchers have proposed temperature-dependent models based on neural networks and support vector machines to more accurately simulate temperature behavior.
3. Nonlinear model:
For nonlinear elements such as diodes and transistors, their characteristics can be described using nonlinear equations, look-up tables, and other methods. In recent years, techniques like polynomial fitting and neural networks have made significant progress in nonlinear modeling. For instance, some researchers have proposed nonlinear models based on radial basis function neural networks, which can provide more accurate representations of nonlinear elements' behavior.
4. Noise model:
Noise is an important performance indicator in microwave circuits. Noise models describe the noise characteristics of components. By introducing parameters like noise figure and noise exponent, the noise characteristics of components can be modeled. Recently, some researchers have also proposed noise models based on stochastic differential equations and stochastic integral equations to achieve more accurate characterization of noise behavior.

5. Surface mount component model:
Surface mount components, such as surface mount resistors, capacitors, and inductors, are widely used in microwave circuits. Due to their small size and fast mounting characteristics, surface mount elements exhibit high parasitic effects. Hence, it is necessary to consider parasitic resistance, parasitic capacitance, parasitic inductance, and other factors when establishing surface mount component models. Generally, multipole network models or transmission line models can be used to describe the characteristics of surface mount elements. In recent years, researchers have proposed surface mount element models based on the finite element method and the finite-difference time-domain method to achieve more accurate characterization of their electromagnetic behavior.
6. Electromagnetic coupling model:
In high-density and highly integrated microwave circuits, the impact of electromagnetic coupling effects on circuit performance cannot be ignored. By introducing parameters such as mutual inductance and coupling capacitance, the electromagnetic coupling effect between circuit elements can be described. Additionally, the influence of circuit board materials, wiring rules, and other factors on the coupling effect must be considered. In recent years, several electromagnetic coupling models based on the time-domain integral equation method and frequency-domain integral equation method have been proposed to describe the electromagnetic coupling characteristics of complex microwave circuits.
7. Bandwidth and group delay model:
Bandwidth and group delay are key performance metrics for frequency-selective circuits like filters and amplifiers. By modeling the impedance characteristics of circuit elements, the bandwidth and group delay response of the circuit can be obtained. Common modeling methods include the transfer matrix method and the scattering parameter method. Recently, some researchers have also proposed bandwidth and group delay models based on matrix decomposition and sparse representation to accurately describe the characteristics of complex microwave circuits.

In low-frequency microwave circuit simulation, accurately modeling the characteristics of circuit elements is vital. By selecting appropriate methods and techniques, precise equivalent circuit models can be established, enhancing the accuracy and reliability of simulations. In practical applications, it is essential to choose the appropriate modeling method based on specific problems and requirements, while considering

3.3.5 Advanced Analysis Function

3.3.5.1 System-Level Simulations

This section provides a detailed overview of the application of system-level simulation in the design of low-frequency microwave circuits. System-level simulation involves simulating and analyzing the entire system, considering the interaction between subsystems and components, to evaluate system performance, identify potential issues, and optimize the design. The main aspects of system-level simulation are as follows:

1. Subsystem Definition and Modeling:

In circuit design, a subsystem refers to a separate component or module with a specific function, such as an amplifier, filter, and mixer. In system-level simulation, the subsystem can be described using an equivalent circuit model or a behavioral model.

(1) Equivalent circuit model:

This is a physical model based on the parameters of actual circuit components, including resistors, capacitors, inductors, diodes, transistors, and other basic components. The equivalent circuit model accurately describes the circuit's performance under different operating conditions, but it involves high computational complexity.

(2) Behavioral model:

This is a mathematical model based on input-output characteristics, implemented through algebraic equations, differential equations, or look-up tables. Behavioral models are less computationally complex but may not capture all the performance characteristics of the circuit.

2. Construction of System-level Signal Flow Graph:

A signal flow graph is a graphical representation used to depict the transmission and interaction of signals between various subsystems in the system. In software tools like ADS, system-level simulation is achieved by connecting the input and output ports of each subsystem to construct the signal flow graph. The construction of the signal flow graph needs to consider the signal transmission characteristics between subsystems, including amplitude, phase, delay, etc. Additionally, the roles of the signal source and the load in the overall system need to be taken into account.

3. Parameter Scanning and System Performance Evaluation:

Parameter scanning involves changing one or more system parameters during the simulation to observe their impact on system performance. System performance evaluation involves assessing the system's operating state under different parameters by calculating various performance indicators such as signal-to-noise ratio, bit error rate, and signal distortion. ADS provides parameter scanning functionality, allowing users to specify parameter ranges, step sizes, and the performance metrics of interest. ADS will automatically perform the parameter scan and output the simulation results.

4. System Optimization and Sensitivity Analysis:

System optimization aims to find the parameter combination that optimizes system performance while satisfying certain constraints. In ADS, system optimization can be achieved using the built-in optimizer. The optimizer searches the parameter space using various algorithms like gradient descent, genetic algorithms, etc., to find the optimal solution. Users need to define objective functions (e.g., minimizing bit error rate, maximizing signal-to-noise ratio) and constraints (e.g., power consumption, size, cost). Sensitivity analysis involves evaluating the system's performance sensitivity to changes in key parameters. It helps identify potential design problems and improve system reliability and stability. In ADS, sensitivity analysis can be performed by calculating partial derivatives of performance indicators with respect to parameter variations. Users need to select the relevant parameters and performance indicators, and ADS will calculate and output the results of sensitivity analysis. Key parameters in low-frequency microwave circuit design may include component values (e.g., resistors, capacitors, inductors), transistor parameters (e.g., cutoff frequency, maximum oscillation frequency), and process parameters (e.g., wire width, layer thickness). Sensitivity analysis helps designers understand the impact of these parameters on system performance, enabling targeted optimization efforts.

3.3.5.2 Noise Analysis

Noise analysis plays a crucial role in the design of low-frequency microwave circuits as it helps identify and mitigate the impact of random electrical signals in the circuit. These signals, commonly known as noise, are typically caused by thermal noise and shot noise from electronic devices. Noise analysis encompasses the following key aspects:

1. Noise Source Identification and Modeling:

The first step is to identify and model the noise sources in low-frequency microwave circuits. The main contributors to noise include resistors, diodes, transistors, and other devices. The noise in these components arises from thermal noise, shot noise, and tunnel noise. Accurate noise analysis requires noise models for these devices, which typically involve parameters such as noise power spectral density, noise figure, and noise temperature.

2. Noise Propagation and Coupling Analysis:

In low-frequency microwave circuits, noise propagates and couples through the circuit, potentially causing signal distortion and degrading system performance. A detailed analysis of noise propagation paths and coupling mechanisms within the circuit is essential. Common analysis methods include the transfer matrix method, signal flow graph method, and network parameter method. These techniques enable understanding of how noise travels and interacts with circuit components.

3. Calculation and Evaluation of Noise Indices:

The next step involves calculating and evaluating noise performance in the circuit. Commonly used noise metrics include noise figure, noise factor, and noise temperature. Noise figure (NF) is a particularly important indicator, representing the ratio of output signal-to-noise ratio to input signal-to-noise ratio. In tools like ADS, these noise metrics can be computed by simulating the circuit in the frequency or time domain.

4. Noise Optimization and Suppression Strategies:

To improve the performance of low-frequency microwave circuits, effective noise optimization and suppression measures are necessary. Common strategies include selecting low-noise devices, optimizing device bias conditions, improving circuit topology and layout, and implementing noise suppression techniques. These efforts aim to minimize the impact of noise on the circuit's functionality and overall performance.

By conducting noise analysis on low-frequency microwave circuits, a comprehensive understanding of noise performance and its influencing factors can be obtained, leading to targeted optimization strategies. Noise analysis finds important applications in wireless communication systems, radar systems, microwave receivers, and other fields. It improves the signal-to-noise ratio, enhances receiver sensitivity, and ultimately enhances overall system performance.

It is important to note that noise analysis is a complex field that requires in-depth knowledge of circuits and a mathematical background. In practical applications, professional electromagnetic simulation software is often used to assist with noise analysis and optimization design. Additionally, different circuit types and devices exhibit varying noise characteristics, necessitating the selection and adjustment of noise analysis methods and strategies based on the specific circumstances.

3.3.5.3 Nonlinear Distortion Analysis

Nonlinear distortion refers to the distortion of the input signal after passing through a circuit that contains nonlinear elements. In the design of low-frequency microwave circuits, nonlinear distortion analysis is particularly important as it directly impacts signal quality and overall system performance. Nonlinear elements in these circuits include nonlinear resistors, diodes, transistors, and others. The characteristic feature of a nonlinear element is the nonlinear relationship between the input and output signals, which can be expressed using a power series expansion:

$$y(t) = \sum_{n=1}^{\infty} a_n x(t)^n \tag{3.77}$$

where $y(t)$ and $x(t)$ denote the output and input signals, respectively, and a_n is the nonlinear coefficient. The modeling of nonlinear components needs to consider the electrical characteristics of the device, packaging effects, temperature changes, and other factors to achieve accurate nonlinear distortion analysis.

Nonlinear distortion measurement provides an index for evaluating the level of nonlinear distortion in a circuit. Key measurements include the total harmonic distortion (THD), intermodulation distortion (IMD), and compression point (CP). THD represents the ratio of harmonic components to the fundamental component in the output signal, serving as a measure of harmonic distortion. It is calculated using the formula:

$$\mathrm{THD} = 20\log_{10}\frac{\sqrt{\sum_{n=2}^{\infty} P_n}}{P_1} \tag{3.78}$$

where P_n is the nth harmonic power and P_1 is the fundamental power.

Intermodulation distortion refers to the generation of intermodulation components in the output signal when multiple input signals with different frequencies are present. Intermodulation distortion can be characterized by the intermodulation order (IMn) and intermodulation ratio (IMR). The intermodulation order is the ratio of the frequency spacing at which intermodulation products are generated to the input signal spacing, while the intermodulation ratio represents the amplitude ratio of the intermodulation products to the input signal. The compression point signifies the point at which the output signal power no longer increases linearly as the input signal power increases. Typically, the 1-dB compression point, which is abbreviated as P1dB, is used, representing the point at which the output signal power increases by less than 1 dB for a 1-dB increase in the input signal power. The compression point is an important measure of the amplifier's linear range.

Harmonic balance (HB), multi-frequency balance (MFB), and waveform relaxation (WR) are commonly used techniques in nonlinear distortion analysis.

The harmonic balance method is a frequency-domain-based nonlinear analysis approach primarily used to calculate the response of a nonlinear circuit under steady-state periodic excitation. The fundamental idea behind the harmonic balance method is to decompose the input signal into harmonic components. Then, the transfer function of each harmonic order in the frequency domain is calculated, and the harmonic components of the output signal are determined through the harmonic balance equation. The harmonic balance method offers computational efficiency but is less suitable for aperiodic excitation and time-varying circuits.

The multi-frequency balance method is an improved harmonic balance method used to analyze nonlinear distortion with multiple input signals. The main idea of the multi-frequency balance method is to decompose the input signal into multiple frequency components, calculate the transfer function of each frequency component in the frequency domain, and solve the intermodulation components of the output signal through the multi-frequency balance equation. This method is applicable to multi-frequency input signals.

The waveform relaxation method is a time-domain-based nonlinear analysis method primarily used to calculate the transient response of nonlinear circuits. The basic idea of the waveform relaxation method is to decompose the input signal into time segments, calculate the response of each time segment in the time domain, and solve the transient waveform of the output signal through the waveform relaxation equation. This method is suitable for aperiodic excitation and time-varying circuits.

The nonlinear distortion optimization strategy includes the following aspects:

1. Selection of devices and materials with lower nonlinearity.

2. Use of linearization techniques such as feedforward linearization and feedback linearization.

3. Optimization of circuit topology and parameters to reduce the effects of nonlinear distortion.

4. Nonlinear distortion compensation using digital signal processing techniques in system-level design.

By employing the above nonlinear distortion analysis methods and optimization strategies, designers can effectively identify and reduce nonlinear distortion during the low-frequency microwave circuit design process, thereby improving system performance and signal quality. In practical applications, nonlinear distortion analysis and optimization are often combined with other analysis methods (such as noise analysis and power integrity analysis) to comprehensively evaluate circuit performance. Additionally, during the nonlinear distortion optimization process, different performance indicators (such as linear range, gain, and noise figure) may need to be traded off to meet the specific requirements of the circuit.

3.3.5.4 Power Integrity Analysis

1. Power network modeling and simulation
In the design of low-frequency microwave circuits, power network modeling serves as the foundation for power integrity analysis. The power supply network can be described using an equivalent circuit model or a behavioral model. An equivalent circuit model represents the physical characteristics of the network by utilizing parameters of actual circuit elements, such as a RLC network. On the other hand, a behavioral model represents the network through mathematical relationships, such as transfer functions or impedance matrices.

Simulations play a vital role in power integrity analysis. When assessing the performance of a power network, it is necessary to simulate it under various operating conditions and input signals. In ADS, simulation can be conducted by setting the voltage source and load conditions of the power network and selecting the appropriate simulation method (e.g., DC analysis, AC analysis, transient analysis, etc.) to perform power integrity simulation.

2. Power supply noise and interference analysis
In low-frequency microwave circuits, power supply noise and interference significantly impact system performance. Power supply noise typically includes internal noise originating from switches and regulators within the power converter, as well as external noise resulting from parasitic capacitance, inductance, and resistance present on the power lines. Power supply interference refers to undesired signals from other systems or devices that can infiltrate the power supply network through electromagnetic coupling, conductance coupling, and other means, thereby affecting circuit performance. To analyze and mitigate power supply noise and interference, the following aspects need to be considered:

- Impedance characteristics of the power supply network
- Electromagnetic shielding design for the power supply
- Effects of parasitic elements and methods of suppression
- Design and implementation of filters

3. Time-domain and frequency-domain response characteristics
Time-domain and frequency-domain response characteristics are two crucial factors in power integrity analysis. Time-domain response characteristics encompass the dynamic changes of the power supply network over time, including power supply voltage waveforms, transient responses, voltage stability, and more. These time-domain response characteristics are critical for evaluating power supply performance under transient load variations and sudden events. On the other hand,

frequency-domain response characteristics reflect the behavior of the power supply network in the frequency domain, including power supply impedance, power supply noise spectrum, frequency response, etc. The frequency-domain response characteristics facilitate the analysis of power supply stability and anti-interference capabilities across different frequencies. In ADS, the time-domain and frequency-domain response characteristics can be obtained through transient analysis and AC analysis functions. Transient analysis calculates the circuit's response in the time domain based on given input signals and initial conditions, while AC analysis determines the circuit's response in the frequency domain within a specified frequency range.

4. Power supply design and optimization

Power supply design and optimization encompass the following aspects:

- Selecting the appropriate power supply topology: Different power supply topologies exhibit distinct performance characteristics. For instance, switching power supplies offer high energy conversion efficiency, whereas linear power supplies feature low-noise levels.
- Ensuring impedance matching in the power supply network: Impedance matching is crucial for minimizing power supply noise and interference. The impedance characteristics of the power supply network are influenced by factors such as power line length, layout, parasitic components, and more.
- Optimizing power supply network layout and parameters of parasitic elements: Regarding power supply network layout, factors such as wiring sequence, line width, line spacing, and other considerations pertaining to power lines should be taken into account. For parasitic elements, the selection and distribution of capacitors, inductors, and resistors must be carefully optimized.
- Designing an effective power management strategy: An effective power management strategy enhances system stability and reliability. This entails aspects such as switching control, voltage regulation, and power supply protection. In low-frequency microwave circuits, the power management strategy should be adaptable to the specific requirements of the system and its operating environment.

3.4 HFSS for Low-Frequency Simulations of Microwave Circuits

3.4.1 Microwave Circuit Component Library and Model Import

Low-frequency microwave circuits find widespread applications in communication, radar, and remote sensing systems. HFSS is a kind of commonly used electromagnetic simulation software that provides an extensive library of low-frequency microwave circuit components. This library includes various types of components, such as transmission line components like microstrips, coaxial lines, and waveguides; coupler components like four-terminal couplers, directional couplers, and mixers; filter components like low-pass filters, band-pass filters, and high-pass filters; amplifier elements like unipolar amplifiers, bipolar amplifiers, and field-effect tube amplifiers; antenna components like dipole antennas, patch antennas, and array antennas; switch and phase modulator components such as single-pole-single-throw (SPST) switches, single-pole-double-throw (SPDT) switches, and PIN diode phase modulators; as well as VSWR and matching network components like Π-type matching networks and T-type matching networks. These components can be parameterized according to specific requirements to meet the diverse needs of designers. With its rich and diverse component library, HFSS enables the selection of suitable components for simulation and optimization based on specific application scenarios.

In the simulation of low-frequency microwave circuits in HFSS, besides selecting the appropriate component library, it is also necessary to configure the parameterization of the components. The variable manager allows setting parameters such as the length, width, and spacing of the microstrip lines for the components, which can be adjusted in subsequent simulations. HFSS also supports parameter scanning function, which enables simulating multiple parameter values simultaneously to evaluate circuit performance under different parameter combinations. Additionally, HFSS supports model import from various file formats, such as CAD files and Gerber files, eliminating the tedious process of manual modeling. When importing a model, attention should be given to adjusting the coordinate system and units to ensure correct representation in the simulation environment. Furthermore, material parameters for the imported model, such as permittivity and permeability, can be set to enhance simulation accuracy.

3.4.2 Geometric Modeling and Material Parameter Setting

3D geometric modeling plays a crucial role in the simulation of low-frequency microwave circuits. An accurate geometric model better represents the actual structure and electromagnetic characteristics of the circuit. HFSS offers a range of modeling tools and operations that facilitate the creation of complex 3D structures. Here are some key points and tips to enhance the effectiveness of 3D geometry modeling:

1. Choose the appropriate coordinate system: Selecting the right coordinate system ensures consistency and simplifies subsequent operations.
2. Utilize basic geometry: Employing basic geometric shapes, such as cubes, cylinders, and spheres, provides a foundation for constructing more complex structures.
3. Employ Boolean operations: Boolean operations like union, subtraction, and intersection enable combining or subtracting different geometric shapes to create complex structures.

4. Utilize parametric modeling: Parameterize the geometric model by defining variables, which allows for easy modification and exploration of different design scenarios.
5. Employ copy and pattern operations: Copying and patterning operations enable efficient replication of geometric elements, reducing modeling time.
6. Import third-party CAD models: HFSS supports the import of CAD models, allowing users to leverage existing designs and streamline the modeling process.

By following these methods, the modeling process becomes more streamlined, saving time and ensuring accuracy. Next, let's discuss how to set material parameters to accurately represent the electromagnetic characteristics of the materials.

Setting material parameters is a crucial step in the simulation of low-frequency microwave circuits. Users can select appropriate linear or nonlinear, isotropic or anisotropic, frequency-dependent or frequency-independent material models to enhance the accuracy and efficiency of the simulation. Some key material parameters to consider are

1. Conductivity of metal materials: Set the conductivity to accurately model the behavior of metal components.
2. Relative permittivity and dielectric loss factor of dielectric materials: Specify these parameters to represent the electrical properties of dielectric materials.
3. Relative permeability and magnetic loss factor of magnetic materials: Set these parameters to capture the magnetic behavior of magnetic materials.
4. Temperature dependence: For temperature-sensitive materials, incorporate temperature-dependent characteristics into the material parameters.

The HFSS provides nonlinear material models for materials with nonlinear properties such as ferromagnetic materials, ferroelectric materials, and semiconductor materials. Anisotropic materials can have electromagnetic parameters defined in different directions. It is also possible to import measured material parameters, improving the simulation accuracy and reducing manual parameter setting errors.

Accurate 3D modeling, combined with appropriate material parameter settings, lays the foundation for boundary condition setup, meshing, optimization, and electromagnetic field-circuit coupling analysis. These further enhance the accuracy and efficiency of the simulation.

3.4.3 Boundary Conditions and Meshing

Boundary conditions play a critical role in electromagnetic field simulations as they define the characteristics of the field distribution on the simulation area's boundaries. Accurate setting of boundary conditions is essential for precise low-frequency microwave circuit simulations. HFSS provides several types of boundary condition, including Perfectly Electric Conductor (PEC), Perfectly Magnetic Conductor (PMC), Absorbing Boundary Condition (ABC), Waveguide Wall, and Radiation Boundary Condition.

- Perfectly Electric Conductor (PEC): The PEC boundary condition assumes infinite conductivity on the boundary surface, resulting in zero tangential component of the electric field strength at the boundary surface.
- Perfectly Magnetic Conductor (PMC): The PMC boundary condition assumes infinite permeability on the boundary surface, leading to zero tangential component of the magnetic field strength at the boundary surface.
- Absorbing Boundary Condition (ABC): Absorbing boundary conditions simulate the propagation of electromagnetic waves from the simulation area to infinity, effectively absorbing the energy of the waves.
- Waveguide Wall: This boundary condition is suitable for simulating transmission line structures like waveguides or microstrips, reducing the simulation area's size.
- Radiation Boundary Condition: The radiation boundary condition simulates the radiation of electromagnetic waves at infinity.

Selecting the appropriate boundary condition type and adjusting the related parameters based on the characteristics and requirements of the low-frequency microwave circuit is crucial for optimizing simulation results. Properly defined boundary conditions ensure the accuracy and reliability of the simulation outcomes.

In electromagnetic field simulations, mesh generation is a fundamental step in discretizing the simulation domain, significantly influencing the accuracy and computational efficiency of the results. HFSS employs the finite element method (FEM) and automatically performs meshing and optimization.

Tetrahedrons serve as the basic unit for mesh generation due to their adaptability to complex geometric structures, adaptive refinement capabilities, and high computational efficiency. Mesh optimization dynamically adjusts the grid density based on the electromagnetic field's distribution characteristics, following the principle of increasing density in areas with significant changes and reducing density in areas with minor changes.

Users can manually define mesh parameters, such as setting local refinement areas, adjusting global density parameters, and specifying special mesh properties. Employing reasonable meshing and optimization methods ensures accurate simulation results and computational efficiency. When simulating low-frequency microwave circuits, it is vital to choose an appropriate meshing and optimization strategy.

It is recommended to use HFSS's adaptive mesh optimization function and follow the default settings. For complex structures or critical performance parameters, consider manually setting local refinement areas or adjusting global mesh parameters. For coupled field problems, carefully consider the meshing strategy to ensure consistency and accuracy. During the simulation, monitor convergence and adjust the grid parameters as necessary.

3.4.4 Post-processing of Simulations Results

In low-frequency microwave circuit simulation, the calculation of circuit parameters (such as S-parameters, impedance, and capacitance) is the key indicator to evaluate the circuit performance. The S-parameter describes the characteristics of the circuit under high-frequency conditions, including transmission loss and reflection loss. The S-parameter matrix can be calculated by solving the distribution of electromagnetic field through HFSS. Impedance is a measure of the change in the impedance current of a circuit element in an AC circuit, which can be calculated by solving the distribution of the electromagnetic field through HFSS. Capacitance represents the ability of a circuit element to store charge in an AC circuit, which can also be calculated by solving the distribution of electromagnetic fields through HFSS. In addition, admittance parameter (Y-parameter), system impedance (Z_{sys}), characteristic impedance (Z_0), efficiency (η), and power loss (P_{loss}) are also important parameters to evaluate the performance of the circuit. They can be calculated by setting the port mode and mode coefficients in HFSS. The accuracy of the simulation can be ensured by properly setting the boundary conditions, the operating frequency, and the waveguide modes. To sum up, HFSS can be used to calculate various circuit parameters of low-frequency microwave circuits and evaluate the circuit performance.

In the design of low-frequency microwave circuits, visualizing the electromagnetic field distribution is crucial for understanding the interaction mechanism between the electromagnetic field and the circuit. This includes visualizing the electric field intensity, magnetic field intensity, and current density. HFSS software provides the capability to intuitively display the changing electromagnetic field distribution through three-dimensional vector diagrams. Users can select the electric or magnetic field strength and adjust the vector length, arrow size, and color. Contour maps can also be used to display the distribution of field intensity on two-dimensional sections. Users can choose the field quantity type and customize the color, line type, and line width of the contour. Additionally, color-filled maps can represent the field strength using different colors, allowing users to select the field quantity type and adjust the color distribution scale.

By analyzing the visualization results of the electromagnetic field distribution, we can draw the following conclusions. Firstly, the distribution of electric field intensity reflects the charge distribution in the circuit, and areas with high electric field intensity may negatively impact circuit performance. Secondly, in transmission line microwave circuits, the distribution of electric field intensity can provide insights into the characteristic impedance and transmission loss of the transmission line. For low-frequency microwave circuits, especially broadband circuits, the distribution of electric field intensity is closely related to the transmission characteristics. Therefore, visual analysis of electric field intensity distribution is of significant importance. Additionally, the distribution of magnetic field intensity reflects the current transmission in the microwave circuit, and regions with higher magnetic field intensity may experience higher transmission loss. In low-frequency microwave circuits, the distribution of magnetic field intensity is closely related to the electromagnetic shielding performance of the circuit.

Lastly, the distribution of current density represents the current transmission in microwave circuits, and regions with higher current density may have higher transmission loss, requiring optimization. In low-frequency microwave circuits, the current density distribution is closely related to the circuit's parasitic parameters, such as parasitic inductance and parasitic capacitance.

3.4.5 Optimization Design and Parameter Scanning

In the design of low-frequency microwave circuits, optimization plays a crucial role in maximizing or minimizing specific performance metrics while satisfying engineering requirements and constraints. Designers must determine the optimization objectives and constraints. The optimization objective can be a single objective or multiple objectives involving circuit performance measures, such as minimizing reflection loss (S_{11}) and insertion loss (S_{21}). Constraints include power limits, voltage and current limits, size limits, material constraints, and more. These constraints can be equality constraints or inequality constraints. To solve the optimization problem, designers need to select the appropriate optimization algorithm, such as gradient descent, Newton's method, stochastic gradient descent, simulated annealing algorithm, genetic algorithm, or particle swarm optimization algorithm. Each algorithm has its own characteristics, advantages, and disadvantages, so it's important to choose the appropriate algorithm based on the specific problem and requirements.

Parameter scanning and sensitivity analysis are essential in the simulation of low-frequency microwave circuits. Parameter scanning observes the changes in system performance by varying circuit components and structural parameters to optimize circuit performance. Sensitivity analysis helps designers identify key parameters that have a significant impact on system performance and enables them to optimize the design accordingly. The principles and methods of parameter scanning include determining the parameters to be scanned and their ranges, setting the step size for parameter variations and the scanning range, performing the parameter scanning simulation, and analyzing the simulation results to find the optimal parameter values. HFSS provides a parametric design function that allows users to set the parameters and ranges to be scanned, automating the scanning process and simplifying user operations. Sensitivity analysis involves determining the parameters to be analyzed, calculating the changes in system performance indices caused by small perturbations in each parameter, evaluating the impact of each parameter on system performance, and optimizing the parameters with high sensitivity. HFSS provides a sensitivity analysis function to evaluate parameters and optimize key parameters based on the analysis results, thereby improving the design efficiency.

3.4.6 *Some Problems in Low-Frequency Simulations of Microwave Circuits*

3.4.6.1 Addressing Low-Frequency Singularity

The issue of low-frequency singularity is common and challenging in the simulations of low-frequency microwave circuits. It refers to the coupling of electrostatic and magnetostatic fields at low frequencies, leading to a significant increase in the matrix condition number and affecting the stability and accuracy of the simulation. The causes of this problem include the coupling of electrostatic and magnetostatic fields, the small induced current at low frequencies, and the ratio of mesh size to wavelength. To tackle this problem, several methods can be employed, such as the hybrid field method, magnetic potential-vector potential coupling method, non-uniform mesh generation, preconditioning technology, multi-scale method, and multi-grid technology. These methods help reduce the matrix condition number, improve computational stability and accuracy, and decrease computational complexity. Depending on the specific problem's characteristics, the appropriate method can be flexibly chosen to ensure the reliability of the simulation results.

3.4.6.2 Calculation Method for Low-Frequency Radiation Loss

Radiation loss is a significant performance metric in low-frequency microwave circuits, representing the energy dissipated by the outward radiated electromagnetic wave from the microwave circuit structure. In HFSS, there are two commonly used methods to calculate low-frequency radiation loss: the surface current method and the Poynting vector method. The surface current method estimates the radiation loss by calculating the current distribution on the circuit structure's surface. On the other hand, the Poynting vector method directly calculates the Poynting vectors of the electric and magnetic fields to determine the radiation loss. In the surface current method, HFSS first calculates the complex current distribution on the circuit structure's surface at a specified frequency and then solves the radiation field and radiation loss based on this distribution. The procedure for the Poynting vector method involves calculating the electric and magnetic field components and subsequently determining the Poynting vector. Finally, the radiation loss is obtained by integrating the Poynting vector over a closed surface surrounding the entire microwave circuit structure. When selecting the calculation method, considerations should include the circuit structure's characteristics and the calculation requirements. The surface current method is more complex but provides a more direct reflection of the current distribution's influence. The Poynting vector method is relatively simpler but may be subject to numerical errors in some cases. For complex circuit structures, various methods can be used to calculate and compare the results to assess the accuracy of the calculation.

3.4.6.3 Low-Frequency Electromagnetic Shielding Analysis

Low-frequency electromagnetic shielding is a technology used to protect electronic equipment from electromagnetic interference or damage. It involves isolating sensitive components, reducing the influence of external electromagnetic fields, and preventing electromagnetic wave leakage in low-frequency microwave circuits. The effectiveness of low-frequency electromagnetic shielding is measured by the shielding efficiency (SE), which is calculated using the formula: $\mathrm{SE}(dB) = 10 \times \log_{10}\left(\frac{\mathrm{P_{in}}}{\mathrm{P_{out}}}\right)$, where P_{in} is the electromagnetic field power outside the shield, and P_{out} is the electromagnetic field power inside the shield. Electromagnetic shielding is based on the principles of reflection, absorption, and multiple shielding. Reflection exploits the interaction between free electrons on the surface of the conductor and electromagnetic waves to reflect the waves. Absorption utilizes magnetic materials to convert the energy of electromagnetic waves into heat energy. Multiple shielding reduces the influence of external electromagnetic waves by arranging multiple layers of shielding materials within the shielding structure. The process of low-frequency electromagnetic shielding analysis in the HFSS includes geometric modeling, material parameter setting, boundary condition setting, mesh generation, electromagnetic field solution, and shielding efficiency calculation. After analyzing the results, the shield's structure and materials can be optimized to enhance the shielding effect according to the requirements.

3.4.6.4 Thermal Stability Analysis of Low-Frequency Microwave Circuits

Thermal stability analysis is crucial for ensuring the reliability and performance of low-frequency microwave circuits. In practical applications, factors such as current density, electromagnetic field distribution, and material properties have an impact on circuit-generated heat and operating temperatures. To analyze thermal stability, it is necessary to establish the heat conduction equation while considering the coupling effect between the electromagnetic and thermal fields. The heat conduction equation describes the propagation of heat within the microwave circuit, which can be simplified to a steady-state equation. Coupling analysis of the electromagnetic and thermal fields involves calculating the electromagnetic field distribution using HFSS, determining Joule heat based on the current density distribution, and solving the heat conduction equation to obtain the temperature distribution. This process includes updating material property parameters and recalculating the electromagnetic field distribution until convergence is achieved. Common numerical calculation methods for thermal stability analysis include the finite element method, finite-difference method, and finite volume method. HFSS typically utilizes the finite element method to solve the heat conduction equation and performs multi-field coupling analysis with other electromagnetic field simulation software and thermal analysis software. It is essential to select reasonable thermal conductivity and specific heat capacity parameters, consider boundary conditions during the heat conduction process, and choose appropriate convergence criteria and time steps. Thermal stability analysis can optimize circuit performance, reduce the risk of failure, and enhance circuit reliability and lifespan.

3.4.6.5 Coupling Analysis of Electromagnetic Field and Mechanical Vibration

In low-frequency microwave circuits, the coupling effect between the electromagnetic field and mechanical vibration significantly impacts the circuit performance and can lead to failures or degradation of vibration-sensitive microwave devices. Therefore, analyzing and controlling the coupling relationship between the electromagnetic field and mechanical vibration is of great significance in the design and simulation of low-frequency microwave circuits. This section introduces the method of coupling analysis, including the coupling mechanism and the establishment and solution of the coupling equation. The coupling equations involve using Maxwell's equations to describe the distribution and propagation of the electromagnetic field and linear vibration equations to describe mechanical vibration. The coupling relationship can be described by introducing a coupling coefficient. Analytical methods, finite element methods, or multi-scale methods can be employed to solve these coupled equations. In HFSS, coupling analysis can be performed by creating a model, defining material properties, setting boundary conditions and excitation sources, establishing coupling relationships, and solving the coupled equations. The circuit design can be optimized and applied to real-world scenarios through result analysis and experimental verification. Practical applications involve reducing the coupling effect, utilizing the coupling effect, conducting sensitivity analysis, and performing verification experiments to solve problems and optimize designs.

3.4.6.6 Fault Diagnosis and Fault-Tolerant Design

With the widespread application of low-frequency microwave circuits in communication, radar, and navigation, ensuring circuit reliability and stability has become a key concern. This section introduces the fundamental principles, methods, and techniques of fault diagnosis and fault-tolerant design for low-frequency microwave circuits. Fault diagnosis aims to locate faulty components and identify the causes, which can be achieved through parameter identification, signal analysis, and electromagnetic field distribution analysis. Fault-tolerant design improves circuit reliability and robustness through redundancy design, adaptive regulation, fault-tolerant coding, and fault detection and isolation. As an type of electromagnetic simulation software, HFSS can be utilized for fault diagnosis and fault-tolerant design. HFSS provides functions for circuit parameter calculation, signal response analysis, and visualization of electromagnetic field distribution, offering a theoretical basis for fault diagnosis and fault-tolerant design. For the fault diagnosis, HFSS can pinpoint faulty components and analyze fault causes based on actual test data. Regarding fault-tolerant design, HFSS can verify redundant designs, optimize adaptive regulation functions, and evaluate the performance of fault-tolerant coding. Additionally, HFSS can be used to design and analyze fault detection and isolation mechanisms, mitigating the risk of fault propagation.

3.5 Co-simulations for Low-Frequency Microwave Circuits

3.5.1 *Concept and Method*

Low-frequency microwave circuit co-simulations are a simulation method that integrates multiple simulation tools, methods, and techniques to solve the design and performance problems of low-frequency microwave circuits [55, 56]. This method can improve the accuracy of simulation by taking into account the various parameters of the circuit (such as inductance, capacitance, and impedance) and the interaction between them. The principle of co-simulations includes circuit theory, electromagnetic field theory, and signal processing technology. The application of circuit theory in co-simulations mainly includes the parameter extraction of circuit elements and the establishment of circuit topology. The circuit parameters can be obtained analytically or numerically. For example, the characteristic impedance (Z_0) and propagation constant (β) of a transmission line can be calculated by Eqs. 3.79 and 3.80:

$$Z_0 = \sqrt{\frac{L}{C}} \tag{3.79}$$

$$\beta = \frac{2\pi}{\lambda} = \omega\sqrt{LC} \tag{3.80}$$

where L and C denote the inductance and capacitance per unit length, respectively, ω is the angular frequency, and λ is the wavelength.

Electromagnetic field theory is mainly used to analyze the coupling effect between circuit elements and the interaction between the circuit and external electromagnetic environment. In the co-simulations of low-frequency microwave circuits, the electromagnetic field problem is usually solved by the finite element method, the finite-difference time-domain method, or the integral equation method. For example, the electromagnetic field distribution on a microstrip line can be solved by the finite element method:

$$\nabla \times \left(\mu^{-1}\nabla \times \mathbf{E}\right) - \omega^2\epsilon\mathbf{E} = -i\omega\mathbf{J} \tag{3.81}$$

Signal processing technology is mainly used to analyze the frequency response, time-domain response, and noise characteristics of the circuit in co-simulations. In frequency-domain analysis, the commonly used methods are harmonic balance, vector network analysis, and frequency response analysis. In time-domain analysis, the commonly used methods are transient analysis and eye diagram analysis.

In the co-simulations of low-frequency and microwave circuits, it is necessary to establish a circuit model that includes the required circuit components and connection topologies. Circuit models can be categorized into two types: component-based models and physics-based models. A component-based model consists of a series of standard components (such as resistors, capacitors, and inductors), which facilitates circuit simulation and optimization. On the other hand, physics-based models consist of actual microwave structures (such as microstrip lines and waveguides), which closely resemble practical application scenarios but have higher computational complexity.

When building a circuit model, the following points need to be considered:

1. Select appropriate circuit components and connections to ensure that the circuit structure meets the design requirements.
2. Assign suitable parameter values to circuit elements, including impedances, capacitances, inductances, etc.
3. For complex microwave structures, equivalent circuits or sub-circuits can be used to reduce simulation complexity.
4. For regions with strong electromagnetic field coupling, subdivision of sub-regions or hierarchical design methods can be employed.

In the co-simulations of low-frequency microwave circuits, the following types of simulation parameters need to be set:

1. Frequency range: Determine the start frequency, end frequency, and frequency step for the simulation.
2. Time-domain parameters: Set the total simulation time, time step, and sampling rate.
3. Boundary conditions: Choose appropriate boundary conditions, such as conducting walls, open regions, and periodic boundaries.
4. Mesh parameters: Define the type, size, and density of the mesh to ensure accurate simulation results.
5. Solver options: Select the appropriate solver, such as the finite element method, finite-difference time-domain method, or integral equation method.
6. Optimization and sensitivity analysis: Specify optimization objectives, constraints, and algorithms for optimization tasks.

The steps for performing the simulations are as follows:

1. Import or create a circuit model.
2. Set the simulation parameters.
3. Run the simulation.
4. Analyze the simulation results, including S-parameters, impedance, electromagnetic field distribution, etc.
5. Optimize and modify the circuit based on the analysis results.

3.5.2 Co-simulations with ADS for Low-Frequency Microwave Circuits

3.5.2.1 Keysight Momentum Co-simulations with ADS

The co-simulation capabilities of Keysight Momentum and ADS allow engineers to perform electromagnetic field and circuit simulations in a unified design environment. Using the electromagnetic field analysis results provided by the Momentum, ADS can provide more accurate estimates of circuit parameters, thus providing engineers with more reliable simulation results. In the design of low-frequency microwave circuits, the interconnection, parasitic effects, and electromagnetic coupling between circuit elements have an important impact on the performance of the circuit. Through the coupling analysis of the electromagnetic field and circuit, the comprehensive evaluation of microwave circuit performance is realized.

Specifically, Momentum simulates the three-dimensional electromagnetic field of the circuit structure and calculates the electromagnetic coupling, radiation loss, and parasitic parameters between the circuit components. Then, the electromagnetic field information is imported into ADS in the form of S-parameters, Y-parameters, or Z-parameters. ADS uses this electromagnetic field information to analyze the circuit and calculate the circuit performance indicators, such as gain, intermodulation distortion, phase noise, and so on.

The following are the specific implementation method and operation steps for the co-simulations of low-frequency microwave circuits with ADS:

1. Create circuit designs in ADS, including components, interconnects, ports, etc. Save the circuit design as a sub-design.
2. Insert a Momentum emulator in the ADS main design. In order to realize the coupling analysis of electromagnetic field and circuit, the Momentum simulator is connected to the port of the sub-design.
3. Import the circuit design into Momentum. The Momentum automatically converts circuit components into 3D models for electromagnetic field simulation. Users can set simulation parameters such as mesh density, boundary conditions, and frequency range in the Momentum.
4. Run the Momentum simulation. The Momentum will carry out three-dimensional electromagnetic field analysis of the circuit structure and calculate the electromagnetic field coupling, radiation loss, and parasitic parameters between the circuit components.
5. Export the Momentum simulation results to an S-parameter, Y-parameter, or Z-parameter file. These parameter files contain information about the electromagnetic field coupling between circuit elements.
6. In the ADS master design, connect the output port of the Momentum emulator to the input port of the sub-design. Import the Momentum-generated parameter file and associate it with the port of the child design.

7. Set simulation parameters in ADS, such as frequency range, temperature, and material parameters. Run the circuit simulation to calculate the circuit performance, such as gain, intermodulation distortion, phase noise, and so on.
8. Analyze the simulation results. Users can view the waveform diagram, eye diagram, and Smith diagram of the circuit performance in ADS to optimize the circuit design.
9. Co-simulations can be performed iteratively if necessary. By modifying the circuit design or simulation parameters, the user can further improve the circuit performance.

Through the above steps, the Momentum and ADS can realize the co-simulations of low-frequency microwave circuits. Using this co-simulation method, engineers can conduct electromagnetic field and circuit simulations in a unified design environment, which provides much convenience for circuit optimization and performance evaluation.

3.5.2.2 Co-simulations of NI AWR Design Environment with ADS

NI AWR Design Environment is a kind of design software developed by Applied Wave Research (AWR) Corporation acquired by National Instruments (NI) that integrates microwave circuit design, electromagnetic field simulation, system-level simulation, and other functions. The software offers a rich component library, advanced optimization algorithms, and powerful post-processing capabilities, making it widely used in the design and simulation of microwave circuits, RF circuits, and wireless communication systems. When co-simulating with ADS, NI AWR Design Environment leverages ADS's powerful circuit analysis capabilities along with its own electromagnetic field simulation and system-level simulation capabilities, providing a comprehensive simulation solution.

The principle of co-simulations between NI AWR Design Environment and ADS is based on data exchange and model sharing. Users can import the circuit design from ADS into NI AWR Design Environment and use its simulation functions to simulate electromagnetic fields, thermal fields, structural fields, and other domains. Simultaneously, users can also import the simulation results from NI AWR Design Environment back into ADS for further circuit analysis and optimization.

The following are the specific implementation steps for co-simulations of low-frequency microwave circuits using ADS and NI AWR Design Environment:

1. Preparation: Ensure that ADS and NI AWR Design Environment are correctly installed and configured.
2. Data import: Create a circuit design in ADS and save it as a sub-design. Import the sub-design into NI AWR Design Environment.
3. Circuit simulation: Set the circuit simulation parameters in NI AWR Design Environment and run the simulation. The software will automatically call the circuit analysis function of ADS for simulations.

4. Electromagnetic field simulation: Set the electromagnetic field simulation parameters in NI AWR Design Environment as required and run the simulation.
5. System-level simulation: If a system-level simulation is needed, set the system-level simulation parameters in NI AWR Design Environment and run the simulation.
6. Data export: Export the simulation results from NI AWR Design Environment and import them into ADS for further circuit analysis and optimization.
7. Verification and optimization: Verify the simulation results in ADS and optimize the circuit as needed.

When performing co-simulations with NI AWR Design Environment and ADS, there are primarily two approaches:

1. Serial co-simulations: In this approach, users first complete the circuit design in ADS, import the design into NI AWR Design Environment for electromagnetic field simulation, and then import the simulation results back into ADS for further circuit analysis and optimization.
2. Parallel co-simulations: In this approach, users perform circuit design and simulation simultaneously in both ADS and NI AWR Design Environment. The two software tools achieve co-simulations through data exchange and model sharing. This method significantly improves simulation efficiency, especially for complex circuits and large systems.

During the co-simulations of NI AWR Design Environment and ADS, users should consider the following points: ensure smooth data exchange between the two software tools, pay attention to compatibility when importing circuit designs, set appropriate electromagnetic field simulation parameters, define clear interfaces, and make full use of ADS's post-processing capabilities. Additionally, selecting the appropriate co-simulations method will achieve the best balance between simulation efficiency and accuracy.

The co-simulations of NI AWR Design Environment and ADS offer significant advantages for the design of low-frequency microwave circuits. Through data exchange and model sharing between the two software tools, users can achieve integrated simulations encompassing circuit design, electromagnetic field simulation, thermal field simulation, structural field simulation, and other domains. This approach helps improve circuit performance, shorten design cycles, and reduce costs.

3.5.2.3 Co-simulations of PathWave RF Synthesis with ADS

PathWave RF Synthesis, which is also known as Genesys, is a kind of design software for integrated circuits and microwave circuits and it is equipped with a comprehensive component library and powerful simulation capabilities. During the co-simulations of Genesys and ADS, data exchange is facilitated through the establishment of a data communication interface. Specifically, Genesys takes charge of the design and preliminary simulation of low-frequency microwave circuits and subsequently imports the design data into ADS for further electromagnetic field simulation and optimization. Once the ADS simulation is completed, the results are fed back to Genesys for post-processing and analysis. This seamless integration of Genesys and ADS provides users with a unified platform for low-frequency microwave circuit simulations.

The following are the specific implementation steps for co-simulations of low-frequency microwave circuits using Agilent Genesys and ADS:

1. Create a circuit design in Genesys and perform a preliminary simulation. Utilize Genesys' component library to design the circuit and adjust component parameters as needed.
2. Import the Genesys circuit design into ADS. Import circuit design files generated by Genesys (e.g., S-parameter files, Touchstone files) into ADS and convert them to ADS-compatible formats.
3. Set simulation parameters in ADS. Configure appropriate simulation parameters based on the circuit's characteristics and requirements, such as the electromagnetic field simulation method (e.g., FEM, FDTD), grid density, and boundary conditions.
4. Run the ADS simulation. ADS will simulate the electromagnetic field of the imported circuit and generate simulation results based on the user-defined parameters.
5. Feedback ADS simulation results to Genesys. Import ADS-generated simulation results (e.g., S-parameters, electromagnetic field distributions) back into Genesys for comparison and analysis with the preliminary simulation results.
6. Perform post-processing and analysis in Genesys. Utilize the ADS simulation results to further optimize and adjust the circuit within the Genesys environment.
7. Validate and optimize the design. Iterate the circuit optimization process based on the simulation results until the performance requirements are met.

Throughout the entire process, Genesys and ADS seamlessly integrate, ensuring an efficient circuit design and simulation. The design environment of PathWave encompasses circuits, systems, and electromagnetic simulators that facilitate RF system simulation and construction using actual components. It enables the creation of high-performance circuits, identification and resolution of circuit degradations, translation of ideal electrical designs into physical implementations, automated custom filter topology creation, analysis, tuning, and optimization through co-simulations, and prevention of costly downstream errors. PathWave circuit design incorporates automatic synthesis and optimization of matching networks. The design can then be transferred to PathWave Advanced Design System (ADS) for inclusion in more complex designs.

3.5.2.4 Co-simulations of Empire XPU with ADS

Empire XPU is a kind of simulation software for 3D electromagnetic fields based on the finite-difference time-domain (FDTD) method. It offers fast, accurate, and efficient simulation capabilities, making it suitable for design analysis in the fields of microwave, radio frequency, and photonics. When co-simulating with ADS, Empire XPU can fully leverage the powerful circuit analysis function of ADS and its own electromagnetic field simulation capabilities to provide comprehensive and accurate simulation results for low-frequency microwave circuit designs.

The co-simulations of Empire XPU and ADS follow the following implementation method and operation steps:

1. Create a circuit design in ADS and save it as a sub-design. Ensure that all microwave components are properly configured and parameterized.
2. Export the circuit design to file formats supported by Empire XPU, such as IE3D and DXF. Take care to preserve the information of conductor layers, dielectric layers, and circuit elements during the export process.
3. Import the circuit file into Empire XPU and set the simulation parameters, including mesh density, boundary conditions, frequency range, etc. Verify that all conductors, dielectric layers, and material properties are correctly assigned.
4. Run Empire XPU for electromagnetic field simulation. During the simulation process, you can observe the real-time electromagnetic field distribution, S-parameters, and other information.
5. Export the simulation results from Empire XPU to file formats supported by ADS, such as S2P and S3P. Ensure that the information of S-parameters and electromagnetic field distribution is retained during the export.
6. Import the simulation results from Empire XPU into ADS and apply them to the circuit design. In the circuit simulator of ADS, you can further analyze the circuit by setting parameters such as the reflection coefficient and transmission coefficient.
7. If necessary, adjust the circuit parameters in ADS and repeat Steps 2 through 6 until a satisfactory simulation result is achieved. During each iteration, closely observe the electromagnetic field distribution and S-parameters to optimize the circuit performance.
8. Based on the results of co-simulations, optimize the circuit in ADS. Adjust the parameters, layout, etc., of the microwave components to improve circuit performance. Pay close attention to electromagnetic field distribution, S-parameters, circuit loss, and other indicators during the optimization process.
9. Once the optimization is complete, export the final circuit design as a production file, such as a Gerber file and ODB++. Additionally, generate a report that includes the circuit design and simulation results for reference by the design team.
10. For complex low-frequency microwave circuit designs, consider utilizing multiple iterations of Empire XPU with ADS. During each iteration, the circuit parameters are automatically updated to improve the simulation accuracy.

By following these steps, Empire XPU and ADS enable efficient and accurate co-simulations for low-frequency microwave circuits.

3.5.3 *Co-simulations with HFSS for Low-Frequency Microwave Circuits*

3.5.3.1 Co-simulations of NI AWR Design Environment with HFSS

NI AWR Design Environment is an advanced design automation (EDA) tool that integrates circuit, system, and electromagnetic field simulation capabilities for RF, microwave, and communication system designs. It can co-simulate with HFSS to enable multi-domain analysis for low-frequency microwave circuits. The connection between these two software tools is established through a standard interface called AWR's EM Socket, which allows users to access a wide range of electromagnetic (EM) tools within the Microwave Office design environment.

The co-simulations of NI AWR Design Environment and HFSS facilitate the exchange of data and interfaces between the two software tools, enabling accurate multi-domain simulations. This method provides a reliable basis for circuit design, optimization, and fault diagnosis. The implementation method and steps for co-simulations are as follows:

1. Create the circuit diagram in the NI AWR Design Environment and add the necessary components.
2. Set simulation parameters such as frequency range and scanning step.
3. Build the 3D model in HFSS and configure boundary conditions, mesh parameters, etc.
4. Establish the connection between NI AWR Design Environment and HFSS for co-simulations. This can be achieved through different approaches:

   – Shared data file: Simulations data from both tools is saved in a shared data file, allowing real-time reading and updating of simulation data by both software.
   – Scripts and API interfaces: Write scripts or use API interfaces to exchange data and synchronize simulation tasks between NI AWR Design Environment and HFSS. Custom scripts or API functions enable seamless data transmission and synchronization between the two tools.

5. Analyze the simulation results, optimize and adjust the circuit. NI AWR Design Environment offers comprehensive post-processing functions, including data visualization, waveform analysis, and parameter optimization. Users can utilize these functions to analyze simulation results, identify performance bottlenecks, and further optimize the circuit.
6. Verify the simulation results. After optimizing the circuit, it is recommended to use test instruments to validate the accuracy of the simulation results. NI AWR Design Environment supports data interaction with various test instruments, facilitating the comparison between actual test results and simulation results.

By following these steps, the co-simulations of NI AWR Design Environment and HFSS enable accurate multi-domain analysis in low-frequency microwave circuits. It provides designers with a comprehensive platform for circuit design, optimization, and verification.

3.5.3.2 Co-simulation of Microwave Office with HFSS

The co-simulations of Microwave Office and HFSS combine RF microwave circuit simulation with high-frequency electromagnetic field simulation, enhancing the realism and reliability of simulation results during circuit design and optimization. This co-simulation approach leverages the strengths of both tools to capture the low-frequency and high-frequency characteristics of the circuits.

Microwave Office excels at simulating low-frequency circuits and calculating their electrical characteristics, such as capacitance, inductance, and impedance. On the other hand, HFSS specializes in high-frequency electromagnetic field simulation, providing insights into parameters like electromagnetic wave transmission, power transmission, and scattering. By combining the results from both simulations, the overall performance of the circuit at low and high frequencies can be accurately evaluated.

Implementation Method:

1. There are two methods to achieve Microwave Office + HFSS co-simulations: Static coupling and dynamic coupling.

2. Static Coupling: This method involves importing circuit components and layout information from Microwave Office into HFSS to create a 3D model of the circuit. High-frequency simulation is then performed in HFSS, and the results are imported back to Microwave Office for low-frequency simulation. Static coupling offers faster calculations but requires detailed circuit element and layout descriptions.

Dynamic Coupling: In this approach, circuit information is transmitted to HFSS in real time during low-frequency simulation in Microwave Office. High-frequency simulation is performed simultaneously in HFSS, and the results are sent back to Microwave Office for low-frequency simulations. Dynamic coupling provides more accurate results but has a slower calculation speed.

Operation Steps:

1. Build the circuit model: Create the circuit model in Microwave Office, including circuit components, layout, and parameter settings.

2. Export the model: Export the circuit model from Microwave Office to HFSS, ensuring the accuracy of the exported model by selecting the appropriate export option.

3. Perform high-frequency simulation: Conduct high-frequency simulation in HFSS to obtain the circuit's high-frequency characteristics, such as S-parameters, power transmission, and scattering.

4. Import simulation results: Import the simulation results from HFSS into Microwave Office as input parameters for the low-frequency simulations.

5. Perform low-frequency simulation: Use Microwave Office to simulate the low-frequency characteristics of the circuit, including capacitance, inductance, and impedance. Combine the low-frequency simulation results with the high-frequency simulation results to obtain the overall performance of the circuit at both low and high frequencies.

Through co-simulations, important circuit performance metrics like gain, bandwidth, and stability can be assessed at both low and high frequencies. Based on the simulation results, the circuit can be optimized and improved to enhance its performance and stability.

3.5.3.3 Co-simulation of ADS Momentum with HFSS

In the co-simulations of low-frequency microwave circuits, the combination of ADS Momentum and HFSS software platforms can be utilized. The implementation method involves the following steps:

ADS Momentum:

1. Launch ADS Momentum and create a new project within the workspace.
2. Select the "Schematic" mode and create a new circuit diagram.
3. Use the circuit diagram to design and parameterize the layout of the microwave circuit.
4. Switch to the "Layout" mode in ADS Momentum and draw the layout of the microstrip filter.
5. Parameterize the layout and perform circuit simulation in ADS Momentum to obtain the S-parameters of the microstrip filter.
6. Save the simulation results to a file.

HFSS:

1. Open HFSS and create a new project within the workspace.
2. Select the "Modeler" mode and create a 3D model.
3. Build the 3D model in HFSS, including all the components of the microstrip filter such as microstrip lines, capacitors, and inductors.
4. Import the parameters of the microstrip filter from ADS Momentum into HFSS.
5. Modify the microstrip filter in HFSS to align with the circuit characteristics obtained from ADS Momentum.
6. Run the finite element solver in HFSS to solve the electromagnetic characteristics of the microstrip filter.
7. Export and save the simulation results of HFSS to a file.

Co-simulations Procedure:

1. Import the simulation result file from HFSS into ADS Momentum.
2. Use ADS Momentum to simulate the low-frequency circuit of the microstrip filter and obtain its low-frequency characteristics.
3. Combine the simulation results from ADS Momentum and HFSS to obtain the complete low-frequency microwave circuit characteristics.

The above steps describe the specific operational procedure for co-simulations of a low-frequency microwave circuit between ADS Momentum and HFSS, taking a microstrip filter as an example.

3.6 An Example of Co-simulations Based on ADS

3.6.1 Implementing Parametric Modeling in ADS Layout Environment

In electronic design, parametric modeling is an important technique that allows for more flexible and efficient circuit design and simulation processes. The ADS Layout environment provides powerful support for implementing this functionality. This section will introduce how to achieve parametric modeling in the ADS Layout environment and how to use the Subnetwork variable parameterization and perturbation variable parameterization setup methods.

First, open the ADS Layout environment and create a new Layout document. Then, choose the method to create variables and create a variable of type Subnetwork. By doing so, a Subnetwork variable is created with a default parameter set to 10 mil. Next, associate the variable "width" with the device dimensions. Select a model from the ADS model library, such as the microstrip line "MLIN," drag it into the Layout, double-click to edit the variable, set its "W" to "width," and click OK. Then, add ports to both ends of the transmission line, add the substrate, and save the layout. Next, create an EM model for co-simulations. Then, select Symbol and create a look-alike model, click OK, save all windows, and close them. Finally, create a new schematic and place the models created in the Layout into the schematic. Now, the width becomes a variable that can be set, allowing for convenient variable sweeping optimization, and other tasks in the schematic. Secondly, for manually created models, you can create new Layout cells and perform parameterization operations through graphical drawing. For example, parameterize the width of a gap box in Fig. 3.6. Here, let's set the space size to 50 mil. Then, open the parameter setting box through EM > Component > Parameter and set the perturbation parameter. Here, the Nominal Value is set to the same value as the symbol created for the gap width, both set to 50 mil, and the perturbation value is set to 70 mil. After clicking Add, a perturbation setting window will pop up. Select the points that need to be perturbed, and once successfully selected, they will be highlighted. Click Apply, and the graphics will change accordingly. Finally, add ports, create a new Layout, and name it "test." In the newly created layout, click on the library, select "workspace libraries," choose "cell_3," and right-click "Place Component."

At this point, the main steps of parametric modeling have been completed. However, to verify if the perturbation settings are correct, we need to conduct a test. Double-click on the graphic in the Layout, and the left-side window will pop up. Change the space distance to any value, such as 80 mil, click Apply, and you will see that the gap width correctly changes.

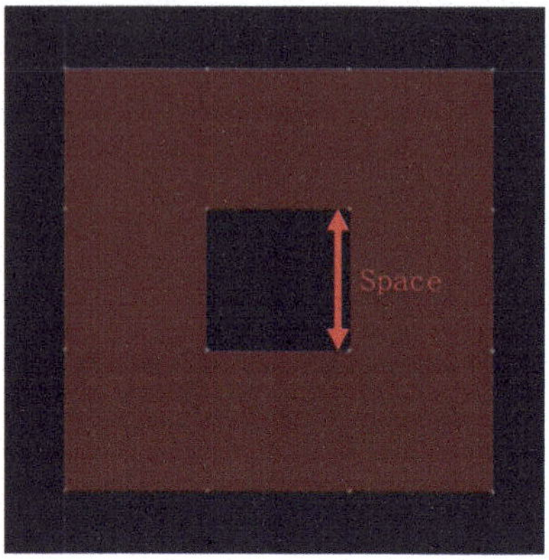

Fig. 3.6 Parameterized gap width, labeled as "Space"

3.6.2 Co-simulations of Field and Circuit Based on RFPro

3.6.2.1 Introduction to RFPro

RFPro is an electromagnetic solver that can be integrated into the Advanced Design System (ADS) or Virtuoso platform. It incorporates classic Momentum and Finite Element Method (FEM) algorithms, primarily targeting radio-frequency (RF) circuit applications. This tool can quickly perform electromagnetic solutions for RF printed circuit boards, various process chips, and multifunctional RF modules. When simulating RF circuits, it's important to pay attention to the parasitic effects of PCB traces and consider the circuit characteristics of RF components (such as LCR elements or S-parameters). If we want to obtain accurate circuit simulation results, we need to perform field-circuit co-simulations. As a brand new platform, RFPro provides a faster field-circuit simulation solution.

3.6.2.2 A Simple RF PCB

Consider a simple RF distribution circuit's PCB as shown in Fig. 3.7, which we import using the ODB++ format. We add excitation sources on both sides of the matching network, where P1 and P2 are the excitation ports, and P3, P4, P5, and P6 are the reference grounds.

3.6.2.3 Set Component Circuit Properties

The imported PCB will contain partial component information. If circuit-field co-simulations is required, these components must include circuit properties. A complete component model needs to include schematic, symbol, and layout as shown in Fig. 3.8. If the imported components don't include these three parts, Layout Symbol

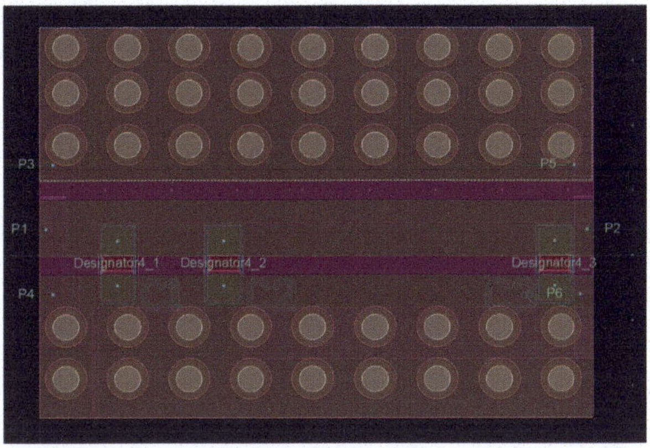

Fig. 3.7 PCB of RF circuit

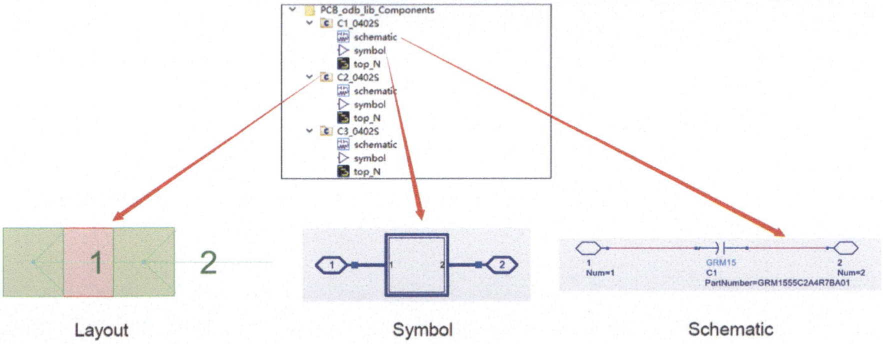

Fig. 3.8 Complete component model, including schematic, symbol, and layout

Schematic needs to be manually completed. By default, the networks of these three aspects are linked digitally.

3.6.2.4 Launch RFPro and Complete Simulations

In the ADS layout design interface, we can launch RFPro to perform subsequent co-simulations of the field and circuit. The component properties that require co-simulations need to be modified. Drag the component to the Component Models, confirm that the component circuit properties are set as Library Cell, and ensure that all three component settings are set as Library Cell. Next, set the ports and frequencies to prepare for simulation. Once the simulation is complete, simply double-click on Generate Test Bench, and the simulation results will be automatically returned to the schematic as shown in Fig. 3.9. Run the simulation again in the schematic, and this time the simulation results will include the co-simulations results of the PCB electromagnetic solution and the circuit components.

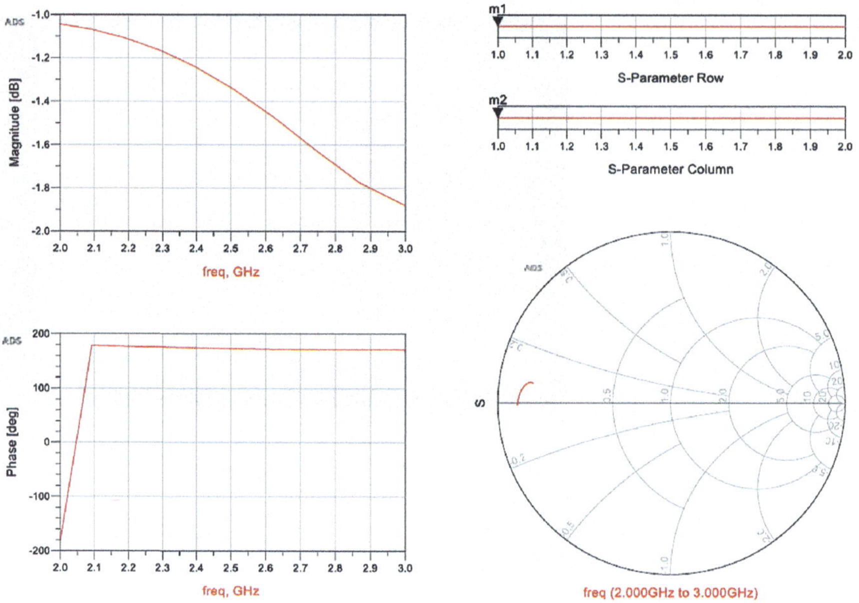

Fig. 3.9 Comprehensive co-Simulation results: frequency response, phase shift, and S-Parameter analysis in RF design

RFPro provides a more convenient method for combined circuit and field simulation. This article summarizes one approach using Library Cells. In this study, ideal LCR components were used. In practical design, these components can be replaced with Process Design Kits (PDK) and measured S-parameter data provided by suppliers, while the methodology for combined simulation remains the same. The complexity of RF circuit design lies in the fact that it involves both circuit design and considerations for electromagnetic fields. Traditional simulation methods often struggle to handle the interaction between these two aspects, but RFPro addresses this issue precisely. By simultaneously addressing both the electromagnetic field and circuit issues in an integrated platform, RFPro can provide more accurate and efficient simulation results.

Problems

P3.1 Design of broadband LC band-pass filter using Genesys and ADS
Genesys offers convenient and efficient filter synthesis capabilities, while ADS provides accurate lumped element modeling and efficient optimization functions. The combination of both tools has resulted in good design and fabrication outcomes.

The key performance specifications for the broadband LC band-pass filter are as follows:

- Passband: 160–660 MHz
- Passband ripple: Less than ±0.5 dB
- In-band reflection: Below −10 dB
- Out-of-band suppression: 20 dB suppression at 110 MHz and 1.86 GHz.

In this design, Genesys is initially employed for filter synthesis to generate a preliminary filter that meets the requirements. Subsequently, ADS is used to fine-tune the preliminary filter to meet the design specifications.

P3.2 HFSS Co-simulation with Microwave Office
ANSYS HFSS has been integrated into AWR's Microwave Office software for high-frequency circuit design, enabling fast and accurate microwave circuit simulations. The practice involves using Microwave Office to connect with HFSS for electromagnetic field analysis and coupling of bond wires in three-dimensional structures. The connection between these two software tools is achieved through AWR's innovative EM Socket open standard interface, allowing AWR users to access other widely used electromagnetic (EM) tools from the Microwave Office design environment.

References

1. Konishi Y (1997) Microwave electronic circuit technology. CRC Press
2. Pozar DM (2011) Microwave engineering. Wiley
3. Rohde UL, Rudolph M (2013) RF/microwave circuit design for wireless applications. Wiley
4. Giannini F, Leuzzi G (2004) Nonlinear microwave circuit design. Wiley
5. Russer P (2003) Electromagnetics, microwave circuit, and antenna design for communications engineering. Artech House
6. Swanson DG, Hoefer WJ (2003) Microwave circuit modeling using electromagnetic field simulation. Artech House
7. Karmel PR, Colef GD, Camisa RL (1998) Introduction to electromagnetic and microwave engineering, vol 53. Wiley
8. Zhang Q-J et al (2021) Simulations and automated modeling of microwave circuits: state-of-the-art and emerging trends. IEEE J. Microw 1(1):494–507
9. Jantunen H et al (2003) Design aspects of microwave components with LTCC technique. J Eur Ceram Soc 23(14):2541–2548
10. Ng CY, Chongcheawchamnan M, Robertson ID (2004) Lumped-distributed hybrids in 3D-MMIC technology. IEE Proc-Microw, Anten Propag 151(4):370–374
11. Yablonovitch E, Baumberg JJ (1993) Photonic band-gap structures. JOSA B 10(2):283–295
12. Oh SS et al (2000) Duplexer using microwave photonic band gap structure. Appl Phys Lett 76(16):2301–2303
13. Dec A, Suyama K (2000) Microwave MEMS-based voltage-controlled oscillators. IEEE Trans Microw Theory Tech 48(11):1943–1949
14. Bahl IJ (2022) Lumped elements for RF and microwave circuits. Artech House
15. Suárez A (2009) Analysis and design of autonomous microwave circuits. Wiley
16. Yeung SH et al (2012) Computational optimization algorithms for antennas and RF/microwave circuit designs: An overview. IEEE Trans Industr Inf 8(2):216–227

17. Vendelin GD et al (2021) Microwave circuit design using linear and nonlinear techniques. Wiley
18. Montgomery CG et al (eds) (1987) Principles of microwave circuits. IET
19. Kurokawa K (ed) (2014) An introduction to the theory of microwave circuits. Elsevier
20. Pozar DM (1998) Microwave filters. In: Microwave engineering, 4th ed. Wiley, pp 381–422
21. Austin ACM, Sarris CD (2013) Efficient analysis of geometrical uncertainty in the FDTD method using polynomial chaos with application to microwave circuits. IEEE Trans Microw Theory Tech 61(12):4293–4301
22. Christoffersen CE, Mughal UA, Steer MB (2000) Object oriented microwave circuit simulation. Int J RF Microwave Comput Aided Eng 10(3):164–182
23. Pedro JC, Maas SA, Grajal J (2018) Nonlinear circuit simulations and modeling: fundamentals for microwave design. Cambridge University Press
24. Snyder RV (2007) Practical aspects of microwave filter development. IEEE Microwave Mag 8(2):42–54
25. Xu L, Mishra SR, Sun W (2009) High-frequency circuit simulator: An advanced three-dimensional finite-element electromagnetic-simulation tool for microwave tubes. IEEE Trans Electron Devices 56(5):1141–1151
26. Gupta KC, Itoh T, Oliner AA (2002) Microwave and RF education-past, present, and future. IEEE Trans Microwave Theory Tech 50(3):1006–1014
27. Steer MB (1991) Simulations of nonlinear microwave circuits-An historical perspective and comparisons. In: IEEE MTT-S international microwave symposium digest, vol 2, pp 1013–1016
28. Monk P (2003) Finite element methods for Maxwell's equations. Oxford University Press
29. Chew WC, Tong MS, Hu B (2008) Integral equation methods for electromagnetic and elastic waves. Morgan & Claypool Publishers
30. Bork AM (1967) Maxwell and the electromagnetic wave equation. Am J Phys 35(9):844–849
31. Ulukus S et al (2015) Energy harvesting wireless communications: a review of recent advances. IEEE J Sel Areas Commun 33(3):360–381
32. Hall DG (1996) Vector-beam solutions of Maxwell's wave equation. Opt Lett 21(1):9–11
33. Grote MJ, Keller JB (1998) Nonreflecting boundary conditions for Maxwell's equations. J Comput Phys 139(2):327–342
34. Weinstein LA (1988) Electromagnetic waves. Radio i svyaz', Moscow
35. Mesa F, Rodriguez-Berral R, Medina F (2018) Unlocking complexity using the ECA: The equivalent circuit model as an efficient and physically insightful tool for microwave engineering. IEEE Microwave Mag 19(4):44–65
36. Caverly RH et al (2000) SPICE modeling of microwave and RF control diodes. In: Proceedings of the 43rd IEEE midwest symposium on circuits and systems, vol 1, pp 554–557
37. Eisenstadt W, Stengel R, Thompson B (2006) Microwave differential circuit design using mixed-mode s-parameters. Artech
38. Dobrowolski JA (2016) Scattering parameters in RF and microwave circuit analysis and design. Artech House
39. Reddy JN (2019) Introduction to the finite element method. McGraw-Hill Education
40. Sumithra P, Thiripurasundari D (2017) Review on computational electromagnetics. Adv Electromagnet 6(1):42–55
41. Davidson DB (2010) Computational electromagnetics for RF and microwave engineering. Cambridge University Press
42. Kunz KS, Luebbers RJ (1993) The finite difference time domain method for electromagnetics. CRC Press
43. de Electroniagnetisnio G, de Cieiicias F, de Granada U (1992) Time-domain integral equation methods for transient analysis. IEEE Anten Propag Mag 34(3):15–21
44. de Electromagnetismo G, de Ciencias F, de Granada U (1992) Time-domain integral equation methods for transient analysis. IEEE Anten Propag Mag 34(3):15
45. Zhang Y, Sarkar TK (2009) Parallel solution of integral equation-based EM problems in the frequency domain. Wiley

46. Notaros BM (2008) Higher order frequency-domain computational electromagnetics. IEEE Trans Anten Propag 56(8):2251–2276
47. Poljak D (2007) Advanced modeling in computational electromagnetic compatibility. Wiley
48. Taflove A, Hagness SC, Piket-May M (2005) Computational electromagnetics: the finite-difference time-domain method. The electrical engineering handbook 3:629–670
49. Baumanns S (2012) Coupled electromagnetic field/circuit simulations. Modeling and numerical analysis. Logos Verlag Berlin GmbH
50. Feliziani M, Maradei F (2002) Circuit-oriented FEM: Solution of circuit-field coupled problems by circuit equations. IEEE Trans Magn 38(2):965–968
51. Bangura JF (2011) Directly coupled electromagnetic field-electric circuit model for analysis of a vector-controlled wound field brushless starter generator. IEEE Trans Energy Conv 26(4):1033–1040
52. Zhao N, Liu W (2009) Electromagnetic field and thermal field coupling analysis of brushless DC motor. Small Spec Electr Mach 3:9–11
53. Djurabekova F et al (2011) Atomistic modeling of metal surfaces under electric fields: direct coupling of electric fields to a molecular dynamics algorithm. Phys Rev E 83(2):026704
54. Maas SA (2003) Nonlinear microwave and RF circuits. Artech house
55. Kassim S, Malek F (2010) EM/circuit co-simulations: a highly accurate method for microwave amplifier design. Univ J Comput Sci Engin Technol 2:127–132
56. Lu T, Jin J-M (2016) Electrical-thermal co-simulations for analysis of high-power RF/microwave components. IEEE Trans Electromagn Compat 59(1):93–102

Chapter 4
Co-simulations of Antennas

4.1 Introduction

4.1.1 Significance of Simulations in Antenna Designs

Since the invention of the first-generation (1G) communication technology in 1978, we have witnessed the significant development and advancement of communication technologies, from the basic voice communication of 1G networks to the immersive multimedia communications of fifth-generation (5G) networks with an inclusion of 4K high-definition (HD) live broadcasting [1–3]. In the realm of mobile or wireless communications, antenna design stands out as a pivotal component. Over years, antenna technologies have evolved rapidly, from those of designing omnidirectional antennas in the early stage to those of designing directional antennas, dual-polarized antennas, multi-polarized antennas, multiband antennas, and broadband antennas in recent years. Currently, large-scale array antennas could represent the state-of-the-art ones and they have become increasingly intricate and diverse [4–9].

First and foremost, the demand for faster network connections within a limited bandwidth has led to the prominence of larger-scale Multiple-Input Multiple-Output (MIMO) array antennas [10]. Presently, the industry commonly employs 64-transmit/64-receive array antennas with the number of units reaching 128. In addition, the antennas with 256 and 512 units are being actively deployed [11]. Moreover, the antennas have been in a transition from passive systems to active systems. The integration of radio frequency (RF) devices such as amplifiers and filters into active antenna systems can enhance the system capacity, improve the performances, reduce the operational costs, and minimize the energy losses of entire mobile networks. Using active antennas has become an essential requirement for 5G massive MIMO antenna systems. Additionally, the antenna design is progressing towards a systematization. The use of multi-array, multi-beam, and high-frequency band technologies has imposed a higher demand on the antenna design and compelled designers to design antennas from a system-level perspective. Furthermore,

M. S. Tong and X. Y. Li, *Co-simulations of Microwave Circuits and High-Frequency Electromagnetic Fields*, https://doi.org/10.1007/978-981-99-8307-0_4

the antennas have increasingly become miniaturized and the advent of 5G millimeter-wave communications particularly necessitates the design of smaller antennas with a higher integration.

As the antennas tend to be large and active array systems with high integration, using a powerful and robust simulation software tool is imperative and indispensable to design such antennas. This tool should encompass an entire co-simulation function which can well simulate antenna units, arrays, and RF devices. In addition, since the antennas are the devices or systems of transmitting and receiving electromagnetic waves, the electromagnetic simulations can enable an accurate extraction of model parameters, including those of passive components like the filters within the systems. This facilitates the large-scale field-circuit co-simulations for the antennas, along with an accurate analysis for their heat dissipation and reliability. Because the antennas are integrated into wireless communication systems, their performances will significantly impact the performances of the entire systems, including the operational cost, effectiveness, and reliability, so the optimal design of antennas is very essential. The optimal design of antennas strongly relies on the use of advanced simulation software tools and related technologies which actually utilize the numerical solutions of electromagnetic equations to predict and optimize the performances of antennas. With the powerful tools, the design can become much simpler and can dramatically reduce the costs while enhancing the efficiency and quality [11–15]. In conclusion, it is impossible without using the simulation software tools and related technologies to design the antennas nowadays and this chapter will address the state-of-the-art simulation for antenna designs.

4.1.2 Basic Process of Antenna Simulations

The initial step in antenna design simulation is conducting an analysis of simulation requirements. During this stage, it is essential to define the simulation objectives, which encompass the frequency range, radiation characteristics, polarization mode, and input impedance requirements. Additionally, the selection of appropriate simulation methods and software is a crucial consideration, dependent on the complexity and type of antenna being designed. Simultaneously, it is imperative to analyze system constraints and performance indicators, including antenna size, weight, materials, as well as performance factors like radiation efficiency, bandwidth, and impedance matching. By comprehensively considering these factors, one can ensure that the simulation results align with the performance requirements and system constraints of practical applications, guaranteeing smooth progress throughout the simulation process and accuracy in the results.

The design of antenna structure encompasses three key elements: antenna type selection, antenna geometry design, and antenna material selection and performance analysis [16–19]. Based on the system requirements and performance indicators, diverse antenna types may be chosen, such as microstrip antennas [20], helical antennas [21], and beamforming antennas [22]. For each specific antenna type, it is necessary to design the corresponding geometric structure, which includes determining the appropriate size, shape, layout, and other relevant parameters. Simultaneously, careful consideration and performance analysis of the chosen antenna materials are crucial to ensure optimal performance and reliability. Commonly utilized antenna materials comprise metal, polymer, and ceramic materials. Throughout the design process, it is important to consider the specific application scenarios and design requirements, tailoring the design accordingly.

The primary objective of electromagnetic modeling is to determine the radiation characteristics and electromagnetic parameters of an antenna. This facilitates subsequent simulation calculations and result analysis. During the electromagnetic modeling process, various factors need to be considered, including the antenna's geometric structure, medium properties, and electromagnetic boundary conditions. Electromagnetic theory and numerical methods are employed to model and calculate these factors [23–25].

The electromagnetic modeling of an antenna involves the following steps:

1. Antenna geometric structure modeling: Utilize suitable 3D modeling software to create an accurate 3D model based on the antenna's physical size and geometric structure. Mesh the model appropriately.

2. Modeling of dielectric properties: Consider the impact of the surrounding medium on the antenna. Establish a model for the dielectric constant and dielectric loss, and analyze the frequency response and loss of the medium.

3. Electromagnetic boundary condition modeling: Select suitable boundary conditions to simulate the electromagnetic interaction between the antenna and the surrounding environment. This enhances the accuracy of the calculations.

4. Modeling of antenna radiation pattern: Taking into account the antenna's geometric structure, dielectric properties, and boundary conditions, use electromagnetic theory and numerical methods to establish the antenna's radiation pattern. This includes parameters such as radiation pattern, radiation power, and radiation impedance.

5. Calculation of antenna electromagnetic parameters: Through simulation calculations, determine the electromagnetic parameters of the antenna, such as input impedance, gain, and pattern. Evaluate the antenna's performance and characteristics based on these parameters.

By employing scientific and rational electromagnetic modeling, precise electromagnetic characteristics and performance indicators of an antenna can be obtained. These results can be used for analyzing and optimizing antenna designs.

In the realm of antenna design, the selection of an appropriate simulation method holds the utmost significance. Commonly employed techniques include the finite element method (FEM), finite-difference time-domain method (FDTD), and method of moments (MoM) [26–28]. The finite element method is well-suited for electromagnetic simulation involving intricate geometries and scenarios with multiple media. It enables the calculation of radiation characteristics, input impedance, and radiation field distribution of antennas. On the other hand, the finite-difference time-domain method is a rapid and straightforward approach specifically suitable for assessing the radiation characteristics of broadband antennas. The method of moments, employing the radiation integral equation, facilitates the calculation of antenna radiation properties and is particularly suitable for large-scale and high-frequency antenna simulations. While selecting a simulation method, factors such as simulation accuracy, computational speed, simulation scope, and computational resources must be taken into account. Co-simulation methods, such as combining the finite element method with the finite-difference time-domain method, can also be employed to enhance simulation accuracy and computational efficiency. Additionally, when choosing simulation software, one should consider the software's functionality and user-friendliness.

Simulation software commonly used for antenna design includes HFSS, CST Microwave Studio, FEKO, and others. These software offer powerful simulation capabilities and flexible parameter settings to cater to the diverse simulation requirements of antenna designs. When utilizing simulation software for antenna design, it is important to consult relevant literature and information to improve understanding and application of simulation methods. Additionally, accuracy and efficiency should be prioritized to ensure reliable simulation results. Taking HFSS simulation software as an example, let's outline the basic process and skills involved in using the software:

1. Interface and operation process:

HFSS features a user-friendly interface and streamlined operation process, enabling quick usability. Upon opening the software, users can choose to create a new project or open an existing one. When creating a new project, simulation parameters such as type, coordinate system, and frequency range must be selected, followed by the construction of the antenna model. Once the model is established, simulation calculations and result analysis can be performed.

2. Antenna model construction:

To create an antenna model in HFSS, users can import a CAD file or utilize the built-in geometric modeling tools. After establishing the geometric model, material parameters must be defined, ports set up, boundary conditions added, and electromagnetic modeling conducted.

3. Setting simulation parameters and boundary conditions:

-Prior to simulation calculations, appropriate simulation parameters and boundary conditions need to be set. These parameters include operating frequency, antenna size, material properties, etc., which should align with the specific simulation requirements. Boundary conditions encompass the simulation space size and surrounding conditions, which should be configured based on the desired simulation accuracy and efficiency.

4. Simulations method and solver selection:

HFSS provides multiple simulation methods and solvers that can be selected based on computational resources and simulation requirements. Common simulation methods include time-domain finite element method, frequency domain finite element method, and modal analysis method. Solvers such as direct solvers and iterative solvers are commonly used as well.

5. Visualization of simulation results:

HFSS offers various methods for visualizing simulation results. These include 2D and 3D graphics displays, S-parameter curves, radiation maps, and more. Additionally, data files can be exported for further analysis and processing using external tools.

6. Analysis and optimization of simulation results:
Once the simulation results are obtained, they need to be analyzed and optimized. HFSS provides numerous analysis tools, such as input impedance calculation, radiation characteristics analysis, and bandwidth calculation. Based on the analysis results, the antenna design can be optimized and refined. After completing the simulation calculations, the simulation results must undergo analysis and evaluation to optimize the antenna design. Typically, the subsequent antenna simulation design is carried out within a computer simulation environment. When producing and verifying the prototype, various factors such as materials, manufacturing processes, and test equipment must be taken into account. To create the prototype, suitable conductive and dielectric materials are chosen, and the conductive structure is laid out and processed according to the design requirements. It is then laminated with the dielectric plate. The test verification process includes conducting electrical performance tests and radiation performance tests. To ensure the accuracy and reliability of the test equipment, appropriate test frequency bands and test points are selected. Once the testing is complete, the results are analyzed to evaluate the antenna's performance and the accuracy of the simulation results.

4.2 Principle of Simulations for Antenna Designs

4.2.1 Basic Theory of Antennas

1. Characteristics of Electromagnetic Waves [29]
1.1 Propagation Speed
Electromagnetic waves propagate at the speed of light in a vacuum, which is approximately $c = 2.998 \times 10^8 \text{m/s}$. In other media, the propagation velocity of electromagnetic waves can be calculated using the relative permittivity ε_r and relative permeability μ_r of the medium:

$$v = \frac{c}{\sqrt{\varepsilon_r \mu_r}} \tag{4.1}$$

where v represents the velocity.
1.2 Wavelength and Frequency
The relationship between the wavelength λ and frequency f of an electromagnetic wave can be expressed using the following formula:

$$\lambda = \frac{v}{f} \tag{4.2}$$

where v is the propagation speed of the electromagnetic wave in the medium.

1.3 Polarization
Polarization refers to the orientation and distribution of the electric field vector in a plane perpendicular to the direction of propagation. Electromagnetic waves can exhibit different polarization types, including linear polarization, circular polarization, and elliptical polarization.
1.4 Linear Polarization
Linear polarization is the simplest form of polarization. In the case of linear polarization, the direction of the electric field vector's oscillation remains constant, while the amplitude varies over time. A linearly polarized electromagnetic wave can be expressed as

$$\mathbf{E}(t) = E_0 \cos(\omega t + \phi)\mathbf{e} \tag{4.3}$$

where E_0 represents the amplitude, ω is the angular frequency, ϕ is the phase difference, and $\mathbf{e}$ is a unit vector indicating the polarization direction.
1.5 Circular Polarization
In the case of circular polarization, the electric field vector vibrates in a circular path with a constant amplitude. A circularly polarized electromagnetic wave can be represented as

$$\mathbf{E}(t) = E_0 \cos(\omega t + \phi)\mathbf{e}_x + E_0 \sin(\omega t + \phi)\mathbf{e}_y \tag{4.4}$$

where $\mathbf{e}_x$ and $\mathbf{e}_y$ are mutually perpendicular unit vectors, and E_0 denotes the amplitude.
1.6 Elliptical Polarization
In the case of elliptical polarization, the electric field vector traces an elliptical path, and the amplitude varies with time. An elliptically polarized electromagnetic wave can be expressed as

$$\mathbf{E}(t) = E_{0x} \cos\left(\omega t + \phi_x\right) \mathbf{e}_x + E_{0y} \cos\left(\omega t + \phi_y\right) \mathbf{e}_y \tag{4.5}$$

where E_{0x} and E_{0y} represent the amplitudes, ϕ_x and ϕ_y denote the initial phases, and $\mathbf{e}_x$ and $\mathbf{e}_y$ are mutually perpendicular unit vectors. The choice of polarization type is closely related to practical applications. For instance, linearly polarized electromagnetic waves are suitable for transmission and reception of antennas, while circularly and elliptically polarized electromagnetic waves find applications in rotating polarizers, modulators, and other fields.
1.7 Polarization State Representation
The vector representation of the electric and magnetic field distributions in a general electromagnetic wave can be complex. To facilitate the description, we can employ either the Jones vector or the Stokes vector to represent the polarization state. The Jones vector represents the polarization state of an electric field and consists of two real numbers, representing the components of the electric field along two orthogonal directions. Assuming the vibration direction of the electromagnetic wave aligns with the x-axis, the Jones vector can be written as

$$\mathbf{J} = \begin{bmatrix} E_x(0) \\ E_y(0) \end{bmatrix} = \begin{bmatrix} E_{0x} e^{i\varphi_x} \\ E_{0y} e^{i\varphi_y} \end{bmatrix} \tag{4.6}$$

where E_x and E_y represent the components of the electric field in the x-axis and y-axis directions, respectively. The Stokes vector is another method to represent the polarization state of electromagnetic waves. It comprises four real numbers that describe the intensity of light in different directions and the relative differences in light intensity. The Stokes vector can be expressed as

$$\mathbf{S} = \begin{bmatrix} S_0 \\ S_1 \\ S_2 \\ S_3 \end{bmatrix} \tag{4.7}$$

In this representation, S_0 represents the total light intensity, while S_1, S_2, and S_3 represent the differences in light intensity along the x-axis, y-axis, and diagonal directions, respectively. For a fully polarized electromagnetic wave, the degree of polarization is used to quantify its polarization degree, ranging from 0 to 1. A value of 0 indicates no polarization (i.e., random vibration), while a value of 1 indicates complete polarization (i.e., vibration in a single direction). The degree of polarization can be calculated from the Stokes vector:

$$P = \frac{\sqrt{S_1^2 + S_2^2 + S_3^2}}{S_0} \tag{4.8}$$

1.8 Energy Transmission

The transmission of energy through electromagnetic waves can be described using the Poynting vector **S**, which is defined as

$$\mathbf{S} = \frac{1}{\mu_0} \mathbf{E} \times \mathbf{B} \tag{4.9}$$

where **E** represents the electric field strength, **B** represents the magnetic field strength, and μ_0 denotes the vacuum permeability. The Poynting vector indicates the amount of energy passing through a unit area per unit time and it points in the direction of propagation of the electromagnetic wave.

1.9 Reflection and Refraction

When an electromagnetic wave travels from one medium to another, reflection and refraction phenomena occur. The reflectivity R and transmittance T can be determined using the Fresnel equations. For electromagnetic waves incident normally to the interface between the media, the equations for reflection and transmission are as follows:

$$R_{\perp} = \left| \frac{n_1 - n_2}{n_1 + n_2} \right|^2 \tag{4.10}$$

$$T_{\perp} = \frac{4n_1 n_2}{(n_1 + n_2)^2} \tag{4.11}$$

In the case of oblique incidence, the formula for the reflectance ($R_{\parallel}$) is correct. However, for the calculation of transmittance, what is more commonly used is the ratio of the energy of the refracted light rather than the direct formula for transmittance. Considering energy conservation, the transmittance (T) is also related to the angle of incidence and the angle of refraction, but it requires a more complex expression through the Fresnel equations for accurate calculation

$$R_{\parallel} = \left| \frac{n_2 \cos\theta_i - n_1 \cos\theta_t}{n_2 \cos\theta_i + n_1 \cos\theta_t} \right|^2 \tag{4.12}$$

$$T_{\parallel} = \frac{4n_1 n_2 \cos\theta_i \cos\theta_t}{(n_2 \cos\theta_i + n_1 \cos\theta_t)^2} \tag{4.13}$$

1.10 Attenuation

During the process of transmission, electromagnetic waves encounter various obstacles and media, resulting in the attenuation of the waves. This attenuation leads to a decrease in the energy and signal strength of the electromagnetic waves. Attenuation refers to the phenomenon where the power of an electromagnetic wave decreases as the propagation distance increases. Several factors contribute to attenuation, including free-space path loss, multipath propagation, reflection, refraction, scattering, and absorption. Free space path loss, which accounts for dispersion and geometric diffusion when electromagnetic waves propagate in free space, is the most fundamental among all attenuation factors. The unit of attenuation is the decibel (dB), commonly used to express power ratios. The amount of attenuation can be calculated using the following formula:

$$\text{Attenuation (dB)} = 10 \log_{10} \left(\frac{P_1}{P_2} \right) \tag{4.14}$$

where P_1 represents the power before transmission, and P_2 represents the power after transmission. In practical applications, the attenuation factor is often used to describe the magnitude of attenuation. The attenuation factor is defined as follows:

$$\text{Attenuation factor (multiple)} = \frac{P_1}{P_2} \tag{4.15}$$

The magnitude of attenuation is influenced by factors such as frequency, transmission distance, transmission medium, and obstacles. In practical applications, it is essential to select the appropriate transmission mode and employ signal enhancement methods based on the specific circumstances to ensure transmission quality and stability.

An antenna is an essential electronic component that converts electromagnetic waves from a transmission line, such as a cable, into radiated electromagnetic waves in free space. Its fundamental principle is based on electromagnetic field theory. The radiation field of an antenna can be described by its radiation electric field and radiation magnetic field. The expression for the radiation field is typically formulated in terms of the current distribution and the length of the antenna. For instance, considering a one-dimensional antenna, the expression for its radiation field can be represented as follows:

$$E(\theta) = \frac{j\omega\mu I_0 l}{4\pi}\frac{e^{-jkr}}{r}\sin\left(\frac{\pi l}{\lambda}\cos\theta\right) \tag{4.16}$$

where $E(\theta)$ denotes the electric field strength along the antenna, θ represents the radiation angle, ω denotes the angular frequency, μ represents the medium permeability, I_0 represents the antenna current, l represents the antenna length, k represents the wave number, and r represents the distance from the radiation point to the antenna. Additionally, λ represents the wavelength. By analyzing the expression of the radiation field, we can obtain the basic parameters of the antenna, such as gain, beam width, and directivity. For example, the gain of an antenna can be calculated as follows:

$$G = \frac{4\pi U}{P_{\text{in}}} \tag{4.17}$$

where U is the radiation intensity in a specific direction, and P_{in} represents the input power of the antenna. The antenna plays a crucial role in communication systems, directly impacting their effectiveness and quality. Hence, during antenna design, it is essential to consider parameters such as gain, beam width, and directivity to meet specific communication requirements.

Furthermore, due to the varying structures and operating principles of different antennas, the process of antenna design and optimization is inherently complex. In practical applications, different types of antennas are chosen based on specific communication requirements and environmental conditions. Various antenna options include directional antennas, omnidirectional antennas, and broadband antennas, among others. Directional antennas concentrate radiation energy towards a specific direction, making them suitable for long-distance communication. Omnidirectional antennas, on the other hand, uniformly radiate energy in all directions, making them suitable for short-distance communication. Broadband antennas can operate across multiple frequency bands and are used for multi-band [19] communication. To describe and evaluate antenna performance, several fundamental parameters are employed.

1. Gain

Gain is a crucial parameter that indicates the performance of an antenna. It represents the ratio of the radiation intensity of the antenna in a specific direction to the radiation intensity of an isotropic radiator with the same input power. Gain is typically expressed in decibels (dB). An isotropic radiator is a theoretical antenna that radiates power uniformly in all directions, although it does not exist in practical terms. The gain of an antenna can be calculated using the following formula:

$$G(\text{dBi}) = 10\log_{10}\left(\frac{U}{U_0}\right) \tag{4.18}$$

where U is the radiation intensity of the antenna in a specific direction and U_0 is the radiation intensity of an ideal omnidirectional antenna when the same input power is radiated.

2. Beamwidth

Beamwidth refers to the angular range in which the radiation power in the main radiation direction of the antenna drops to half of its peak value. Beamwidth can be divided into vertical and horizontal components. The vertical and horizontal beamwidths can be calculated using the following formulas:

$$\theta_{\text{HPBW}} \approx \frac{c}{fd} \tag{4.19}$$

where c is the speed of light in air or vacuum, f is the operating frequency of the antenna, and d is the diameter of the antenna or in some contexts, it can be considered as a characteristic size of the antenna aperture.

3. Antenna Directivity

Antenna directivity refers to its ability to radiate electromagnetic energy in a specific direction. It is commonly represented by a radiation pattern, which depicts the power density of the antenna's radiation in different directions. The pattern can be determined using the following formula:

$$D = \frac{4\pi U(\theta, \phi)}{P_{\text{rad}}} \tag{4.20}$$

where P_{rad} is the total power radiated by the antenna, and $U(\theta, \phi)$ is the radiation intensity in a specific direction.

4. Antenna Impedance

The impedance of an antenna refers to its input impedance at a specific frequency. The input impedance can be calculated using the following equation:

$$Z_{\text{in}} = R_{\text{in}} + jX_{\text{in}} \tag{4.21}$$

where R_{in} and X_{in} represent the real and imaginary parts of the input impedance, respectively. The real impedance describes the antenna's resistance to current flow, while the imaginary impedance describes its resistance to voltage. Understanding the input impedance is crucial for designing the matching network and optimizing the antenna system.

5. Voltage Standing Wave Ratio (VSWR)
The Voltage Standing Wave Ratio (VSWR) is the ratio of the maximum reflection coefficient to the minimum reflection coefficient at the input of an antenna. It indicates how well the antenna's input impedance matches the impedance of the transmission line. A lower VSWR signifies a better impedance match, ensuring higher efficiency and stability of the antenna system. Typically, the VSWR should be less than 1.5. The VSWR can be calculated using the following formula:

$$VSWR = \frac{1 + |\Gamma|}{1 - |\Gamma|} \tag{4.22}$$

where $|\Gamma|$ represents the reflection coefficient at the input of the antenna. If reflections occur between the antenna and the transmission line, they can degrade antenna performance and reduce transmission efficiency.

There are various types of antennas, each with its own characteristics and applications. Below are some commonly used antenna types and their respective application areas [30–40]:

1. Half-wave Dipole Antenna

–Principle: The half-wave dipole antenna consists of two equal-length wires, typically fed at the center. Its length is half the operating wavelength ($\lambda/2$).

–Applications: Broadcast reception, wireless communication, RF testing, etc.

2. Passive Microstrip Antenna (Patch Antenna)

–Principle: The passive microstrip antenna is a planar antenna that includes a metal plate (radiating element) separated from the ground by an insulating layer. The size of the radiating element is typically $\lambda/2$.

–Applications: Wireless LAN, satellite communication, radar systems, mobile communication, etc.

3. Active Phased Array Antenna

–Principle: An active phased array antenna consists of multiple independent radiation elements. The beam directivity is adjusted by controlling the phase of each radiation element. The beam directivity angle (θ) is given by $\theta = \sin^{-1}\left[\left(\frac{\lambda}{d}\right)\sin(\Delta\phi)\right]$, where λ is the operating wavelength, d is the array element spacing, and $\Delta\phi$ is the phase difference between adjacent elements.

–Applications: Radar systems, satellite communications, aerospace communications, etc.

4. Log-periodic Antenna

–Principle: A log-periodic antenna is composed of multiple dipoles arranged at logarithmic intervals. This antenna type exhibits broadband characteristics, meaning its performance parameters remain consistent over its operating frequency range.

–Applications: TV broadcast reception, broadband communication systems, etc.

5. Helical Antenna

–Principle: The helical antenna is an axially symmetrical antenna that consists of a spiral wire. It can generate circularly polarized waves and offers good antenna directivity and broadband characteristics.

–Applications: Satellite communications, Global Positioning System (GPS), radio navigation, etc.

6. Time-domain pulse antenna (Ultra-WideBand Antenna)

–Principle: The time-domain pulse antenna operates with ultra-wideband characteristics, usually in conjunction with short-time pulse signals. It offers excellent pulse fidelity, broadband performance, and maintains high gain and directivity over a wide frequency range.

–Applications: UWB communication systems, radar systems, geological exploration, etc.

7. Slot Antenna

–Principle: A slot antenna is a type of microwave antenna that features one or more slots on a conductive plane. The length of the slot is usually $\lambda/2$, and the size and shape of the slot determine the antenna's radiation characteristics.

–Applications: Satellite communications, radio broadcasting, radar systems, etc.

8. Yagi-Uda Antenna

–Principle: The Yagi-Uda antenna is a directional antenna comprising a primary radiator, a reflector, and one or more directors. Adjusting the spacing and length between radiators allows for higher gain and directivity.

–Applications: Radio broadcasting, television reception, amateur radio communication, etc.

9. Load Capacitance Antenna (Electrically Small Antenna)

–Principle: A loaded capacitive antenna is significantly smaller than the operating wavelength. Adding a capacitive load to the top or base of the antenna improves radiation efficiency and bandwidth.

–Applications: Internet of Things (IoT) devices, wireless sensor networks, mobile communications, etc.

The above are just some of the common antenna types and their application areas. In fact, there are many other types of antennas and they are designed according to different needs and scenarios.

4.2.2 Radiation Characteristics of Antennas

The radiation pattern of an antenna refers to the distribution of electromagnetic waves radiated by the antenna. In three-dimensional space, the radiation pattern of an antenna is usually represented by a radiation pattern. A radiation pattern is a polar plot in which the horizontal direction represents azimuth, the vertical direction represents pitch, and the magnitude in each direction represents the radiated power density in that direction. The radiation pattern can be expressed in terms of the magnitude of the electric or magnetic field, or in terms of the power density. The radiation pattern of the antenna is related to the antenna structure, operating frequency, input excitation, and other factors. When designing an antenna, it is necessary to simulate the antenna to get the radiation pattern of the antenna for optimization. In order to conveniently calculate and analyze the radiation pattern of the antenna, we usually use the ideal point source antenna model. An ideal point source antenna is an ideal antenna model, which can be regarded as an invisible point source and radiates electromagnetic waves uniformly to the surrounding space. The radiation field distribution of an ideal point source antenna can be expressed as

$$E(r) = \frac{e^{-jk_0 r}}{r} \tag{4.23}$$

where E represents the electric field strength, r is the distance, θ and ϕ are the pitch and azimuth angles, respectively, and k_0 is the wave number in free space. The wave number is related to the frequency f by $k_0 = \frac{2\pi f}{C}$, where C is the speed of light in vacuum. The radiation pattern of an antenna describes the distribution of radiation power density in different directions, often described by the power density function $P(\theta, \phi)$.

For an ideal point source antenna, the power density function can be expressed as

$$P(r) = \frac{|E(r)|^2}{2Z_0} = \frac{1}{2Z_0 r^2} \tag{4.24}$$

where Z_0 represents the characteristic impedance of free space, which has a value of 377Ω. In the spherical coordinate system, the radiation field distribution of an antenna can be represented using spherical harmonic functions. Spherical harmonics are mathematical tools used to solve spherically symmetric problems and can be seen as a special type of Fourier series that expands an arbitrary function on the sphere into a series of orthogonal functions. The general form of spherical harmonics is given by

$$Y_{lm}(\theta, \phi) = \sqrt{\frac{(2l+1)}{4\pi} \frac{(l-m)!}{(l+m)!}} e^{im\phi} P_l^m(\cos\theta) \tag{4.25}$$

where l and m are integers, $P_l^m(\cos\theta)$ is a Legendre polynomial, and $Y_l^m(\theta, \phi)$ satisfies

$$\int_0^{2\pi} \int_0^{\pi} |Y_{lm}(\theta, \phi)|^2 \sin\theta d\theta d\phi = 1 \tag{4.26}$$

Due to the orthogonality of spherical harmonics, they can be used to expand the radiation field distribution of an antenna as follows:

$$E(\theta, \phi) = \sum_{l=0}^{\infty} \sum_{m=-l}^{l} a_{lm} Y_{lm}(\theta, \phi) \tag{4.27}$$

where a_{lm} is a complex coefficient that represents the radiated power of the antenna in the l and m directions. The radiation power density of the antenna can be expressed as

$$P(\theta, \phi) = \frac{1}{2Z_0} \left| \sum_{l=0}^{\infty} \sum_{m=-l}^{l} a_{lm} Y_{lm}(\theta, \phi) \right|^2 \tag{4.28}$$

The beamwidth of an antenna refers to the angle range in which the radiationpower density of the main lobe drops by 3dB, which is commonly expressed as the 3dB beamwidth. For the radiation field expanded in spherical harmonics, the approximate 3dB beamwidth can be calculated as

$$\theta_{3dB} \approx \frac{60°}{\sqrt{l_1}}, \quad \phi_{3dB} \approx \frac{120°}{\sqrt{l_1}} \tag{4.29}$$

where l_1 represents the smallest value of l, which can be obtained by solving the following equation:

$$\frac{\sum_{l=0}^{l_1} \sum_{m=-l}^{l} a_{lm}^2}{\sum_{l=0}^{\infty} \sum_{m=-l}^{l} a_{lm}^2} = 0.5 \tag{4.30}$$

The radiation pattern of a real antenna differs from that of an ideal point source antenna due to factors such as the non-ideal nature of the antenna structure, ground effects, and multipath propagation. To analyze the radiation pattern of an actual antenna, antenna simulation software is commonly used for calculation and analysis.

Antenna gain represents the ratio of power density radiated by the antenna to the power density radiated by an ideal point source. It is typically measured in dBi (decibels relative to an isotropic radiator) and quantifies the antenna's ability to transmit power. Higher gain results in increased transmission power and extends the communication range. Antenna directivity refers to the variation in power density radiated by the antenna across different directions. It is influenced by the antenna's radiation pattern, with higher directivity indicating a more focused radiation power and greater communication distance. The gain of an antenna can be calculated using the following formula:

$$G = 10 \log_{10} \frac{P_{rad}}{P_{in}} \tag{4.31}$$

where P_{rad} denotes the radiated power of the antenna, and P_{in} represents the input power of the antenna. The radiation pattern of the antenna can be described by the power density function $P(\theta, \phi)$. The radiated power of the antenna can be expressed as

$$P_{rad} = \int_0^{2\pi} \int_0^{\pi} P(\theta, \phi) r^2 \sin\theta d\theta d\phi \tag{4.32}$$

The input power of the antenna can be expressed as

$$P_{in} = \frac{1}{2} \frac{|V_{in}|^2}{R_{in}} \tag{4.33}$$

where V_{in} represents the input voltage of the antenna, and R_{in} represents the input impedance of the antenna.

Antenna directivity refers to the variation in the power density of antenna radiation in different directions, commonly represented by radiation pattern. A radiation pattern is a two-dimensional polar map that depicts the azimuth angle on the horizontal axis, the pitch angle on the vertical axis, and the power density in each direction. It can be described in terms of the amplitude of the electric or magnetic field, or the power density itself. Power density is often used to represent the radiation pattern as it directly reflects the transmitting capability of the antenna. Mathematically, the power density can be expressed as

$$P(\theta, \phi) = \frac{1}{2Z_0} |E(\theta, \phi)|^2 \tag{4.34}$$

where Z_0 represents the impedance in free space, and $E(\theta, \phi)$ denotes the radiated electric field component of the antenna in spherical coordinates. The directivity of an antenna primarily relies on its radiation pattern. In the case of an ideal point source antenna, its radiation pattern resembles a spherical distribution centered on the antenna, indicating that the radiation power density is the same in all directions. Thus, the radiation pattern of an ideal point source antenna takes the form of a circle. However, a real antenna exhibits a significantly different radiation pattern compared to an ideal point source antenna. Various factors, such as antenna structure, material, excitation mode, and operating frequency, affect the radiation pattern of a real antenna. Hence, the radiation pattern of an actual antenna can exhibit various shapes and characteristics. Antenna directivity is commonly quantified by the Half Power Beam Width (HPBW), which represents the angular range where the radiation power density of the main lobe decreases by 3 dB. The HPBW for a specific antenna can be determined using simulation software. In general, a smaller HPBW indicates better antenna directivity. An antenna's radiation pattern typically consists of one or more main lobes and several side lobes. The main lobe corresponds to the region with the highest radiation power density in a particular direction, serving as the primary radiation direction of the antenna. Side lobes, on the other hand, refer to radiation power density in directions adjacent to the main lobe. When designing an antenna, optimizing the direction and width of the main lobe becomes one of the crucial parameters to consider. Sidelobes represent radiation directions other than the main lobe in the antenna's radiation pattern, typically exhibiting lower radiation power density. The size and number of sidelobes significantly impact antenna performance. A large number of sidelobes can adversely affect signal transmission quality, directivity, and gain of the antenna. Therefore, antenna design must take into account directivity to achieve superior radiation performance.

Polarization refers to the characteristics of the direction of electromagnetic wave vibration, which is an important factor in antenna design. The polarization characteristic of antenna is usually expressed by polarization vector, which includes polarization direction and polarization degree. The polarization direction refers to the direction of the electromagnetic wave vibration direction relative to the antenna. The polarization direction of the antenna can be divided into horizontal polarization, vertical polarization, oblique polarization, circular polarization, and other types. For horizontally polarized antenna, the vibration direction of electromagnetic wave is perpendicular to the ground plane and parallel to the ground, which is suitable for communication of parallel transmission between cities and buildings. Vertically polarized antenna, the vibration direction of electromagnetic wave is perpendicular to the ground plane, which is suitable for ground communication. The inclined polarized antenna is suitable for communication in mountainous areas, forest areas, and other complex terrains because the vibration direction of the electromagnetic wave forms a certain angle with the ground plane. Circularly polarized antenna, the vibration direction of electromagnetic wave moves along a circular orbit, which is suitable for radio navigation, radar, and other applications. The polarization degree refers to the degree of deviation of the vibration direction of the electromagnetic wave with respect to the polarization direction of the antenna. The polarization degree can be divided into two types: linear polarization and circular polarization. For linearly polarized antennas, the vibration direction of electromagnetic waves moves along a straight line with a small deviation angle, which is suitable for transmitting high-speed data, voice, and other applications. For circularly polarized antennas, the vibration direction of electromagnetic waves moves along a circular orbit with a large deviation angle, which is suitable for radio navigation, radar, and other applications. Because the polarization characteristics of the antenna are different from those of the transmission medium, the polarization loss of the signal will occur in the transmission process. Polarization loss can result in signal attenuation and distortion, degrading the performance of wireless communication systems. The polarization loss can be reduced by polarization matching. Polarization matching refers to matching the polarization characteristics of the antenna with the polarization characteristics of the transmission medium to reduce the polarization loss of the signal. Commonly used polarization matching methods include: selecting the appropriate antenna polarization, adjusting the polarization angle of the antenna, and using the polarization converter.

Impedance matching is the process of matching the input port impedance of an antenna to the input impedance of a transmission line or receiver. The purpose of impedance matching is to maximize the energy transfer when transmitting signals from the antenna to the receiver or from the receiver to the antenna, so as to improve the system performance. Impedance matching is very important in wireless communication systems, because impedance mismatch will lead to signal reflection and attenuation, thus reducing the quality and reliability of signal transmission.

The complex impedance Z_{in} represents the input impedance of the antenna and can be expressed as follows:

$$Z_{\mathrm{in}} = R_{\mathrm{in}} + jX_{\mathrm{in}} \tag{4.35}$$

where R_{in} denotes the real part of the impedance, representing the resistance at the antenna's input. X_{in} represents the imaginary part of the impedance, which corresponds to the impedance of the reflection device connected at the antenna's input, such as inductance or capacitance.

To achieve impedance matching, it is necessary to minimize the reflection coefficient Γ, which quantifies the ratio of the reflected signal to the forward signal:

$$\Gamma = \frac{Z_L - Z_0}{Z_L + Z_0} \tag{4.36}$$

In this equation, Z_L refers to the impedance of the load, while Z_0 represents the characteristic impedance of the transmission line.

A common approach to impedance matching is through an antenna matching network. The antenna matching network usually includes some passive devices, such as inductors, capacitors, and transformers, which are used to adjust the input impedance of the antenna to match the input impedance of the transmission line or receiver. The design of antenna matching network needs to consider the operating frequency of the antenna, the input impedance, the characteristic impedance of the transmission line, and other factors, which can be simulated and optimized by circuit simulation software. Another common method of impedance matching is to use an attenuator to adjust the power of the input signal to a level that matches the transmission line or receiver input impedance. Attenuators are typically implemented using variable resistors or variable capacitors. Impedance matching is very important in wireless communication systems. It can improve the transmission performance of the system, avoid signal reflection and loss, and thus improve the communication distance and reliability. In impedance matching, the commonly used parameters include matching coefficient, matching loss, and bandwidth. The matching factor is the ratio of the actual load impedance to the input impedance of the transmission line or receiver and is used to describe how well the impedance is matched. The matching coefficient ranges from 0 to 1. When the matching coefficient is 1, the impedance matching is perfect and the reflection coefficient is 0. Matching loss is the loss of energy due to impedance mismatch. The magnitude of the matching loss is related to the matching coefficient and the reflection coefficient, and can be calculated by the following formula:

$$L_m = 10 \log_{10}\left(\frac{1}{1 - |\Gamma|^2}\right) \text{ dB} \tag{4.37}$$

where L_m represents the matching loss and Γ denotes the reflection coefficient.

The antenna's bandwidth refers to its capability to function effectively within a specific range of frequencies. It depends on factors such as impedance matching, radiation efficiency, and structural parameters of the antenna. As the operating frequency of an antenna changes, its impedance characteristics also vary, thereby impacting its performance. Hence, considering the antenna's bandwidth is crucial to ensure optimal functioning within a designated frequency range. Specifically, the antenna's bandwidth is defined as the frequency range where the difference between its gain and that of an ideal point source antenna is less than 3 dB at the operating frequency. The ideal point source antenna possesses identical radiation characteristics but does not encounter any losses or matching issues. Consequently, the antenna's bandwidth directly affects its performance across diverse frequencies. The antenna's bandwidth can be determined by calculating its standing wave ratio (SWR). SWR represents the ratio of the antenna's input port impedance to the characteristic impedance of the transmission line, indicating the reflection loss at the antenna's input port. Assuming the antenna's operating frequency is denoted as f_c, with f_1 and f_2 representing its lower and upper endpoint frequencies, the antenna's bandwidth (BW) is defined as follows:

$$BW = \frac{f_2 - f_1}{f_c} \tag{4.38}$$

In general, the wider the bandwidth of the antenna, the wider the applicable frequency range, but its gain and directivity will be affected accordingly. Therefore, when designing the antenna, it is necessary to determine the bandwidth of the antenna according to the requirements of the actual application scenarios.

The efficiency of an antenna refers to the ratio of the electromagnetic wave power radiated by the antenna to the total power input to the antenna. The efficiency of the antenna is related to the structure, material, impedance matching, radiation pattern, and the surrounding environment of the antenna. The higher the efficiency of the antenna, the greater the radiation power of the antenna, and the farther the communication distance. In practical applications, the efficiency of an antenna is a crucial performance metric, particularly in radio communications, satellite communications, radar systems, and other related fields. The antenna's efficiency has a significant impact on the overall system performance. Thus, when designing an antenna, it is essential to maximize its efficiency to ensure optimal performance. The efficiency of an antenna can be quantified using the following formula:

$$\eta = \frac{P_{\text{rad}}}{P_{\text{in}}} \tag{4.39}$$

where P_{rad} represents the power of electromagnetic waves radiated by the antenna, and P_{in} represents the total power input to the antenna. The total power of the antenna encompasses radiation power, scattering power, and absorption power. The radiation power of the antenna can be calculated using the following formula:

$$P_{\text{rad}} = \frac{1}{2}\int_S |E|^2 dS \tag{4.40}$$

In this equation, S denotes the surface area over which the antenna radiates, and E represents the electric field strength of the radiated waves. On the other hand, the input power of the antenna can be expressed as

$$P_{\text{in}} = \frac{1}{2} \operatorname{Re}\{Z_{\text{in}}\} |I_{\text{in}}|^2 \tag{4.41}$$

where Z_{in} represents the input impedance of the antenna, and I_{in} denotes the input current.

The efficiency of the antenna can be described by the radiation efficiency and the impedance matching efficiency of the antenna. The radiation efficiency of an antenna refers to the ratio of the radiation power of the antenna to the total power input to the antenna, where the total power input to the antenna includes radiation power, scattering power, and absorption power. Therefore, in the antenna design, it is necessary to improve the efficiency of the antenna as much as possible on the basis of ensuring the impedance matching and radiation efficiency of the antenna. The efficiency of the antenna can be achieved by optimizing the antenna structure, selecting the appropriate antenna material, adjusting the impedance matching of the antenna input port, and other means. In practical applications, in order to improve the efficiency of the antenna, multi-layer antenna structure and radiation enhancement technology are usually used to enhance the radiation effect of the antenna.

4.2.3 Electromagnetic Problems in Antenna Designs

Static electric field and static magnetic field are two basic types of electromagnetic field, which are related to static charge and current, respectively. The electrostatic and magnetostatic fields are usually not a major concern in antenna design because the antenna operates at high frequencies and the charges and currents are usually variable.

Grounding is one of the key issues in antenna design. Good grounding can reduce electromagnetic interference and improve antenna performance. The grounding resistance is used to evaluate the grounding quality, which can be calculated by the formula. Grounding resistance is affected by many factors, such as shape, size, depth, soil resistivity, and water content. The grounding resistance can be determined by experiment or simulation and adjusted according to the requirements. Good grounding not only improves antenna performance but also reduces the impact of electromagnetic radiation on the environment. Static electric and magnetic fields are generated by cables and power lines and can affect antenna performance, so these factors must be considered. Electromagnetic field simulation is a common tool to predict and optimize antenna performance. It can analyze radiation patterns, input impedance, gain, and other parameters, and study electrostatic and magnetostatic fields. Simulations can help to analyze the antenna performance and optimize the design, such as analyzing the distribution of grounding resistance and grounding charge density to determine the optimization scheme or analyzing the distribution of electrostatic field and magnetostatic field to reduce the influence on the antenna performance.

In antenna design, the consideration of electromagnetic wave propagation and reflection is crucial. When electromagnetic waves propagate between different media, phenomena such as refraction and reflection occur, significantly influencing the performance of the antenna. The propagation of electromagnetic waves refers to the transmission of energy through space via electromagnetic fields. At the boundary between media, electromagnetic waves undergo refraction and reflection.

For a general plane wave, its propagation direction is determined by the wave vector $\mathbf{k}$, with a magnitude of $k = 2\pi/\lambda$, where λ represents the wavelength. In empty space, electromagnetic waves travel at the speed of light, denoted as $c = 3 \times 10^8$ m/s. However, in different media, the propagation speed is altered to $v = c/n$, where n represents the refractive index of the medium. The process of refraction and reflection of electromagnetic waves at the boundary of media can be accurately described using Fresnel's formulas. For a general incident wave, it can be divided into two electric field components: $E_{i,\mathrm{n}}$, which is perpendicular to the incident plane, and $E_{i,\mathrm{t}}$, which is parallel to the incident plane. By defining the reflection coefficients r_n and r_t, as well as the refraction coefficients t_n and t_t, we can obtain the relationships as follows:

$$r_\mathrm{n} = \frac{E_{r,\mathrm{n}}}{E_{i,\mathrm{n}}} = \frac{n_i \cos\theta_i - n_t \cos\theta_t}{n_i \cos\theta_i + n_t \cos\theta_t} \tag{4.42}$$

$$r_\mathrm{t} = \frac{E_{r,\mathrm{t}}}{E_{i,\mathrm{t}}} = \frac{n_t \cos\theta_i - n_i \cos\theta_t}{n_t \cos\theta_i + n_i \cos\theta_t} \tag{4.43}$$

$$t_\mathrm{n} = \frac{E_{t,\mathrm{n}}}{E_{i,\mathrm{n}}} = \frac{2n_i \cos\theta_i}{n_i \cos\theta_i + n_t \cos\theta_t} \tag{4.44}$$

$$t_\mathrm{t} = \frac{E_{t,\mathrm{t}}}{E_{i,\mathrm{t}}} = \frac{2n_i \cos\theta_i}{n_t \cos\theta_i + n_i \cos\theta_t} \tag{4.45}$$

where n_i and n_t are the refractive indices of the incident and refractive media, respectively, and θ_i and θ_t are the angles of incidence and refraction, respectively. When the electromagnetic wave encounters the medium boundary, part of the energy will be reflected back, while the other part of the energy will pass through the medium boundary and enter the new medium. The proportion of reflection and refraction depends on the wavelength of the incident wave, the direction of polarization, the angle of incidence, and the refractive index of the medium. The reflection coefficient Rrepresents the ratio of the reflected energy to the incident energy and can be expressed as

$$R = \left(\frac{E_\mathrm{r}}{E_\mathrm{i}}\right)^2 = \left(\frac{Z_2 - Z_1}{Z_2 + Z_1}\right)^2 \tag{4.46}$$

where E_i and E_r are the electric field strengths of the incident and reflected waves, respectively, and Z_1 and Z_2 are the characteristic impedances of the media on both sides, respectively. For a general electromagnetic wave, its characteristic impedance is

$$Z = \frac{E}{H} = \sqrt{\frac{\mu}{\epsilon}} \frac{1}{\cos\theta} \tag{4.47}$$

where μ and ϵ are the permeability and permittivity of the medium, respectively, and θ is the angle of incidence at the surface of the medium. The reflection coefficient and the refraction coefficient can be calculated by the Fresnel formula, and the specific formula is the same as the above. The reflection and refraction coefficients of electromagnetic wave at the boundary of medium can be calculated by the Fresnel formula. For a general incident wave, it is divided into an electric field component $E_{i,\mathrm{n}}$ perpendicular to the incident plane and an electric field component $E_{i,\mathrm{t}}$ parallel to the incident plane. They can be used to define the reflection coefficients r_n and r_t, and the refraction coefficients t_n and t_t, namely

$$r_\mathrm{n} = \frac{E_{r,\mathrm{n}}}{E_{i,\mathrm{n}}} = \frac{n_i \cos\theta_i - n_t \cos\theta_t}{n_i \cos\theta_i + n_t \cos\theta_t} \tag{4.48}$$

$$r_\mathrm{t} = \frac{E_{r,\mathrm{t}}}{E_{i,\mathrm{t}}} = \frac{n_t \cos\theta_i - n_i \cos\theta_t}{n_t \cos\theta_i + n_i \cos\theta_t} \tag{4.49}$$

$$t_\mathrm{n} = \frac{E_{t,\mathrm{n}}}{E_{i,\mathrm{n}}} = \frac{2n_i \cos\theta_i}{n_i \cos\theta_i + n_t \cos\theta_t} \tag{4.50}$$

$$t_\mathrm{t} = \frac{E_{t,\mathrm{t}}}{E_{i,\mathrm{t}}} = \frac{2n_i \cos\theta_i}{n_t \cos\theta_i + n_i \cos\theta_t} \tag{4.51}$$

where n_i and n_t are the refractive indices of the incident and refractive media, respectively, and θ_i and θ_t are the angles of incidence and refraction, respectively. The reflection coefficient R represents the ratio of the reflected energy to the incident energy and can be expressed as

$$R = \left(\frac{E_\mathrm{r}}{E_\mathrm{i}}\right)^2 = r_\mathrm{n}^2 + r_\mathrm{t}^2 \tag{4.52}$$

Likewise, the refractive index Trepresents the ratio of the refracted energy to the incident energy and can be expressed as

$$T = \left(\frac{E_\mathrm{t}}{E_\mathrm{i}}\right)^2 = \frac{n_t \cos\theta_t}{n_i \cos\theta_i}(t_\mathrm{n}^2 + t_\mathrm{t}^2) \tag{4.53}$$

It should be noted that the values of reflection coefficient and refraction coefficient are related to the polarization direction of the incident wave and the frequency of the electromagnetic wave.

In antenna design, the interaction of the electromagnetic field with the antenna results in an increase or decrease in the electromagnetic field and a change in the impedance of the antenna. Direct coupling means that the electromagnetic field directly acts on the antenna conductor, resulting in the change of antenna parameters. To reduce the effects of direct coupling, you can change the antenna geometry, use shielding materials, reduce the distance between the antennas, or adjust the antenna impedance using elements such as couplers and tuners. Direct coupling can also cause energy transfer, mode mixing, and cross-polarization. Inductive coupling means that the electromagnetic field induces the current in the antenna and changes the antenna parameters. The influence of inductive coupling can be supplemented by adding inductors or capacitors in the antenna loop to reduce the inductive effect of electromagnetic field on the antenna. In the antenna design, the frequency, power, antenna geometry, and other factors of the electromagnetic field need to be considered, which can be analyzed and optimized by using electromagnetic field simulation software. The effect of inductive coupling can be reduced by changing the antenna structure or placement.

Electromagnetic compatibility (EMC) refers to the ability of electronic devices to operate together in the same electromagnetic environment without interfering with each other. EMC and EMI issues are the factors that must be considered in antenna design. The principle of electromagnetic compatibility is to ensure the coexistence of equipment in a common electromagnetic environment without mutual interference through effective design and control. In antenna design, in order to meet the requirements of electromagnetic compatibility, it is necessary to consider whether the electromagnetic wave radiated by the antenna will interfere with other equipment, such as mutual interference with the same frequency. The methods to evaluate the degree of electromagnetic interference include field strength measurement, spectrum analysis and electromagnetic compatibility simulation. In order to control electromagnetic interference, measures such as filters, shielding, grounding, and power control can be taken.

4.2.4 Optimization Algorithms in Antenna Designs

In antenna designs, optimization algorithms are used to determine the optimal design parameters of the antenna to meet the performance requirements. These algorithms can be divided into two types: mathematical optimization algorithms and simulation-based optimization algorithms [41–43]. Mathematical optimization algorithms include gradient descent method, Newton method, quasi-Newton method, conjugate gradient method, and genetic algorithm. The gradient descent method adjusts the parameters by iteration to minimize the objective function, the Newton method adjusts the parameters using the second derivative of the objective function, the quasi-Newton method adjusts parameters by approximating the Hessian matrix, the conjugate gradient method optimizes the parameters by adjusting the conjugate gradient direction, and the genetic algorithm optimizes the parameters based on the principle of biological evolution. Simulations-based optimization algorithms, including global optimization algorithms and local optimization algorithms, are used to establish electromagnetic field simulation model and optimize parameters on the model. The global optimization algorithm searches for the optimal solution through global search. Genetic algorithm, particle swarm optimization algorithm, and ant colony algorithm are commonly used. Local optimization algorithms adjust the parameters in the design space step by step to obtain the local optimal solution. Gradient descent method, Newton method, quasi-Newton method and conjugate gradient method are commonly used. The application examples of optimization algorithms include optimization of radiation characteristics based on genetic algorithm, optimization of antenna matching based on simulation, optimization of beam width based on the quasi-Newton method, and optimization of mechanical structure based on conjugate gradient method. The advantage of optimization algorithm is to find the optimal solution or approximate optimal solution quickly and improve the antenna performance and design efficiency, but it needs fine adjustment to ensure convergence and robustness and consider the cost of computing resources and time. In a word, optimization algorithm is an important tool in antenna design, which is selected and adjusted according to design requirements and computing resources to achieve the best design effect.

4.2.5 Error Analysis and Calibration

In the simulation process of antenna designs, there are many error sources and influencing factors, including simulation model error, antenna material error, manufacturing error, and measurement error. In order to reduce the error, calibration methods such as comparison method, parameter fitting method, and optimization algorithm can be used. The comparison method corrects and calibrates the simulation model by comparing the simulation results with the measured results, the parameter fitting method improves the precision and accuracy of the model by fitting the parameters of the simulation model with the measured data, and the optimization algorithm optimizes the simulation model to make the simulation results closer to the actual situation. In practice, it is necessary to refine the simulation model, collect accurate measurement data, and combine a variety of calibration methods to obtain accurate and reliable simulation results.

4.3 ADS for Antenna Simulations

In the ADS, in addition to its applications in designing RF and microwave circuits, communication systems, and digital signal processing (DSP) systems, it can also be used for designing planar structure antennas. The following example illustrates the steps to design a UHF RFID tag antenna using the RF simulation software ADS. To achieve maximum power transmission, both antenna design and impedance matching with the connected tag chip need to be considered.

1. Launch the ADS software and create a new project file. Click on "File" → "New Project" to set the project file name and storage path. Click on "Length Unit" to set the length unit as millimeters. Then create a new ADS LAYOUT file.

2. Draw the antenna in the layout, considering whether it is a single-sided or double-sided antenna.

3. Set the material properties for the tag antenna through layer definition. Separate settings are required for the Substrate Layer and Metallization Layer.

3.1 Setting for Substrate Layer: From the main menu, go to "Momentum" → "substrate" → "create" to enter the substrate Layer definition dialog window.

3.2 Setting for Metallization Layer: From the main menu, go to "Momentum" → "Substrate" → "Create/Modify" to enter the metallization Layer definition dialog window. Fill in the conductivity in the "Conductivity" field and the metal thickness in the "Thickness" field.

4. Port definition: The port impedance setting for the tag antenna is different from a regular antenna as it needs to match the impedance of the tag chip it is connected to. In ADS, select and add two ports. Then, select the ports with the mouse and edit them using "Momentum" → "Port Editor". Since the GND was canceled in the previous layer definition, the Single Port option is not available. Therefore, a solution using an Internal port with a Ground Reference Port is used.

5. Mesh partitioning and S-parameter simulation settings: Before performing the Momentum simulation, set the mesh partitioning in "Momentum" → "Mesh". The mesh settings determine the simulation accuracy. Generally, a higher mesh frequency and a larger number of cells per wavelength result in higher accuracy. However, this comes at the cost of increased simulation time. Sometimes, it is necessary to compromise accuracy for shorter simulation time. Then, select "Simulations" → "S-parameters" from the main menu to open a dialog box for setting up S-parameter simulation. Choose the desired sweep type, click on "Update", and then "Simulate" to start the simulation and obtain the S-parameter simulation results.

To observe the far-field 2D radiation pattern, in ADS, go to "Momentum" → "Post-Processing" → "Radiation Pattern" and make the settings as shown in Fig. 4.4. Set the frequency to the resonant frequency at the center of the antenna. After clicking on "Compute", the far-field 2D or 3D radiation pattern is obtained. This allows observing antenna polarization characteristics, antenna gain, antenna efficiency, and radiation pattern.

7. Optimization of antenna parameters: If the simulation results are not satisfactory, ADS Layout offers optimization capabilities for the antenna. From the main menu, go to "Momentum" → "Optimization" → "Parameters" to enter the dialog box for setting the optimization parameters of the antenna size. After setting the optimization parameters, go to "Momentum" → "Optimization" → "Goal" to set the optimization objective. The optimization objective is mainly focused on S_{11} and S_{21} parameters, and parameters such as dielectric constant and substrate thickness cannot be optimized.

4.4 HFSS for Antenna Simulations

In HFSS, antenna modeling plays a crucial role in antenna design and simulation, ensuring the accuracy and reliability of simulation results. This section provides an overview of how to model an antenna in HFSS, covering geometric modeling, material property definition, and modeling.

HFSS offers various tools for geometric modeling, including lines, cylinders, cuboids, spheres, and surfaces. Based on the antenna's shape and size, users can choose the appropriate modeling tool to build the antenna model in the 3D interface. The following steps outline the basic process of modeling an antenna:

1. Launch the HFSS software and create a new project by clicking on "New Project." Then, select "HFSS Design" to create a new design.

2. In the new design, select the "Modeler" interface to access the 3D model interface for the antenna.

3. Utilize the appropriate geometric modeling tool based on the antenna's shape and size. For instance, a straight antenna can be modeled using the straight line tool, while complex-shaped antennas may require advanced modeling tools like surfaces.

4. During the modeling process, ensure that the antenna geometry aligns with the design requirements, such as frequency bandwidth and gain.

Defining accurate material properties is essential in the antenna model, as it significantly affects antenna performance and characteristics. HFSS enables users to select and customize material properties based on the antenna's materials. The following steps outline the process of defining material properties:

1. Select the object or face in the model interface for which you want to define material properties.

2. Use the "Assign Material" command in the model interface and choose the desired material from the "Assign Material" dialog box. You can select predefined materials or create custom ones.

3. Configure the material properties for the selected material. Different materials have distinct properties, such as dielectric constant, electrical conductivity, and magnetic permeability.

4. Complex material properties can be defined by importing external material data files.

Employing accurate modeling is vital to ensure the precision and reliability of the antenna model. Here are some common modeling:

1. Parametric Modeling: Define antenna geometric parameters as variables and generate different antenna geometries by adjusting these variables. This approach enables users to swiftly generate various antenna shapes, expediting the design optimization process.

2. Simplified Modeling: For complex antenna models, simplify the model by reducing its geometric complexity. This method accelerates simulation analysis speed while maintaining accuracy.

3. Symmetry Modeling: Utilize symmetry modeling for antennas with symmetry. Build only a portion of the model and generate the complete model through

symmetry transformation. This technique reduces modeling time and computational requirements during simulation analysis.

4. Reference Plane Modeling: Incorporate a reference plane during antenna modeling to assist with defining geometric parameters and material properties. The reference plane serves as the coordinate system reference for the antenna.

HFSS provides the field calculator for calculating the radiation power density of the antenna, allowing users to determine the antenna's gain. Follow these steps to perform the calculation:

1. In the model interface, select the "Analysis Setup" command to open the analysis setup window.

2. Go to the "Radiation" tab and set the frequency range and step size for the calculation.

3. Click the "Add Solution" button to select the antenna port and calculate the output parameters.

4. Start the simulation calculation by clicking the "Run" button. After completion, you can view the field distribution and radiation pattern data of the antenna in the "Results" interface to assess performance characteristics such as directivity and radiation efficiency.

In HFSS, the setting function plays a crucial role in optimizing the design and analysis process while allowing users to customize the settings according to their specific requirements. Below are the primary setup functions of HFSS in antenna design simulation, refined for clarity and conciseness:

1. Automatic optimization settings

HFSS offers automated optimization of antenna parameters to meet specified performance requirements. Users can define the parameter range, step size, and optimization goals, such as minimum VSWR or maximum gain.

The automatic optimization function significantly enhances the efficiency and accuracy of antenna design, eliminating the tedium and errors associated with manual parameter adjustments. Here are the steps to perform automatic optimization:

(1) Open the optimization setup window by selecting the "Optimization Setup" command in the model interface.

(2) Configure optimization parameters, including optimization type, objective, and constraint conditions.

(3) Select the desired parameter range, step length, starting point, and other settings by clicking the "Add Solution" button.

(4) Initiate the optimization calculation by clicking the "Run" button. Upon completion, review the optimization results in the "Optimization Results" interface.

2. Automated script setup

HFSS supports scripting capabilities to automate various operations, such as modeling, setup, simulation, and optimization. Users can create customized scripts to enhance design efficiency and accuracy.

Leveraging automated scripting eliminates the monotony and potential errors associated with repetitive operations and manual parameter tuning. Follow these steps to utilize the automated script setup:

(1) Open the script editor by selecting the "Script" command in the model interface.

(2) Compose the script by including operations such as modeling, setup, simulation, optimization, etc.

(3) Execute the script to automate the series of operations. During script execution, monitor the process and review the output results.

3. Batch simulation settings

HFSS enables batch simulation analysis for multiple antenna models, providing the ability to output results to a designated folder. Users can configure simulation parameters, analysis types, output formats, and more.

The batch simulation function significantly enhances the efficiency and accuracy of simulation analysis, facilitating the swift acquisition of performance indicators for multiple antenna models. Perform the following steps to set up batch simulation:

(1) Open the batch simulation setting window by clicking "Electronics Desktop (Batch)" in the model interface.

(2) Specify the folder and file type to be simulated, and define simulation parameters and output format.

(3) Commence the batch simulation calculation. Once completed, view the simulation results in the specified folder.

4.5 Co-simulations for Antenna Designs

Antenna design encompasses various factors that should be taken into consideration, such as antenna structure, materials, operating frequency, impedance matching, and radiation performance. Co-simulations play a vital role in leveraging the advantages of different kinds of simulation software, considering the impact of these factors on antenna performance, and obtaining comprehensive and accurate simulation results. By employing co-simulations, the efficiency and accuracy of antenna design are significantly enhanced, leading to an expanded application range and improved performance of antennas. This technology enables the co-simulations of multiple physical fields, including electromagnetic, thermal, mechanical, circuit, sound, and fluid fields, allowing for a comprehensive understanding of their influence on antenna behavior and obtaining precise simulation outcomes.

Furthermore, co-simulation facilitates multi-scale simulation, accommodating different antenna scales and integrating the simulation results accordingly. It also considers the nonlinear effects of antennas, such as harmonic generation, by combining the results from various simulation software, thereby yielding accurate simulation outcomes. Ultimately, co-simulation enables design optimization by swiftly identifying issues and conducting targeted optimizations to enhance the performance and applicability of antennas. Through its application, antenna design can encompass a comprehensive consideration of various factors, conduct co-simulations of multiple physical fields, perform multi-scale simulations, account for the nonlinear effects of antennas, and optimize the overall design. These advancements not only enhance the efficiency and precision of antenna design but also broaden the scope of antenna applications and improve overall antenna performance.

Antenna design simulation encompasses three primary methods: model-based co-simulations, simulation data-based co-simulations, and script-based co-simulations. Model-based co-simulations involve achieving more precise and comprehensive results by converting models and conducting simulations in different kinds of software. This method follows the following steps: first, an antenna model is established and simulated in one type of software; then, a data file is exported. Next, the data is imported into another type of simulation software, where it is converted into a new model file. A new simulation model is established and simulated using this converted file. The data file is exported once again, and finally, the data file is imported back into the initial simulation software to obtain the final result. Model-based co-simulations effectively harness the advantages of diverse simulation software, enabling rapid model transformation and facilitating the co-simulations of multiple physical fields.

Simulation data-based co-simulations employ the integration of simulation results from different software to achieve more comprehensive and accurate outcomes. This method can be implemented through two approaches: format conversion and data interface. Format conversion-based co-simulations require specialized software or programs for format conversion, while data interface-based co-simulations rely on the support of data interfaces in different simulation software.

Script-based co-simulations involve the exchange of data and co-simulations between different simulation software through script writing. This method enhances simulation efficiency, enables comprehensive and accurate results, and offers high flexibility and scalability. Script-based co-simulations find wide application in antenna design and optimization, as well as other fields. It can be combined with various simulation software to meet specific needs and requirements.

4.6 Co-simulations with ADS for Antenna Designs

4.6.1 Co-simulations of CST Studio Suite with ADS

CST Studio Suite is a kind of electromagnetic simulation software that utilizes the finite integral time domain (FITD) method. It is widely employed in the design of various antennas, including linear, nonlinear, array antennas, and other complex structures. With its powerful computational capabilities, CST can simulate RF, microwave, and photonic devices effectively. Co-simulations with ADS enable the analysis and optimization of complete communication links.

The co-simulations of CST and ADS involve the following steps:

1. Antenna Modeling and Simulations: CST Studio Suite offers a range of tools to create antenna geometries. Once the structure is constructed, an appropriate electromagnetic solver, such as the finite integration time domain solver, is selected,

and simulation parameters are set. Running the simulation provides performance parameters like S-parameters, impedance, and radiation patterns.

2. Exporting S-parameters: After completing the antenna simulation in CST, the S-parameters can be exported in Touchstone (.snp) or Citifile (.cit) format using the S-parameter export function.

3. RF Circuit Model Creation: In ADS, an RF circuit model is created, typically consisting of RF components like filters, amplifiers, and mixers. These components can be chosen from the built-in library or imported as third-party models.

4. Importing Antenna Model: The antenna's S-parameter file, exported from CST, is imported into the RF circuit model in ADS. Antenna models are added to the RF circuits using the Data File element in ADS.

5. Co-simulation Analysis: Simulation parameters such as frequency range and power level are configured in ADS. The co-simulation analysis method, such as Harmonic Balance or Circuit Envelope, is selected. Running the simulation provides performance metrics for the complete communication link, including system gain and spurious parameters.

6. Results Optimization and Adjustment: Based on the co-simulation results, the antenna structure or RF circuit can be optimized. Adjustments can be made by modifying antenna geometry parameters in CST or altering RF circuit component parameters in ADS. The simulation analysis is repeated until the performance of the entire communication link meets the design requirements.

7. Verification and Testing: Once the co-simulations are completed, the actual hardware of the antenna and RF circuit can be tested to validate the accuracy of the simulation results. By comparing the simulation and test data, further optimization of the complete communication system can be achieved.

By following the detailed steps of CST Studio Suite and ADS co-simulations outlined above, engineers can identify potential issues between antennas and RF circuits early in the design phase. This helps reduce overall design risks and ensures the integrity and performance optimization of the signal chain. Additionally, the co-simulation approach can shorten the design cycle and enhance design efficiency.

4.6.2 Co-simulations of WIPL-D with ADS

WIPL-D is a kind of professional electromagnetic simulation software, which is used to design antennas, microwave circuits, and RF systems. It can support the simulation of various types of antennas and microwave devices, including antenna arrays, microstrip antennas, laser antennas, and millimeter-wave antennas. Compared with other electromagnetic simulation software, WIPL-D has higher solution speed and higher accuracy. Its solver supports multi-processor, distributed computing, and GPU acceleration, which can greatly reduce the simulation time. WIPL-D's post-processing tools are also very powerful, which can display the simulation results of electric field distribution, power flow density, radiation pattern, and so on. When using WIPL-D and ADS co-simulations for antenna design, the following steps are required:

1. Install WIPL-D and ADS software

First you need to install the WIPL-D and ADS software and make sure that both are working properly. It is recommended to use the latest version of the software for better performance and functionality.

2. Prepare antenna design document

Before starting the simulation, you need to prepare the antenna design file. This can be done in Wipl-D or ADS software. If you use WIPL-D software, you need to create the antenna model in WIPL-D and set the simulation parameters and solver options. If you use ADS software, you need to create the antenna model in ADS and set the simulation parameters and solver options. It is recommended to save the antenna model in HFSS or CST format for sharing between WIPL-D and ADS.

3. Import the WIPL-D model into ADS

In the ADS software, the WIPL-D model can be imported into ADS through the "Import Model" function. This creates an ADS model and copies the geometric information and material properties from the WIPL-D model to the ADS model. Before you import the model, you need to make sure that the WIPL-D model has been geometrically modeled and set up with material properties. After the model is imported, it can be edited and modified in ADS to meet the design requirements.

4. Set simulation parameters

Before the simulation, the simulation parameters need to be set. This can be done in the ADS software. Parameters such as frequency range, electric field excitation, and simulation time need to be specified. You can use the default settings or modify them to meet your design requirements.

5. Run the simulation

After you set the simulation parameters, you can start running the simulation. During the simulation, ADS software will call the solver of WIPL-D to calculate the electromagnetic field distribution and antenna performance. At the end of the simulation, the simulation results can be saved and analyzed in post-processing.

6. Analyze the simulation results

In the simulation results, the S-parameters, radiation pattern, electric field distribution, power flow density, and other information of the antenna can be viewed. Simulation results can be displayed using WIPL-D's post-processing tools. The performance of the antenna design can be evaluated by comparing the simulation results, and necessary modifications and optimizations can be made.

7. Revise and optimize the design

According to the simulation results, the antenna design can be modified and optimized. You can edit and modify the antenna model in WIPL-D or ADS software and rerun the simulation. An optimization algorithm can be used to search for an optimal solution for better performance.

4.6.3 Co-simulations of XFdtd with ADS

XFdtd is a kind of high-frequency electromagnetic simulation software developed by Remcom. It uses the finite-difference time-domain (FDTD) algorithm and can be used in electromagnetic field simulation, antenna design, electromagnetic compatibility (EMC) evaluation, and other fields. In antenna design, XFdtd provides a variety of antenna design tools, including built-in antenna library, fast antenna optimization (FAO), and antenna design wizard. The principle of XFdtd and ADS co-simulations is to interact XFdtd simulation results with ADS circuit simulation results. Specifically, an antenna model is first established in XFdtd for electromagnetic field simulation, and then the obtained electromagnetic field results are imported into ADS as the input of circuit simulation. After circuit simulation in ADS, the circuit simulation results are fed back to XFdtd as the basis for antenna performance evaluation. In this way, the optimal antenna design can be obtained through continuous iterative optimization.

The following describes the specific operation steps in the co-simulations of XFdtd and ADS.

Step 1: Build an antenna model

Build an antenna model in XFdtd. You can use XFdtd's antenna library, customize modeling, or import antenna models from other CAD software. The antenna model should include the complete geometry, material parameters, and excitation scheme.

Step 2: Perform electromagnetic field simulation

Simulate the electromagnetic field in XFdtd, and select the appropriate excitation mode and frequency range to obtain the accurate electromagnetic field distribution.

Step 3: Derive electromagnetic field result

When deriving the electromagnetic field results in XFdtd, you can choose to derive the electric field, magnetic field, power density, and other parameters. The exported result should be in a format that ADS can recognize, such as HFSS format, Touchstone format, and so on.

Step 4: Establish an ADS circuit model

The circuit model is established in ADS, which can include RF front-end circuit, power amplifier, filter, antenna driver circuit, etc. The circuit model should correspond to the antenna model in the XFdtd and contain a device model that ADS can recognize.

Step 5: Import electromagnetic field results

The XFdtd-derived electromagnetic field results are imported into ADS and used as input to the circuit simulation. In order to ensure the accuracy of simulation results, the correct port and excitation mode are selected in ADS.

Step 6: Perform circuit simulation

For circuit simulation in ADS, various simulation tools can be used for circuit performance evaluation, such as S-parameter simulation, power analysis, noise analysis, etc. The simulation results should correspond to the electromagnetic field results in XFdtd to ensure the accuracy of co-simulations.

Step 7: Derive circuit simulation result

Circuit simulation results are derived in ADS and used as feedback inputs to XFdtd.

Step 8: Perform antenna performance evaluation

The ADS-derived circuit simulation results are imported into XFdtd, and the antenna performance is evaluated. Various tools provided by XFdtd can be used for antenna performance analysis and optimization, such as radiation pattern, gain, beamwidth, etc.

Step 9: Iterative optimization

Through the iterative optimization of the above steps, the best antenna design scheme can be obtained. In the optimization process, the antenna model and circuit model can be modified according to the actual needs to get more accurate results.

4.7 Co-simulations with HFSS for Antenna Designs

4.7.1 Co-simulations of Antenna Magus with HFSS

Antenna Magus is a kind of professional antenna design software, which provides a large number of antenna design templates and tools to help engineers design and simulate various types of antennas quickly and accurately. With an easy-to-use interface and powerful features, Antenna Magus can automate many design and simulation tasks, saving time and effort. The specific implementation method and operation steps are as follows:

Step 1: Open the antenna design to be exported in Antenna Magus.

Step 2: Open the HFSS software, select "File"→"Open" in the main interface of HFSS, and select the file just exported to open it.

Step 3: Conduct electromagnetic simulation and analysis in HFSS to obtain information such as performance indicators and electromagnetic field distribution of the antenna.

Using the co-simulation function of Antenna Magus and HFSS, the electromagnetic simulation results of antenna can be obtained quickly, and the optimization and adjustment can be carried out. In Antenna Magus, set the format of the export file. Commonly used export formats include SAT, STEP, IGES, DXF, etc., of which the SAT format is the most commonly used format for HFSS. When setting the export file format, you should pay attention to the following points:

- Select a format that is compatible with the target simulation software to ensure that the exported file can be properly parsed and simulated by HFSS.
- Set the precision and precision unit of the exported file to ensure that the precision and unit of the exported file are consistent with the actual needs.
- Set the dimensions and coordinate system of the exported file to ensure that the dimensions and coordinate system of the exported file are consistent with the actual needs.

Before exporting the antenna design to HFSS for simulation, it is necessary to set some parameters in Antenna Magus to ensure that the exported file can be correctly parsed and simulated in HFSS. It mainly includes the following aspects:

- Antenna size and units: Before deriving the antenna design, you need to determine the size and units of the antenna and set them to the same size and units as HFSS.
- Material properties: In Antenna Magus, you can set the material properties of the antenna, such as permittivity, conductivity, and so on. These properties have an important impact on the electromagnetic performance of the antenna and need to be set correctly at the time of export.
- Mesh Density: Before exporting the antenna design, you need to determine the mesh density of the antenna and set it to the same mesh density as HFSS. This ensures that an accurate electromagnetic field distribution is obtained in HFSS.
- Excitation setting: In Antenna Magus, you can set the excitation mode and parameters of the antenna, such as input impedance, power, etc. These parameters also need to be set correctly at the time of export.

4.7.2 Co-simulations of EMCoS Antenna VirtualLab with HFSS

EMCoS Antenna VirtualLab is a professional antenna design and simulation software, which can simulate various physical processes including antenna radiation, reflection, scattering, scattering matrix, electromagnetic radiation field, and radiation power. The main advantage of EMCoS Antenna Virtual Lab is that it can quickly optimize and simulate complex antenna structures, and the simulation results can be co-simulated with other antenna simulation software. EMCoS Antenna VirtualLab and HFSS are both finite element method (FEM) based software for electromagnetic field calculation, so data exchange and co-simulations can be carried out between them. Specifically, data can be imported and exported between EMCoS Antenna VirtualLab and HFSS, for example, electromagnetic field distribution data in HFSS can be imported into EMCoS Antenna VirtualLab for subsequent antenna radiation simulation and optimization; At the same time, the antenna structure data in EMCoS Antenna Virtual Lab can also be imported into HFSS for simulation and optimization of electromagnetic field distribution. Therefore, the co-simulations of EMCoS Antenna VirtualLab and HFSS can give full play to their respective advantages and improve the accuracy and efficiency of antenna design.

The electromagnetic field distribution data in HFSS is imported into the EMCoS Antenna Virtual Lab for subsequent antenna radiation simulation and optimization. During the co-simulations of EMCoS Antenna VirtualLab and HFSS, the following items shall be noted:

1. For the compatibility of import and export file formats, it is necessary to ensure that the file format can be recognized and read by the software of the other party;

2. For the consistency of simulation setting parameters, it is necessary to ensure that the parameters set in the two types of software are consistent, such as working frequency, material parameters, etc.;

3. Consistency of optimization objectives. It is necessary to ensure that the optimization objectives set in the two types of softwares are consistent, such as gain, radiation efficiency, etc;

4. To ensure the consistency and accuracy of the simulation results, it is necessary to compare the simulation results obtained from the two types of softwares.

4.8 Prospect of Simulations for Antenna Designs

Antenna design simulation has been widely used and developed in the past decades, which provides important support and promotion for the development of electronic communications, radar, satellite communications, and other fields. With the continuous progress and development of science and technology, antenna design simulation is also constantly updated and improved, which is expected to achieve more efficient, more accurate, and more reliable antenna design simulation in the future, and bring more opportunities and challenges for the development of related fields.

Artificial intelligence and machine learning are widely used in antenna design and simulation. These technologies can optimize the antenna design process, realize automatic antenna design, fault diagnosis and prediction, material selection, and performance prediction, optimize the design of antenna array, solve the problem of antenna heat dissipation, and evaluate the impact of radiation environment .With the help of these technologies, engineers can find the optimal design scheme more quickly and improve the overall design efficiency.

The development trend of antenna design simulation in the future includes multi-physical field coupling simulation, digital twin, quantum computing, big data and data mining, cloud computing and distributed computing, virtual reality, and augmented reality. More accurate simulation results and more efficient design process can be achieved by studying emerging technologies such as multi-physics coupling simulation, digital twin, and quantum computing. However, there are some challenges in the field of antenna design simulation. This includes computing resource requirements, multi-physical field coupling, security and privacy protection, interdisciplinary knowledge fusion, integration of artificial intelligence and simulation, reliability and stability, experimental verification, and consistency of simulation results. Solving these challenges will provide more space for the development of antenna design simulation.

4.9 Co-simulations of Microstrip Antennas Based on HFSS and ADS

The ANSYS Electromagnetics Suite provides independent simulation of the electromagnetic performance of components, circuits, and system designs, as well as evaluations of temperature, vibration, and other key mechanical effects. The Advanced Design System (ADS) from Agilent Technologies is a leading electronic design automation software, suitable for radio frequency, microwave, and signal integrity applications. In electromagnetic and radio frequency design, it is often necessary to design antennas using HFSS and then validate the circuits through ADS. At this time, co-simulation is required, and the relay between them is the S-parameters. The schematic used in this section is the microstrip antenna from the article below. It only has one port and is relatively simple (Fig. 4.1).

To perform the simulation, we first design a rectangular microstrip antenna with a center frequency of 2.45 GHz using HFSS and provided its antenna parameters. The substrate used is a 1.6 mm thick FR4 epoxy resin board. The antenna consists of four parts: the FR4 substrate, a 50-ohm feed line, a $1/4\lambda$ impedance matching section, and a radiating patch. Assuming the antenna is in the transmitting state, the signal is loaded into the wave port, passes through a 50-ohm feed line, reaches the $1/4\lambda$ impedance transformer for impedance matching, and then enters the radiating patch to be radiated into space. After completing the antenna design, we export the S-parameters of the antenna. In the HFSS software, right-click on "Results," click on "Solution Data," and enter the results data interface. Go to "Matrix Data" and then "View." Check the box to display all frequency points. In "Export," select "Export Matrix Data." Choose a directory to save the file in the pop-up window. Since this antenna is a single-port circuit, the exported file is named "s1p." Once the S-parameters are exported, we need to create a circuit in ADS and import these S-parameters for simulation. First, draw the schematic diagram. Find the "s1p" file in the components and place it. The placement should resemble Fig. 4.2.

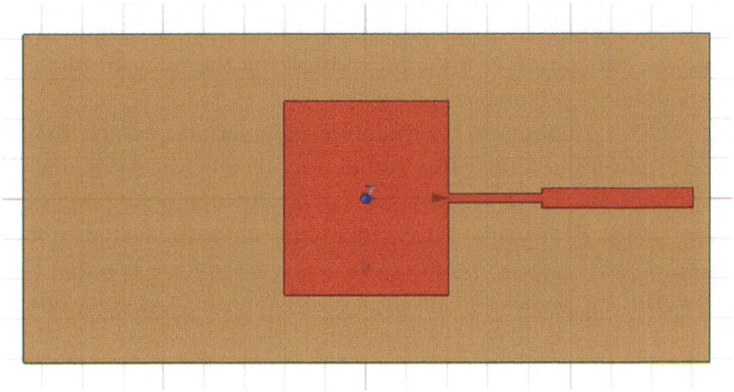

Fig. 4.1 Rectangular microstrip antenna with a center frequency of 2.45 GHz

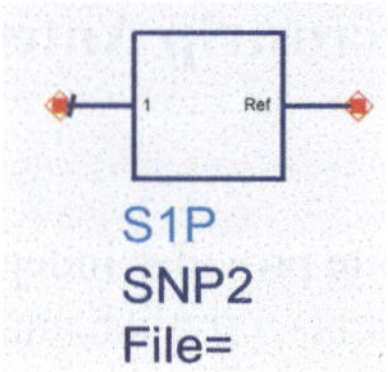

Fig. 4.2 Circuit configuration in ADS

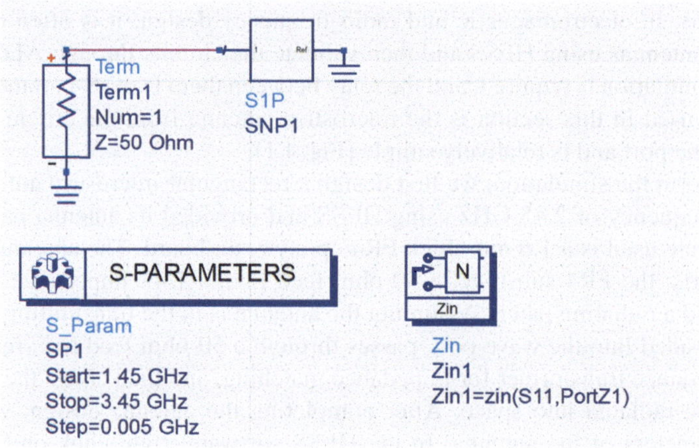

Fig. 4.3 S-parameter simulation settings in ADS

Double-click and select the path of the previously exported S-parameter in the pop-up window. As shown in Fig. 4.3, connect the remaining connections with wires, place the S-parameter simulation component and Z_{in} simulation component, and ground the ref port of the S-parameter device.

After completing all the setups, we can proceed with the simulation. Using ADS, we obtained the simulation results for S_{11}-parameters and input impedance Z_{in}. Similarly, we also obtained the simulation results for these two parameters in HFSS. Upon clicking the simulation button and configuring the relevant plots, the curves appeared (Figs. 4.4, 4.5, 4.6 and 4.7).

Although HFSS is based on interpolation calculations, which may introduce some small deviations, overall, the simulation results of ADS and HFSS are consistent. Through the above process, we can see the complete process of antenna co-simulations using HFSS and ADS, including antenna design, exporting and importing S-parameters, and comparing the final simulation results. This demonstrates the powerful capabilities of these two software tools in electromagnetic and

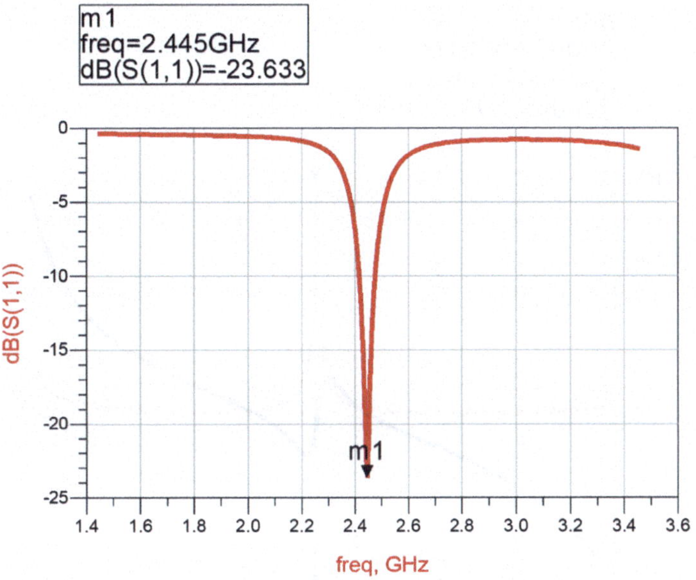

Fig. 4.4 Simulations results of S_{11}-parameter in ADS

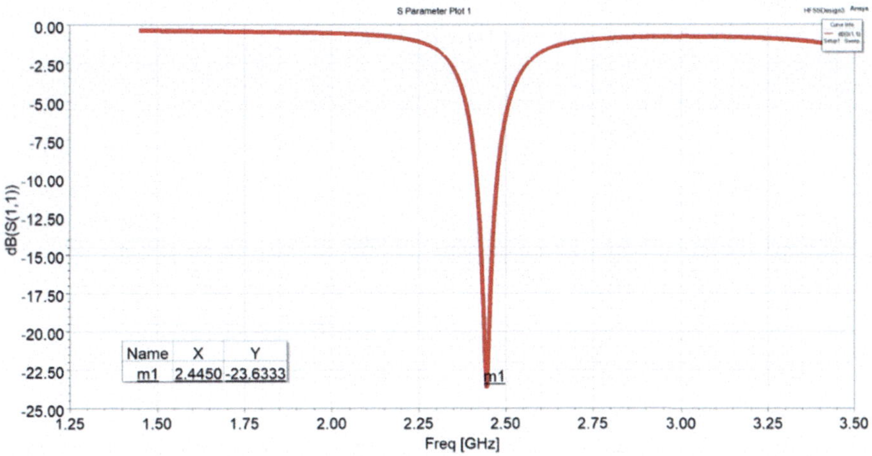

Fig. 4.5 Simulations results of S_{11}-parameter in HFSS

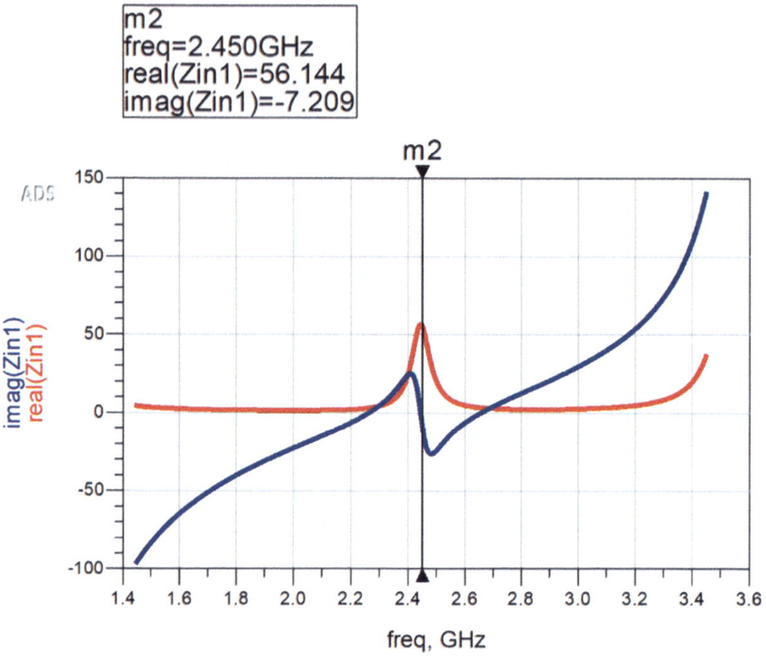

Fig. 4.6 Simulations results of input impedance Z_{in} in ADS

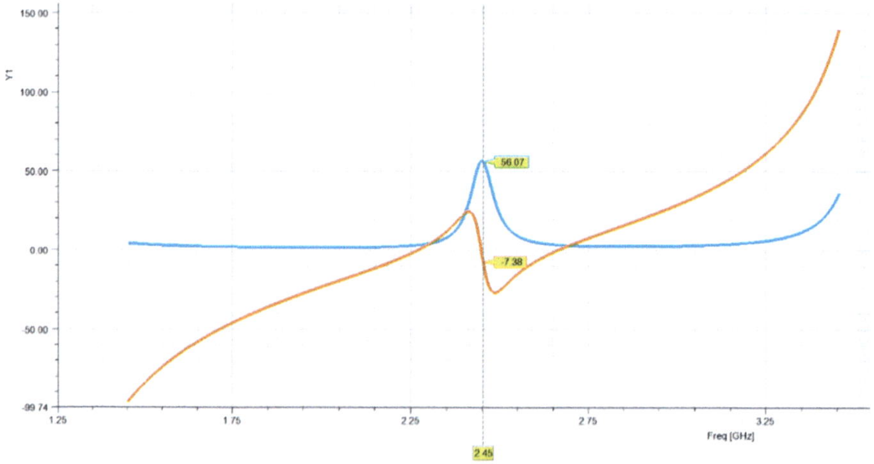

Fig. 4.7 Simulations results of input impedance Z_{in} in HFSS

RF design. In the antenna design process, HFSS provides accurate electromagnetic field simulation, allowing us to consider the electromagnetic performance of the antenna at the early design stage. In the subsequent verification stage, ADS performs circuit simulation using the S-parameters exported from HFSS, further validating the antenna's performance and operating state. This simulation process tightly integrates the physical design of the antenna with circuit verification, improving both design efficiency and accuracy. This is particularly important for applications requiring high-frequency antenna design, such as RF, microwave, and signal integrity.

4.10 Co-simulations of Microstrip Antennas Based on Circuits and HFSS

In this section, we will explore how to perform co-simulations of antennas using Ansys HFSS and Circuit. Firstly, we launch Ansys Electronics Desktop and initiate the Antenna Toolkit. We create a rectangular patch antenna with desired design parameters from the HFSS Antenna Design Kit. HFSS generates the geometric model of the antenna from the toolkit, as shown in Fig. 4.8.

We will use the frequency of this model for matching, using the Smith tool in HFSS Circuit. In the project modeling window, insert HFSS Circuit. Once it appears, select the HFSS Mcad model and copy-paste it into the circuit or drag and drop the HFSS model into the HFSS Circuit. As shown in Fig. 4.9, in the project manager window, the HFSS model is a subcircuit design within the HFSS Circuit, establishing a hierarchical structure.

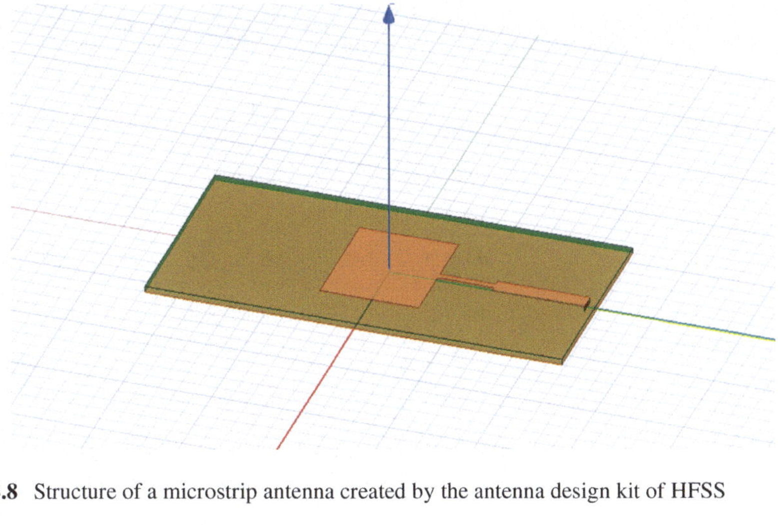

Fig. 4.8 Structure of a microstrip antenna created by the antenna design kit of HFSS

Fig. 4.9 Co-simulation modeling of HFSS and Circuit

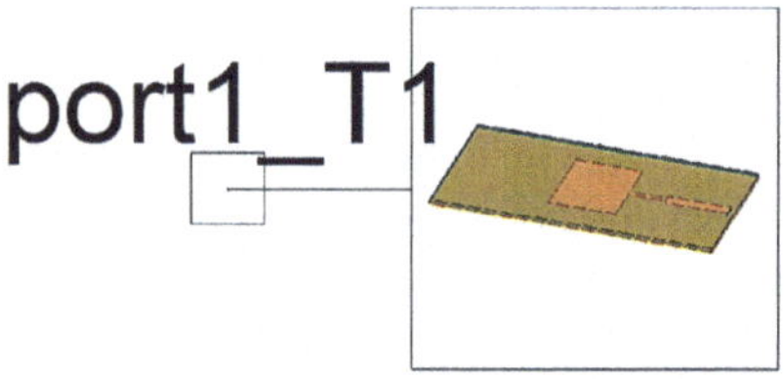

In the 3D modeling window, use Ctrl+D to maximize the circuit schematic block view of the Mcad model in the circuit domain. Next, we are going to insert a port for the model, select a microwave port, and use the Edit Port option by right-clicking on the inserted port. Create a voltage source for this port, as we are using a microwave port implementation, the voltage source should be a sinusoidal voltage source. Make sure to edit the properties of this port, using one volt as the AC voltage, and set the frequency to 3.5 GHz. Add a circuit solution setup for linear network analysis, using the same frequency as the HFSS model, scanning from 1.75 GHz to 5.25 GHz with a step size of 0.05 GHz. Perform the analysis and once completed, view the results. Now, we can see that the matching effect is poor. In the Results tab, we can use the Smith tool to design a matching network. Select 3.5 GHz on the Smith chart to see its location on the chart. We will use this point to create a conjugate match. Click on the button named "Conjugate" and select the position of 3.4 GHz. At this point, a circle will appear representing the conjugate point used for matching on the Smith chart (Fig. 4.10).

Fig. 4.10 Conjugate points used for matching on Smith Chart

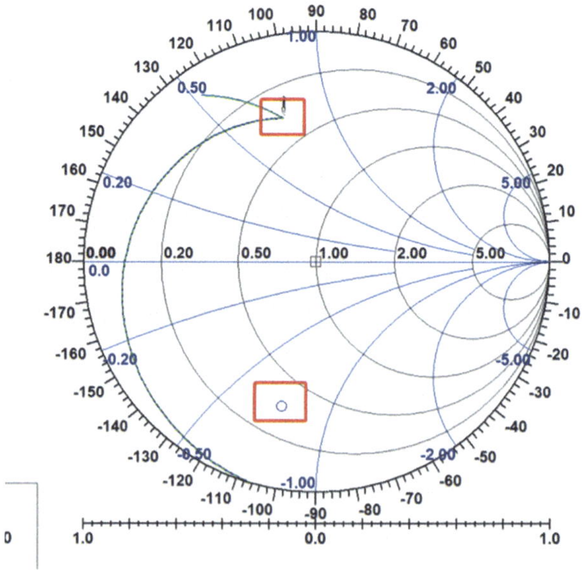

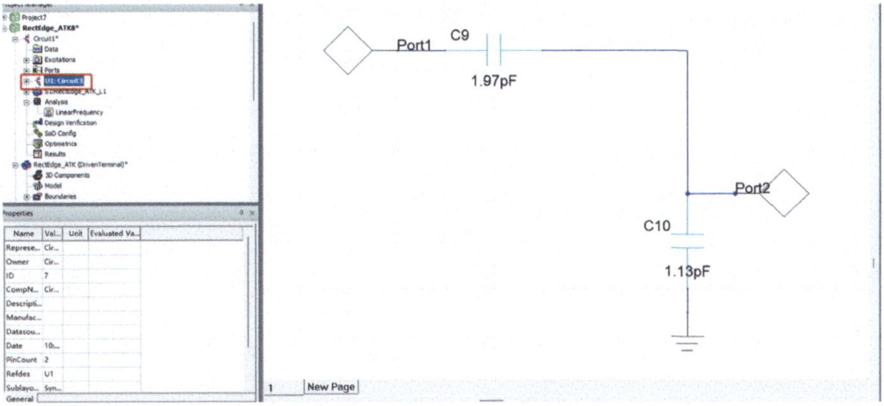

Fig. 4.11 Circuit design for matching

Now select the Match tab and choose the "New Match" button to create some matching network configurations, then select the center of the Smith chart. The number of elements in the component scheme now depends on your design goals. For this model, we will use a series C and a parallel C. As you can see, these values are very small. Then export this network, drag and connect this matching network between the port and the HFSS model, and reanalyze it (Fig. 4.11).

Upon reviewing the results again, we can observe the variations in return loss and input impedance with and without a matching network. It can be seen that the return loss of the folded antenna model has improved compared to the matching circuit created by the Smith tool in the HFSS circuit. Additionally, the gain of the radiation pattern has also been enhanced. These improvements are the outcome of the combined simulation using HFSS electromagnetic simulation and HFSS circuit. When creating the matching network in the HFSS circuit using the Smith tool, our strategy in antenna design was to minimize the number of elements for achieving a match. In this case, we employed a series capacitor and a parallel capacitor. These values were chosen to be very small, showcasing the meticulousness and precision in our design. By exporting this network, we were able to connect the matching network between the port and the HFSS model. Upon reanalysis, significant changes in return loss and input impedance were observed with and without the matching network. This variation once again demonstrates the importance of matching networks in antenna design (Fig. 4.12).

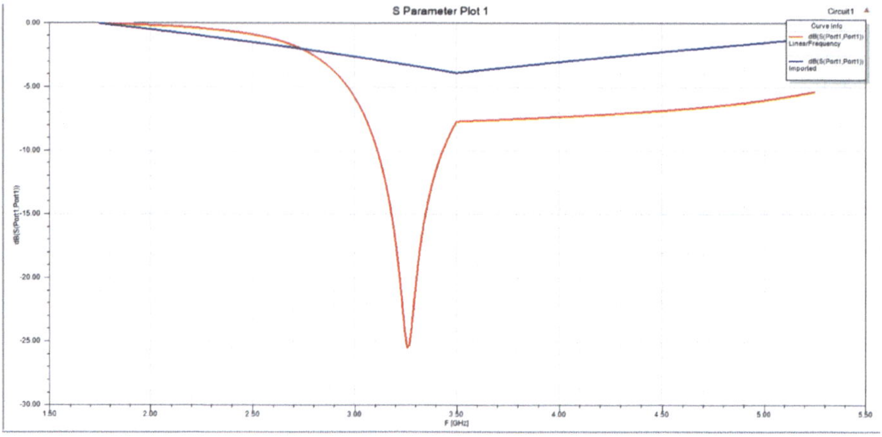

Fig. 4.12 Results of optimized return loss

By observing the results, we can see that the folded antenna model with the implementation of a matching network shows an improvement in return loss. These results were obtained through the combined simulation of HFSS electromagnetic simulation and HFSS circuit simulation. This combined simulation allows us to consider the operating conditions of the circuit while making full use of the powerful electromagnetic field analysis capabilities of HFSS, thereby achieving the goal of optimizing antenna performance. In summary, this section demonstrates how to perform co-simulations of antennas using Ansys HFSS and Circuit. Through detailed steps and explanations, we can understand that this approach provides a comprehensive and detailed simulation environment for antenna design, helping us optimize antenna performance more effectively.

Problems

P4.1 ADS Co-simulation with CST

In antenna design processes, it is often necessary to incorporate matching networks, which require tuning the component values to achieve the desired outcome. Performing a far-field simulation in CST can be time-consuming, especially for complex models, and it can heavily burden computer resources. However, using ADS as an aid allows for obtaining the trends of S_{11} and Smith chart variations in a much shorter time, thereby significantly reducing the development cycle.

In this exercise, CST is used to simulate the S_{11} parameters of an antenna, taking a dipole antenna as an example. The S_{11} parameter data in s1p format is exported from CST for use in ADS. In ADS, a data module is added through Data Items, and the S_{11} data is imported. A 50-ohm port and ground are included in ADS, along with an S-parameter sweep module to scan and observe the S_{11} results. The S_{11} and Smith chart results in ADS are compared with those obtained in CST. A matching network is added between the S1p data module and the 50-ohm port, such as a pi-shaped matching circuit. The matching results are instantaneously simulated and observed, and the values of the matching components can be swept to observe the trends.

P4.2 Co-simulation of Antenna Magus and HFSS
Antenna Magus not only provides the functionality of the Antenna Design Kit but also offers an expanded antenna library, enabling immediate access to preliminary model results and production-related reports. It also possesses powerful features such as generating documents for FEKO or CST. In this exercise, Antenna Magus is used in CST to generate an antenna model, which is then imported into HFSS for co-simulation. The simulation results from both tools are compared.

References

1. Zhou L et al (2019) 5G multimedia communications: theory, technology, and application. IEEE Multimedia 26(1):8–9
2. Whai-En C, Liu CH (2019) Performance enhancement of virtualized media gateway with dpdk for 5g multimedia communications. In: 2019 International conference on intelligent computing and its emerging applications (ICEA). IEEE
3. Milovanovic DA, Bojkovic ZS, Kukolj DD (2021) Machine learning in 5G multimedia communications: open research challenges and applications. In: Research anthology on developing and optimizing 5G networks and the impact on society, pp 204–225
4. Zhu X, Murch RD (2002) Performance analysis of maximum likelihood detection in a MIMO antenna system. IEEE Trans Commun 50(2):187–191
5. Quan XL, Li R (2012) A broadband dual-polarized omnidirectional antenna for base stations. IEEE Trans Antennas Propag 61(2):943–947
6. Abbosh AM (2008) Directive antenna for ultrawideband medical imaging systems. Int J Antennas Propag
7. Zhang Q, Gao Y (2018) A compact broadband dual-polarized antenna array for base stations. IEEE Antennas Wirel Propag Lett 17(6):1073–1076
8. Forward RL (1971) Multidirectional, multipolarization antennas for scalar and tensor gravitational radiation. Gen Relativ Gravit 2:149–159
9. Mailloux RJ (2017) Phased array antenna handbook. Artech house
10. Zhu X, Murch RD (2002) Performance analysis of maximum likelihood detection in a MIMO antenna system. IEEE Trans Commun 50(2):187–191
11. Balanis CA (1992) Antenna theory: a review. Proc IEEE 80(1):7–23
12. Koziel S, Ogurtsov S (2014) Antenna design by simulation-driven optimization. Springer International Publishing, Cham, Switzerland
13. Remski R (2000) Analysis of photonic bandgap surfaces using Ansoft HFSS. Microwave J Euroglobal Edn 43(9):190–199
14. Cendes Z (2016) The development of HFSS. 2016 USNC-URSI Radio Science Meeting. IEEE
15. Zhao K, Petersson LR (2018) Overview of hybrid solver in HFSS. In: 2018 IEEE international symposium on antennas and propagation & USNC/URSI national radio science meeting. IEEE
16. Chen S, Shanghui Y (1986) A guide-weight criterion method for the optimal design of antenna structures. Eng Optim 10(3):199–216
17. Alam MS et al (2013) Development of electromagnetic band gap structures in the perspective of microstrip antenna design. Int J Antennas Propag

18. Khan MU et al (2015) Microstrip patch antenna miniaturisation techniques: a review. IET Microw Antennas Propag 9(9):913–922
19. DuHamel R, Isbell D (1966) Broadband logarithmically periodic antenna structures. 1958 IRE International Convention Record 5
20. Pozar DM (1992) Microstrip antennas. Proc IEEE 80(1):79–91
21. Kraus JD (1949) The helical antenna. Proc IRE 37(3):263–272
22. Roy Choudhury, R. (2006) Utilizing beamforming antennas for wireless multihop networks
23. Bastos JPA, Sadowski N (2003) Electromagnetic modeling by finite element methods. CRC Press
24. Sevgi L (2014) Electromagnetic modeling and simulation. Wiley
25. Kolundžija BM, Djordjević AR (2002) Electromagnetic modeling of composite metallic and dielectric structures. Artech House
26. Reddy JN (2019) Introduction to the finite element method. McGraw-Hill Education
27. Taflove A, Hagness SC, Piket-May M (2005) Computational electromagnetics: the finite-difference time-domain method. Electric Eng Handbook 3:629–670
28. Hall AR (2004) Generalized method of moments. OUP Oxford
29. Kong JA (1975) Theory of electromagnetic waves. New York
30. Raaijmakers AJE et al (2016) The fractionated dipole antenna: A new antenna for body imaging at 7 Tesla. Magn Reson Med 75(3):1366–1374
31. Ta SX, Park I, Ziolkowski RW (2015) Crossed dipole antennas: a review. IEEE Antennas Propag Mag 57(5):107–122
32. Khan MU, Sharawi MS, Mittra R (2015) Microstrip patch antenna miniaturisation techniques: a review. IET Microw Antennas Propag 9(9):913–922
33. Waterhouse R (1995) Small microstrip patch antenna. Electron Lett 31(8):604–605
34. Jeon S-I, Kim Y-W, Oh D-G (2000) A new active phased array antenna for mobile direct broadcasting satellite reception. IEEE Trans Broadcast 46(1):34–40
35. Isbell D (1960) Log periodic dipole arrays. IRE Trans Antennas Propag 8(3):260–267
36. Kraus JD (1949) The helical antenna. Proc IRE 37(3):263–272
37. Yekeh Yazdandoost K, Kohno R (2004) Ultra wideband antenna. IEEE Commun Mag 42(6):S29–S32
38. Peroulis D, Sarabandi K, Katehi LP (2005) Design of reconfigurable slot antennas. IEEE Trans Antennas Propag 53(2):645–654
39. Grajek PR, Schoenlinner B, Rebeiz GM (2004) A 24-GHz high-gain Yagi-Uda antenna array. IEEE Trans Antennas Propag 52(5):1257–1261
40. Patel RH, Desai A, Upadhyaya T (2015) A discussion on electrically small antenna property. Microw Opt Technol Lett 57(10):2386–2388
41. Darvish A, Ebrahimzadeh A (2018) Improved fruit-fly optimization algorithm and its applications in antenna arrays synthesis. IEEE Trans Antennas Propag 66(4):1756–1766
42. Yeung SH et al (2012) Computational optimization algorithms for antennas and RF/microwave circuit designs: an overview. IEEE Trans Ind Inf 8(2):216–227
43. Roy GG et al (2010) Design of non-uniform circular antenna arrays using a modified invasive weed optimization algorithm. IEEE Trans Antennas Propag 59(1):110–118

Chapter 5
Co-simulations of Electromagnetic Compatibility

5.1 Introduction

Electromagnetic Compatibility (EMC) refers to the ability of an electronic device or system to function properly in an electromagnetic environment without causing adverse electromagnetic effects on other electronic devices or systems [1, 2]. The EMC encompasses two main aspects, i.e., Electromagnetic Interference (EMI) and Electromagnetic Susceptibility (EMS) [2, 3]. The EMI is a measurement of radiated electromagnetic energy from a device or system that will interfere with the normal operation of other electronic devices or systems. In contrast, EMS is the capability of an electronic device that resists the EMI from other electronic devices and systems [4, 5].

The EMI can be categorized into the intentional interference and unintentional interference based on the purpose [3]. The intentional interference or jamming is an actively applied electromagnetic signal to interfere with other electronic devices and systems and has been usually employed for military or security purposes like electronic warfare [4]. The unintentional interference primarily refers to an undesired interference caused by the operations of various devices and systems. Some unintentional interferences could also be produced by unartificial or natural factors like lighting. The unintentional interference can be further classified into the following categories [3, 5, 6]: (a) Conducted EMI which is the interference transmitted through conductive media such as wires and cables; (b) Radiated EMI which is the interference in the form of electromagnetic waves propagating through space; (c) Low-frequency EMI which is the interference caused by the electromagnetic wave with a frequency below 30 MHz; (c) High-frequency EMI which is the interference caused by the electromagnetic wave with a frequency above 30 MHz.

The EMS reflects or measures the ability of an electronic device that prevents itself from the EMI of external devices or systems and it is related to the EMC design for the device. The good EMC design is very essential to enhance the EMS of electronic devices or systems and the simplest EMC design is the addition of metallic

M. S. Tong and X. Y. Li, *Co-simulations of Microwave Circuits and High-Frequency Electromagnetic Fields*, https://doi.org/10.1007/978-981-99-8307-0_5

shielding covers on the surfaces of protected devices or components. Similar to the EMI, the EMS can also be divided into the following categories in general [7–10, 28]: (a) Conducted Susceptibility which is the resisting ability of an electronic device or component to the electromagnetic interference transmitted through conductive media such as wires and cables; (b) Radiated Susceptibility which is the resisting ability of an electronic device or component to the electromagnetic interference transmitted through free space; (c) Low-Frequency Susceptibility which is the resisting ability of an electronic device or component to the electromagnetic interference with a frequency below 30 MHz; (d) High-Frequency Susceptibility which is the resisting ability of an electronic device or component to the electromagnetic interference with a frequency above 30 MHz.

The good EMS of electronic components, devices, and systems relies on the good EMC design for them. As the components, devices, and systems have become more and more complicated and diverse, the EMC designs for them have also become more and more difficult. The traditional simple designs have no longer been suitable for them and novel and better designs are very desirable. The EMC designs require a full understanding of the interactions between the incident electromagnetic fields and interfered components or devices or systems, and this can only be achieved by seeking accurate numerical solutions of governing electromagnetic equations. The powerful software tools play such a role and they have been indispensable for the simulations of EMC designs. This chapter will introduce the simulation methods of various software tools for EMC designs and related analyses and optimizations of simulation results.

5.2 Fundamental of EMC Designs

5.2.1 Quantification of EMC Performances

To quantify the EMC performance of a device or system, it is necessary to evaluate its electromagnetic radiation and electromagnetic susceptibility. Evaluation methods include experimental testing and numerical simulation. Experimental testing is typically conducted in specialized EMC laboratories, while numerical simulation involves the use of EMC simulation software for calculation and analysis. By solving Maxwell's equations and applying boundary conditions, equations describing the propagation characteristics of the electromagnetic field can be obtained.

Interference from electromagnetic sources can propagate through various means. Interference propagated through wires can be divided into differential mode interference and common mode interference [29–31]. Differential mode interference voltage U_{dg} appears between the input and output leads of a current loop and is connected in series with the operating power supply voltage U_s as shown in Fig. 5.1. Differential mode interference generates a differential mode current I_{dg} within the loop and produces an interference voltage signal between the loop and the input terminal

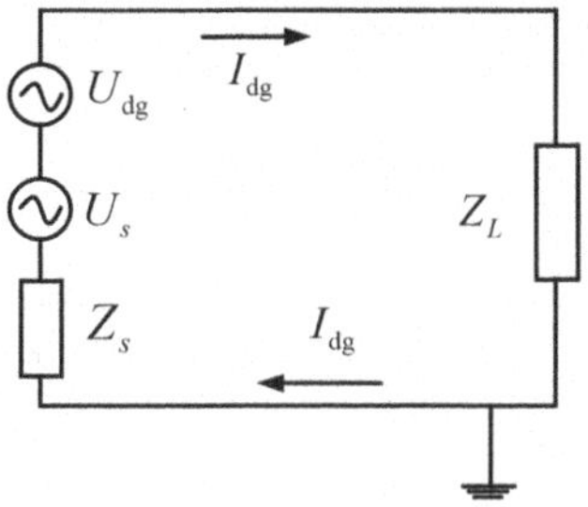

Fig. 5.1 Definition of differential mode voltage

(represented by Z_L) of the affected system. This interference voltage signal is superimposed on the useful signal of the loop. Differential mode interference is primarily caused by magnetic coupling or common mode to differential mode conversion. Since it is connected in series with the useful signal, it can lead to problems such as measurement errors and malfunctions. When the internal resistance of the power supply $Z_S \ll Z_L$, the interference voltage U_{dg} induced in the loop is effectively considered as the interference voltage affecting the system (in series with the useful signal).

Common mode interference U_{cg} occurs between the signal core and the neutral conductor (e.g., the measurement ground). It results from the introduction of a disturbing voltage source U_{cg} between the neutral and the return, such as a transient ground potential increase, as illustrated in Fig. 5.2. The common mode voltage is not an interfering voltage in series with the desired signal, but a large common mode voltage may cause surface flashover between the signal line and the instrument case or line neutral, potentially leading to permanent damage. When multiple connections exist between the signal circuit and the ground return, the common mode voltage will drive different currents in the input and output leads (I_{cg1} and I_{cg2} in Fig. 5.2), resulting in different voltage drops across the impedance. Consequently, the input and output leads have different voltages with respect to ground, thereby generating a differential mode interference voltage at the receiver input (represented by Z_L). This phenomenon is known as common mode to differential mode conversion. The different impedances convert the common mode voltage partially or completely into a differential mode voltage, equal in magnitude to the difference between the ground potentials of the input and output leads.

To enhance the EMC performance of equipment, several measures are typically employed, including shielding, grounding, and filtering [32, 50, 51]. Shielding involves surrounding the equipment with metallic or magnetic materials to minimize the propagation of electromagnetic interference. Grounding refers to connecting the metal components of the equipment to the ground wire in order to reduce voltage generated by interference currents. Filtering aims to decrease the amplitude of interference signals by installing filters at the device's input and output. This effect can be quantified using the following formula:

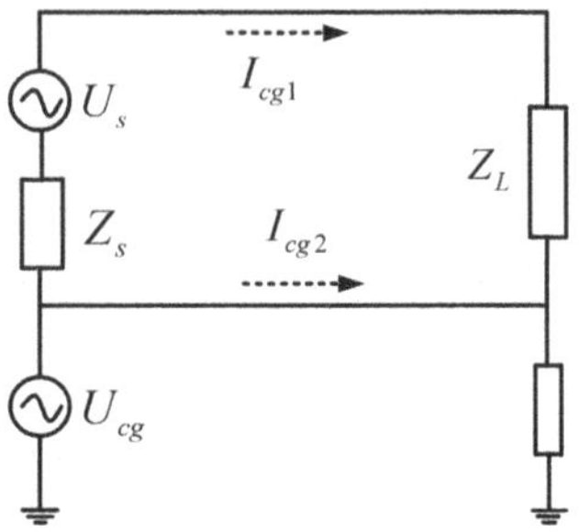

Fig. 5.2 Definition of common mode voltage

$$SE = 10 \times \log_{10}\left(\frac{P_{\text{in}}}{P_{\text{out}}}\right) \tag{5.1}$$

where SE represents the Shielding Effectiveness, P_{in} denotes the input power, and P_{out} represents the output power. Evaluating and optimizing EMC performance is an integrated process that spans multiple domains and technologies. With the rapid development of wireless communication, Internet of Things, and other emerging technologies, the issue of electromagnetic compatibility (EMC) is becoming increasingly prominent. The research on EMC design and analysis technology will face greater challenges and require foresight.

5.2.2 EMC Analysis Methods

EMC analysis involves the application of electromagnetic field theory and electromagnetic compatibility principles to analyze and optimize the design of electronic equipment and systems. It encompasses the following aspects [23, 24, 52]:

1. Electromagnetic Field Theory

Electromagnetic field theory forms the foundation of EMC analysis and focuses on studying the generation, propagation, and interaction of electromagnetic fields. In EMC analysis, it is crucial to understand the fundamental concepts and calculation methods of electromagnetic field theory, including electromagnetic field equations, Maxwell's equations, radiation field calculation methods, and transmission line models [25, 26].

The transmission line model is a fundamental model used to describe electromagnetic fields on transmission lines and analyze electromagnetic compatibility issues. The electromagnetic field on a transmission line can be classified into TEM mode, TE mode, and TM mode. In TEM mode, both the electric and magnetic fields are aligned with the direction of the transmission line, having equal magnitudes and directions, representing an uncoupled electromagnetic field. In TE or TM mode, either the electric or magnetic field is perpendicular to the direction of the transmission line, resulting in a coupled electromagnetic field.

The calculation method for radiation fields is essential for analyzing electromagnetic radiation and wave propagation. Radiation field calculations can be performed using analytical, numerical, or experimental methods. The analytical method involves solving mathematical formulas to obtain the distribution of electromagnetic fields, such as charge models and radiation-based models. The numerical method utilizes computer simulations to analyze the distribution of electromagnetic fields, employing techniques like finite element method and finite-difference time-domain method. The experimental method directly measures electromagnetic fields using equipment like antenna test systems and frequency sweepers [27, 28].

2. Electromagnetic Compatibility Principles

The principles of electromagnetic compatibility define the requirements that electronic equipment must meet in the electromagnetic environment. These requirements encompass EMC standards and EMC design principles.

EMC standards specify the electromagnetic compatibility requirements that electronic equipment must adhere to in the electromagnetic environment. They form the basis for ensuring the electromagnetic compatibility of electronic equipment. Common EMC standards include the International Electrotechnical Commission (IEC), Federal Communications Commission (FCC), European Telecommunications Standards Institute (ETSI), and others. These standards specify limitations on electromagnetic field strength, frequency, waveform, etc., to ensure the proper operation and interoperability of equipment in the electromagnetic environment.

EMC design principles outline the fundamental guidelines to follow when designing electronic equipment in the electromagnetic environment. These principles include the isolation principle, suppression principle, and shielding principle. The isolation principle emphasizes isolating the electromagnetic interference source within the electronic equipment from the target being interfered with, preventing the interference source from affecting the target. The suppression principle involves implementing various measures to limit the electromagnetic radiation and conduction of interference sources, thus reducing the interference level. The shielding principle entails enclosing the equipment with a shield to restrict the radiation and conduction of electromagnetic interference sources within the equipment, thereby preventing the interference source from affecting external equipment.

3. EMC Analysis Methods

EMC analysis methods comprise modeling analysis, simulation analysis, and experimental verification. The modeling analysis method involves establishing electromagnetic field models of electronic equipment and systems on computers and analyzing electromagnetic compatibility problems by solving electromagnetic field equations. The simulation analysis method utilizes electromagnetic field simulation software to numerically simulate and analyze electromagnetic fields, enabling the evaluation of electromagnetic compatibility levels. The experimental verification method entails quantitative analysis and evaluation of electromagnetic compatibility through experimental measurements.

Common EMC design optimization measures include optimizing circuit board layouts, selecting appropriate filters and suppression devices, and optimizing grounding and shielding techniques. In the optimization design process, factors such as circuit layout, signal transmission paths, electromagnetic radiation, and conduction need to be comprehensively considered to achieve optimal EMC design outcomes. EMC test methods include radiation testing and conducted testing. Radiation testing involves measuring the electromagnetic radiation power and radiation field intensity of electronic equipment within a specified frequency range, thereby evaluating the equipment's radiation compatibility. Conducted testing measures conducted electromagnetic interference and anti-interference capabilities of electronic equipment within a specified frequency range, assessing the equipment's conducted compatibility.

5.2.3 EMC Analyses and Applications

The primary objective of EMC analysis is to ensure the reliable operation of electronic equipment and systems in the presence of electromagnetic fields, while preventing undesirable interference and damage to the surrounding environment and other equipment. Specifically, EMC analysis aims to achieve the following:

1. Electromagnetic compatibility design:

EMC analysis enables the design and optimization of electronic equipment and systems to enhance their compatibility within the electromagnetic environment. This involves analyzing and optimizing circuitry, signal lines, power lines, antennas, enclosures, and other components to reduce the levels of radiated and conducted interference.

2. Electromagnetic compatibility testing:

EMC analysis facilitates the assessment of the compatibility level of electronic equipment and systems through experimental testing. This includes evaluating electromagnetic emissions and conducting interference to determine their electromagnetic compatibility within the given environment.

3. EMC troubleshooting:

EMC analysis assists engineers in identifying and resolving EMC issues, such as circuit interference, signal crosstalk, power harmonics, and more. Through analysis and simulation of the identified problem, the root cause of the fault can be determined quickly, enabling appropriate measures to be taken for repair. By addressing these objectives, EMC analysis plays a crucial role in ensuring the reliable and interference-free operation of electronic equipment and systems in diverse electromagnetic environments. EMC analysis finds extensive applications in various fields, including electronic communication, military, aerospace, medical devices, automotive, industrial automation, and more [31–33]. Commonly employed methods in EMC analysis include:

1. Electromagnetic field simulation:

This involves utilizing electromagnetic field theory and computer simulation technology to analyze and predict the distribution of electromagnetic fields in electronic equipment and systems. Popular electromagnetic field simulation software includes Ansys HFSS, CST Microwave Studio, FEKO, among others.

2. Circuit simulation:

Circuit simulation entails using circuit analysis software to assess the interference and anti-interference performance of circuits. Commonly employed circuit simulation software includes PSPICE, ADS, and others.

3. Radiation measurement:

Radiation measurement involves the actual measurement and evaluation of the electromagnetic radiation emitted by electronic equipment and systems using radiation measurement instruments. Commonly used radiation measurement instruments include antennas, spectrum analyzers, magnetic field detectors, and more.

4. Conducted measurement:

Conducted measurement refers to the practical measurement and evaluation of conducted interference in electronic equipment and systems using conducted measurement instruments. Differential mode probes, common mode probes, current probes, etc., are commonly employed conducted measurement instruments.

5. Electromagnetic compatibility testing:

Electromagnetic compatibility testing employs laboratory equipment and standardized test methods to assess and verify the electromagnetic compatibility of electronic equipment and systems. Radiation tests, conduction tests, and other EMC test methods are commonly used.

5.2.4 Typical Problems and Solutions

Common issues in EMC analysis primarily include radiated interference, conducted interference, and grounding problems [31, 34, 35].

(1) Radiated interference:

Radiated interference refers to the electromagnetic radiation generated by electronic equipment, which can cause interference and damage to the surrounding environment and other equipment. To address radiated interference, two main approaches are commonly employed: reducing the radiation power of the source and enhancing interference suppression capability. Methods to improve interference suppression capability involve implementing interference suppression measures during the design stage, such as reducing circuit area, optimizing interface design, and incorporating filters. Isolation measures can be taken during the layout stage, such as reducing distances and increasing shielding. At the system level, shielding measures like metal shielding boxes and covers can be employed. To decrease the radiation power of the source, strategies include reducing circuit area, minimizing the size of power supply loops, and optimizing layout. Additionally, selecting appropriate electronic components with low noise, low radiation, and low scattering characteristics is essential in mitigating radiated interference [37].

(2) Conducted interference:

Conducted interference occurs when common mode or differential mode interference affects signal transmission through shared conductors or interfaces, resulting in interference and damage. Addressing conducted interference involves two primary approaches: reducing the signal voltage of conducted interference and enhancing anti-interference capability. Techniques to reduce the signal voltage of conducted interference include reducing the circuit area, increasing the number of windings in signal lines, employing shielded wires or sleeves, and optimizing interface design. Enhancing anti-interference capability encompasses implementing interference suppression measures during the design stage, such as adopting differential signal transmission and incorporating filters. Isolation measures can be implemented during the layout stage, including increasing isolation distance and utilizing shielding. At the system level, shielding measures such as metal shielding boxes and covers can be employed.

(3) Grounding problems:

Grounding problems arise during the design and implementation of the ground wire system in electronic equipment. They mainly involve single-point grounding and multi-point grounding. Single-point grounding consolidates all ground wires at one location, which reduces interference between ground wires. However, single-point grounding can result in electromagnetic interference caused by large currents flowing through the ground wire. Therefore, methods to mitigate this interference include increasing the area of the ground wire, reducing ground wire resistance, and employing metal shielding. Multi-point grounding, on the other hand, connects different ground wires to the ground separately, avoiding the issue of large currents associated with single-point grounding. However, multi-point grounding can lead to interference problems between ground wires. To address this, measures such as increasing the grounding area, reducing grounding resistance, and using anti-interference grounding wires can be implemented.

5.3 EMC Simulations in ADS

5.3.1 ADS Implementation of EMC Simulations

The ADS is a kind of Electronic Design Automation (EDA) software developed by Keysight Technologies, widely utilized in microwave circuits, RF circuits, communication systems, and various other fields. ADS offers robust simulation and analysis capabilities, encompassing circuit simulation, system simulation, electromagnetic (EM) simulation, and other simulation methods. Notably, the EM simulation functionality in ADS supports diverse algorithms based on the finite element method (FEM), method of moments (MoM), and finite-difference time-domain (FDTD). This allows for high-frequency electromagnetic field simulation and electromagnetic compatibility (EMC) analysis [36, 38].

The underlying principle of ADS lies in the interaction between circuit components, circuit structures, and physical simulation models, enabling comprehensive support for circuit design and performance analysis. Through ADS, users can engage in circuit design, optimization, simulation, and verification processes to enhance design efficiency and quality.

5.3.2 EMC Analysis Module in ADS

ADS is a kind of exceptional EDA software offering a wide range of circuit simulation and system simulation tools. Within its comprehensive suite, ADS includes specific modules dedicated to EMC simulation analysis, namely EMPro, EMX, and EMDS. These tools encompass diverse simulation methods and analysis functions tailored to address various electromagnetic field simulation and electromagnetic compatibility challenges [39].

EMPro, a component of ADS, is a remarkable 3D electromagnetic field simulation tool based on the finite element method (FEM). It empowers users with comprehensive modeling and simulation capabilities, encompassing tasks like model creation, material definition, boundary condition configuration, simulation parameter adjustment, and more.

When creating models in EMPro, users have the flexibility to employ various methods such as part and assembly-based modeling, importing existing CAD models, or even drawing models manually. Furthermore, users can refine and modify models to suit specific simulation requirements.

In the process of configuring simulation parameters, users are required to set parameters such as the frequency of the electromagnetic field, size of the calculation domain, material properties, boundary conditions, error limits, and other relevant factors. Boundary conditions, being a crucial element impacting simulation accuracy, include considerations for incident waves, radiation waves, and scattering waves of the electromagnetic field. To enhance simulation accuracy, users can choose appropriate boundary conditions and optimize them accordingly.

EMPro is equipped with a diverse range of result analysis tools, enabling users to examine electric field distribution and magnetic field distribution, perform S-parameter analysis, impedance analysis, power transmission analysis, and more. By analyzing and optimizing simulation results, users can enhance both simulation accuracy and computational efficiency. Additionally, EMPro seamlessly integrates with other ADS tools, such as enabling S-parameter simulation analysis and system-level simulation analysis through SystemVue.

EMDS, an integral part of ADS, serves as a dedicated tool for electromagnetic compatibility analysis. Its primary function is to provide a comprehensive time-domain simulation and analysis environment, specifically tailored for analyzing electromagnetic radiation, interference, and anti-interference problems.

During the model creation process in EMDS, users can choose from various modeling methods, such as importing existing CAD models or creating models manually. Users also have the ability to edit and refine models to meet the specific requirements of their simulations. In the configuration of simulation parameters, users need to define essential parameters such as the frequency of the electromagnetic field, size of the calculation domain, material properties, boundary conditions, error limits, and more. Unlike other simulation tools, EMDS specifically focuses on time-domain simulations, requiring users to set the time step and time length for the simulation. EMDS encompasses a rich set of result analysis tools, including time-domain electromagnetic field distribution analysis, frequency-domain power spectrum analysis, time-domain impulse response analysis, and more. By thoroughly analyzing and optimizing the simulation results, users can enhance both the accuracy and computational efficiency of their simulations.

Electromagnetic compatibility (EMC) refers to the ability of electronic devices or systems to coexist and interact within their environment. EMC research primarily focuses on two aspects: electromagnetic radiation and electromagnetic interference. Simulations analysis and optimization design are crucial in ensuring the EMC performance of electronic equipment. Advanced Design System (ADS) offers robust simulation analysis capabilities, including circuit simulation, system simulation, and EM simulation methods. Notably, the EM simulation function in ADS supports various algorithms such as finite element method (FEM), method of moments (MoM), and finite-difference time-domain (FDTD), enabling high-frequency electromagnetic field simulation and EMC analysis. The EMC simulation method and process in ADS typically involve the following steps:

1. Modeling and geometric design

EM simulation in ADS necessitates initial modeling and geometric design. Tools like EMPro and EMX are employed to model and design the physical structure intended for simulation. During the modeling process, factors such as the geometry of the structure, material properties, size parameters, and circuit connections need to be considered. The precision and accuracy of the model significantly impact subsequent simulation calculations and analysis results.

2. Setting simulation parameters

Before conducting EMC simulation, relevant simulation parameters must be defined. These parameters include the frequency of the electromagnetic field, boundary conditions, material properties, etc. The choice of electromagnetic field frequency should align with the actual application. Common boundary conditions include perfect electric conductor (PEC) boundary, perfect magnetic conductor (PMC) boundary, and impedance boundary. Material properties play a crucial role as well, as the permittivity, permeability, and conductivity of different materials affect the distribution and propagation of electromagnetic fields.

3. Performing simulation calculations

EM simulation is executed based on the chosen algorithm. In ADS, users can select algorithms like FEM, MoM, FDTD, etc., for simulation purposes. During the calculation process, parameters can be optimized based on the simulation results, improving both simulation accuracy and computational efficiency. If multiple frequency points need simulation, the batch mode can be utilized to calculate multiple points simultaneously, enhancing simulation efficiency.

4. Analyzing simulation results

After completing the simulation calculations, the obtained results should be analyzed and processed. This analysis encompasses studying the characteristics of the electromagnetic field, including its distribution, propagation, and radiation, as well as evaluating circuit traits such as transmission characteristics, power transmission, reflection coefficient, and scattering coefficient. ADS provides analysis tools such as the Smith diagram, power transmission diagram, and electric field distribution diagram for result interpretation. Additionally, in EMC simulation, it is essential to analyze electromagnetic interference and anti-interference measures. Addressing electromagnetic interference entails analyzing the radiation characteristics of the electromagnetic field, determining radiation source and field parameters, and subsequently identifying interference mitigation measures. For anti-interference purposes, it involves analyzing the coupling characteristics of the electromagnetic field, determining coupling paths and mechanisms, and devising anti-interference measures. While analyzing the results, attention should be given to result accuracy and reliability. Verification and cross-referencing with measured results are necessary. If discrepancies arise, further analysis of the reasons, optimization of simulation models and parameter settings, and enhancement of result accuracy should be pursued.

5.4 EMC Simulations in HFSS

5.4.1 HFSS Implementation of EMC Simulations

In EMC analysis, HFSS plays a vital role in assisting engineers in swiftly evaluating design schemes for EMC, identifying potential EMC issues, and offering guidance for design optimization. Specifically, HFSS enables the simulation of electromagnetic field distribution and characteristics during the early stages of circuit design, facilitating the evaluation and optimization of electromagnetic compatibility in the design scheme. Moreover, in the middle and later stages of circuit manufacturing, HFSS can simulate electromagnetic field distribution and characteristics within the circuit, evaluate potential electromagnetic compatibility problems that may arise during the manufacturing process, and provide solutions.

In practical applications, the workflow of HFSS typically involves the following steps: model preparation, simulation type definition, simulation scenario setup, simulation calculation, simulation result analysis, and design optimization. Model preparation entails establishing the structure model to be simulated in HFSS and configuring the necessary simulation parameters, such as model size, material properties, and boundary conditions. The definition of simulation type involves selecting appropriate analysis modules and methods to conduct interference analysis, radiation interference analysis, conduction coupling analysis, radiation coupling analysis, or field coupling simulation as required. Setting the simulation scenario involves configuring boundary conditions, excitation signals, and simulation time, such as setting the excitation signal and simulation range of an antenna. Simulations calculation involves utilizing the HFSS solver to compute the electromagnetic field distribution and electromagnetic characteristics of each circuit element, such as solving for electromagnetic field distribution and frequency response. Analysis of simulation results entails examining the simulation results, including locating and analyzing interference sources, affected circuits, coupling paths, etc., and evaluating electromagnetic compatibility indicators. Design optimization involves making structural adjustments based on the analysis results to improve electromagnetic compatibility, such as repositioning components or optimizing filter design.

5.4.2 *EMC Analysis Modules in HFSS*

EMC analysis in HFSS encompasses Conducted Emission, Radiated Emission, Conducted Susceptibility, Radiated Susceptibility, and Field Coupling [40–44]. Among these modules, the most commonly used ones in electromagnetic compatibility (EMC) analysis are Conducted Interference Analysis, Radiated Interference Analysis, and Conductive Coupling Analysis.

Conducted Interference Analysis focuses on the impact of interference voltages and currents transmitted through wires on other circuits. It primarily occurs due to high-speed signals and fast switches in circuits, which can generate high-frequency noise and electromagnetic waves, thus interfering with signals in other circuits. HFSS offers various tools and methods for evaluating and mitigating the effects of conducted interference.

The workflow for conducted interference analysis is as follows:

1. Model Preparation:

Establish the HFSS model of the structure to be simulated and define the necessary parameters for simulation. This may include transmission lines, interfaces, filters, isolators, and other components.

2. Simulations Type Definition:

Select the conducted interference analysis type and set the frequency range and simulation time for analysis.

3. Simulations Scenario Setup:

Configure the boundary conditions, excitation signals, and simulation time for the analysis. HFSS supports different excitation signals, such as sine waves and pulses, allowing users to choose based on their specific needs.

4. Perform Simulations Calculation:

Utilize the HFSS solver to calculate the electromagnetic field distribution and characteristics of each circuit element. In conducted interference analysis, it is often necessary to calculate the interference currents and voltages in the conduction path.

5. Analyze the Simulations Results:

Analyze the simulation results, including identifying and analyzing interference sources, affected circuits, coupling paths, and evaluating electromagnetic compatibility indicators.

Radiated Interference Analysis focuses on the impact of electromagnetic radiation generated by circuits on surrounding circuits and equipment. High-speed circuits and electronic devices can produce strong electromagnetic radiation due to rapid signal transitions. These emissions can interfere with signals in neighboring equipment and circuits, potentially affecting system performance and stability. Therefore, radiated interference analysis plays a crucial role in EMC analysis. Conductive Coupling Analysis examines the effect of conduction pathways on signal transmission. In a circuit, conductive coupling typically arises from electromagnetic interference from adjacent circuits when signals are transmitted through wires, cables, boards, or other media. Conductive coupling can lead to signal distortion, interference, and other issues, necessitating its analysis in EMC analysis.

5.4.3 EMC Simulation Method and Process in HFSS

HFSS is a type of highly precise electromagnetic field simulation software widely used in various aspects of EMC analysis, including conducted interference, radiated interference, conducted coupling, radiated coupling, and field coupling. In EMC simulations, the following steps are generally involved:

1. Model Preparation:

Establish the HFSS model of the structure to be simulated, including the circuits, signal lines, ground lines, and other components to be analyzed, as well as the model of the surrounding environment. The model should accurately reflect the real-world situation as it greatly influences the simulation results.

2. Simulation Type Definition:

Select the appropriate simulation type based on the specific requirements, such as conducted interference, radiated interference, conducted coupling, radiated coupling, or field coupling. Different simulation types require setting different simulation scenarios and boundary conditions.

3. Simulation Scenario Setup:

Configure the boundary conditions, excitation signals, and simulation time for the chosen simulation scenario. The simulation scenario should closely resemble the real-world conditions to enhance the accuracy of the simulation results. For instance, in conducted interference simulation, it is necessary to set the excitation signal, interface impedance of the circuit, and conducted parameters of the surrounding environment. In radiated interference simulation, the radiation characteristics of the circuit and the calculation area of the radiation field need to be set.

4. Simulation Calculation:

Utilize the HFSS solver to calculate the electromagnetic field distribution and electromagnetic characteristics of each circuit element. Since simulation computations can be time and resource-intensive, they are typically performed on high-performance computers or using distributed computing methods to expedite the process. During the calculation, different solver options and error control parameters can be adjusted to strike a balance between calculation speed and simulation accuracy.

5. Analysis of Simulations Results:

Analyze the simulation results, including identifying and analyzing interference sources, affected circuits, coupling paths, and evaluating Electromagnetic Compatibility (EMC) indicators. Spectrum analysis, time-domain analysis, voltage, and current analysis are commonly employed to conduct in-depth analysis of the simulation results. Additionally, the structure can be optimized based on the simulation results to enhance electromagnetic compatibility.

6. Optimization Design:

Based on the analysis results, perform structural optimization to improve electromagnetic compatibility. For example, optimizing circuit layout, introducing filter devices, and refining interface design can help reduce the coupling between interference sources and affected circuits, thereby enhancing the EMC performance of the system.

It is important to note that the accuracy and reliability of EMC simulation are influenced by various factors, such as model accuracy, simulation scenarios, solver options, error control parameters, and among others.

5.5 Co-simulations of EMC with ADS

5.5.1 Co-simulations of CST Studio Suite with ADS

CST Studio Suite is a kind of professional electromagnetic simulation software that enables three-dimensional electromagnetic field simulation, high-frequency circuit simulation, electromagnetic compatibility analysis, and more. It is characterized by its simplicity in model creation, fast calculations, high precision, and powerful functionality. The software supports multiple simulation modes, such as time domain, frequency domain, and mode simulations, allowing users to select the most suitable mode for their specific simulation requirements. In EMC co-simulations, it is common to combine electromagnetic field simulation software with circuit simulation software to achieve interactive analysis between the electromagnetic field and the circuit. Co-simulations enhance the accuracy of simulation analysis, produce more realistic and reliable results, reduce the need for repetitive model creation, improve simulation efficiency, and save time and cost.

The specific procedure for CST Studio Suite and ADS co-simulations is as follows:

1. Establish an electromagnetic field simulation model:

Create the electromagnetic field simulation model in CST Studio Suite, considering factors such as circuit board layout, electronic device placement, and interface line locations. This model enables simulation and analysis of the electromagnetic performance of electronic equipment.

2. Export the electromagnetic field simulation model:

After establishing the electromagnetic field simulation model, export it to ADS using an appropriate file format for subsequent simulation analysis in ADS.

3. Establish a circuit simulation model:

Create the circuit simulation model in ADS, including the circuit board layout, electronic device connections, signal source placement, etc. This model allows simulation and analysis of the circuit performance of electronic equipment.

4. Perform the co-simulations:

With the electromagnetic field simulation model and the circuit simulation model in place, carry out the co-simulations. Combine the two simulation models to analyze the interaction between the electromagnetic field and the circuit. This involves adding the CST Studio Suite simulation model in ADS and conducting simulation analysis.

5. Analyze the simulation results:

After completing the co-simulations, analyze the simulation results. These results provide insights into the operating conditions of electronic equipment in the electromagnetic environment and the mutual interference between different signal sources. By analyzing the simulation results, the design of electronic equipment can be optimized, and the EMC performance can be improved.

CST Studio Suite and ADS can be Co-simulated using various methods, including:

1. Data transfer mode:

Export the electromagnetic field simulation model file from CST Studio Suite to ADS for simulation analysis. Various file formats, such as IGES, STEP, and SAT, can be chosen when exporting the model.

2. Bidirectional co-simulations mode:

Establish a bi-directional connection between CST Studio Suite and ADS to enable interactive analysis between the electromagnetic field and the circuit. In this mode, the two software tools can share simulation models and parameters, leading to more accurate simulation analysis.

3. API mode:

Utilize the API interfaces provided by CST Studio Suite and ADS for co-simulations. This allows for flexible simulation analysis, catering to different simulation requirements.

5.5.2 Co-simulations of AWR Microwave Office with ADS

AWR Microwave Office is widely used for microwave circuit designs and simulations. It features a powerful simulation engine and a comprehensive component library, enabling efficient and accurate design, simulation, and optimization of microwave circuits. In terms of electromagnetic compatibility (EMC) simulation, AWR Microwave Office provides engineers with robust functions and tools to analyze and address EMC issues in circuits.

The EMC simulation capabilities of AWR Microwave Office encompass the following:

1. Routing simulations of printed circuit boards (PCBs):

AWR Microwave Office can perform three-dimensional electromagnetic simulations of PCB routing and analyze electromagnetic interference and radiation problems within the circuit. By simulating the electric field, magnetic field, and power distribution in PCB traces, circuit layout, and routing can be optimized to minimize electromagnetic radiation and interference.

2. Antenna simulations:

AWR Microwave Office allows for three-dimensional simulation of antennas, facilitating analysis of radiation characteristics and matching networks. Through simulations, important parameters such as gain, radiation pattern, and impedance matching can be obtained. This enables optimization of antenna designs to enhance performance and efficiency.

3. RF front-end simulations:

AWR Microwave Office can simulate RF front-end circuits and analyze their electromagnetic compatibility. By conducting simulations, engineers can gain insights into power distribution, impedance matching, noise, and other key aspects of RF front-end circuits. This information aids in optimizing circuit designs to improve performance and stability.

4. System simulations:

AWR Microwave Office enables system-level simulations to analyze overall electromagnetic compatibility. Through simulations, engineers can study the electric field, magnetic field, and power distribution within the system. This knowledge can be utilized to optimize system designs and mitigate electromagnetic radiation and interference.

For EMC co-simulations analysis, AWR Microwave Office can leverage the EMPro (Electromagnetic Professional) module provided by ADS (Advanced Design System). The EMPro module is a powerful electromagnetic simulation tool within ADS. It supports three-dimensional structure simulations and analysis of electromagnetic field distribution, characteristics, and parameters. In EMC co-simulations analysis, the EMPro module of AWR Microwave Office and ADS can exchange data through an interface, allowing for seamless co-simulations analysis. In the context of EMC co-simulations analysis between AWR Microwave Office and ADS, the following steps can be taken:

1. Design and optimize the circuit in AWR Microwave Office, obtaining its electrical performance.

2. Export the circuit in a compatible format (e.g., CITI file) from AWR Microwave Office and import it into ADS for EMC simulation.

3. Set up the simulation model in ADS, including the circuit's structure, materials, and boundary conditions. Conduct electromagnetic simulations to obtain the electromagnetic field distribution and characteristics.

4. Export the simulation results as S-parameter files and import them back into AWR Microwave Office.

5. Perform circuit simulations in AWR Microwave Office and compare/analyze the results with those obtained from ADS. Identify any EMC issues and develop optimization strategies.

6. Based on the simulation results, optimize the circuit design and repeat the simulation process until satisfactory electromagnetic compatibility and performance requirements are met.

7. Finally, export the PCB layout file of the circuit for actual testing and verification.

It is crucial to ensure compatibility between the versions and interfaces of both software tools during the EMC co-simulations analysis of AWR Microwave Office and ADS. Incompatible versions or interfaces may lead to errors and inaccuracies in data transmission and exchange. Additionally, attention should be given to establishing and setting up the simulation model, as well as analyzing and interpreting the simulation results. This ensures that the final circuit design meets the desired electromagnetic compatibility and performance requirements.

The EMC co-simulations analysis of AWR Microwave Office and ADS can be carried out through various methods. One common approach involves utilizing the EMPro module of ADS for electromagnetic simulation and subsequently importing the results into AWR Microwave Office for circuit simulation and optimization. Furthermore, AWR Microwave Office and ADS can exchange data through other interfaces and file formats like IBIS, VHDL-AMS, SPICE, etc. Engineers can choose the appropriate interface and file format based on specific requirements and software version compatibility to facilitate the EMC co-simulations analysis between AWR Microwave Office and ADS.

5.5.3 Co-simulations of Simbeor with ADS

Simbeor is a powerful software tool for high-speed digital signal simulations, widely used in circuit design, signal integrity analysis, and EMC simulation. It supports co-simulations with ADS (Advanced Design System), enabling more accurate EMC simulation analysis. In this section, we will provide a detailed introduction to Simbeor software in EMC simulation, as well as the principles and implementation methods of co-simulations analysis with ADS.

In the field of EMC simulation, Simbeor software assists engineers in analyzing the characteristic parameters of high-speed digital signal lines on PCB boards, such as transmission lines, microstrip lines, differential pairs, and power buses. It predicts the EMC performance and transmission characteristics of these lines in different electromagnetic environments. Simbeor software calculates parameters such as impedance, transmission rate, transmission distance, loss, clock jitter, and noise, helping engineers design better high-speed digital signal lines. Additionally, it simulates signal propagation and reflection, and analyzes signal integrity issues in various environments, providing corresponding solutions.

The principle of co-simulation analysis with ADS involves importing the model to be co-simulated into ADS software and creating a new project in ADS. A new circuit diagram is then generated, and Simbeor software is imported as a sub-circuit within the diagram. During the creation of the sub-circuit, input, and output ports, as well as simulation parameters, must be set. Subsequently, other circuit models are added to the diagram, and the signal lines requiring co-simulations are connected to the Simbeor sub-circuit. Simulations parameters are set, and the simulation is executed.

The simulation results from Simbeor software are outputted to ADS software for further analysis.

The implementation method and steps for co-simulations of Simbeor software and ADS are as follows:

1. Preparations: Complete the design of the signal circuit to be Co-simulated and import the circuit parameters into Simbeor software for simulation analysis. Save the simulation results locally.
2. Create a new project in ADS software and generate a new circuit diagram.
3. Import Simbeor software as a sub-circuit into the circuit diagram. While creating the sub-circuit, specify the input and output ports as well as simulation parameters.
4. Add other circuit models to be simulated and analyzed in the circuit diagram, connecting the signal lines requiring co-simulations to the Simbeor sub-circuit.
5. Set simulation parameters, including simulation type and duration, and run the simulation.
6. Export the simulation results to ADS software for further simulation and analysis.

It is important to note that accurate parameter settings and proper signal line connections are crucial for the co-simulations of Simbeor software and ADS. When building the Simbeor sub-circuit, ensure the input and output ports, as well as simulation parameters, are set according to the specific parameters of the signal line. When connecting the signal line to the Simbeor sub-circuit, ensure the physical connection matches the simulation configuration to guarantee the accuracy of the simulation results.

Co-simulations between Simbeor and ADS enhance the accuracy and reliability of EMC simulation analysis, particularly in the analysis of high-speed digital signal lines, signal integrity, PCB design, and EMC design optimization. It enables engineers to accurately predict EMI issues of electronic equipment in different electromagnetic environments, thus improving the performance and reliability of such equipment.

5.6 Co-simulations of EMC with HFSS

5.6.1 Co-simulations of Ansys SIwave with HFSS

Ansys SIwave and Ansys HFSS are two kinds of powerful electromagnetic simulation software tools that effectively address electromagnetic compatibility issues. Ansys SIwave is a kind of specialized software for signal integrity (SI), power integrity (PI), and electromagnetic compatibility (EMC) analysis. It is primarily used in high-speed circuit board (PCB) and packaging subsystem designs to predict, diagnose, and resolve signal integrity and power integrity problems in complex designs, thereby ensuring the performance and reliability of electronic products.

Before conducting EMC co-simulations, the PCB model to be analyzed must be prepared. PCB models can be exported from EDA software like Altium Designer and Cadence Allegro and imported into Ansys SIwave and HFSS as required. Additionally, material parameters and boundary conditions should be set based on the actual scenario.

In SIwave, the imported PCB model is analyzed for signal integrity, power integrity, and electromagnetic emissions using the following steps:

1. Set electromagnetic radiation parameters: Access SIwave's Engineering Settings, select the appropriate solver (such as FEM or MoM), define the frequency range and sweep step, and choose between Radiated Field or Near-Field Calculation.

2. Signal integrity analysis: Add the signal lines or differential pairs to be analyzed, set the ports and time-domain excitation signals, and perform the simulation. The simulation results allow the evaluation of signal integrity by observing parameters such as time-domain waveforms, transmission characteristics, and crosstalk of the signal lines.

3. Power integrity analysis: Add the power and ground grids to be analyzed, set the port, current, and voltage parameters, and conduct the simulation. The simulation results enable observation of power supply voltage distribution, power supply ground loop impedance, and the effect of decoupling capacitors, thereby evaluating power supply integrity.

In HFSS, the imported PCB model is analyzed for electromagnetic fields using the following steps:

1. Set boundary conditions and material parameters: Set the boundary conditions for importing the PCB model, such as radiation boundaries and waveguide ports, based on the actual scenario. Define the electromagnetic material parameters such as permittivity, permeability, and conductivity.

2. Set mesh and solver parameters: Access the "Analysis Settings" in HFSS, select the appropriate mesh type and solver (e.g., finite element solver), and configure parameters like frequency range and sweep step.

3. Electromagnetic field analysis: Examine simulation results for electromagnetic field distribution, S-parameters, electromagnetic radiation, and electromagnetic sensitivity to evaluate electromagnetic compatibility performance.

By performing co-simulations between SIwave and HFSS, a comprehensive evaluation of signal integrity, power integrity, and electromagnetic compatibility can be achieved through the following steps:

1. Export the SIwave simulation model: Export the simulation results of electromagnetic radiation from SIwave in a compatible format, such as CITI or MDIF.

2. Import the SIwave simulation model into HFSS: In HFSS, import the exported SIwave simulation model using the Import function and configure the corresponding boundary conditions, material parameters, and solver parameters.

3. Conduct co-simulations analysis: In HFSS, with the electromagnetic radiation simulation results from SIwave, analyze the electromagnetic field. The combined performance of signal integrity, power integrity, and electromagnetic compatibility can be observed in the simulation results.

Co-simulations of SIwave and HFSS enable a comprehensive assessment of signal integrity, power integrity, and electromagnetic compatibility, facilitating better design and optimization decisions.

5.6.2 Co-simulations of Altium Designer with HFSS

Altium Designer, as a kind of comprehensive circuit design software, offers powerful EMC simulation capabilities. It enables simulations for radiated and conducted noise, signal integrity, power integrity, and wiring. Radiated and conducted noise simulation is a crucial aspect of EMC design, allowing designers to quickly identify and resolve noise-related issues in the circuit, such as crosstalk, reflection, attenuation, power echo, and power jitter. Additionally, Altium Designer facilitates cabling simulations to help analyze signal integrity and power integrity problems associated with the wiring.

The co-simulations between Altium Designer and HFSS involve a common mode simulation based on a Simulink model. Specifically, Altium Designer converts the circuit design into a Simulink model, which is then transferred to HFSS for electromagnetic simulation. During the simulation process, Altium Designer and HFSS exchange data and control through the Simulink model, enabling co-simulations. The Simulink model serves as a graphical programming tool for establishing mathematical and simulation models of the circuit.

By converting the circuit design into a Simulink model, the circuit model can be easily transferred to HFSS for electromagnetic simulation. Throughout the simulation process, Altium Designer and HFSS exchange data and control via the Simulink model to achieve co-simulations. Specifically, after Altium Designer converts the circuit model into a Simulink model, it transmits the model to HFSS. HFSS then performs an electromagnetic simulation and returns the simulation results. Finally, Altium Designer feeds the simulation results back to the Simulink model for analysis and processing.

The co-simulations of Altium Designer and HFSS can be accomplished through the following steps:

1. Circuit design and EMC simulation: Use Altium Designer to design the circuit and conduct EMC simulation to obtain the simulation results. Altium Designer provides tools like the Signal Integrity Wizard and PDN Analyzer for simulating radiated and conducted noise, as well as a cabling simulation tool for analyzing cabling.

2. Convert the circuit design to a Simulink model: In Altium Designer, convert the circuit design to a Simulink model. This conversion process allows for selecting different parts of the circuit model to be converted, enabling partial or complete model conversion.

3. Transfer the Simulink model to HFSS for simulation: Transfer the Simulink model to HFSS for simulation. In HFSS, the Simulink model can be simulated, and simulation results can be obtained. During the simulation, set simulation parameters and conditions such as frequency and power.

4. Feed the simulation results back to the Simulink model: Feed the simulation results obtained from HFSS back to the Simulink model for analysis and processing. Within the Simulink model, simulation analysis tools can be used to analyze the results, including spectrum analysis, time-domain analysis, power analysis, etc.

5. Optimize the circuit design and perform additional simulations: Based on the simulation results within the Simulink model, optimize the circuit design by adjusting component parameters, optimizing wiring, etc. Then, repeat Steps 3–5 until satisfactory simulation results are achieved.

The co-simulation approach of Altium Designer and HFSS is relatively straightforward and involves circuit design and simulation, Simulink model conversion, Simulink model transfer to HFSS for simulation, feedback of simulation results to the Simulink model for analysis and processing, optimization of circuit design, and subsequent re-simulations. It is important to note that simulation parameters and conditions need to be appropriately set to ensure an accurate representation of the circuit's actual operation.

5.6.3 Co-simulations of HyperLynx with HFSS

HyperLynx, a kind of professional EMC simulation software developed by Mentor Graphics, is widely utilized for EMC simulation. It can be integrated with HFSS software through co-simulations, enabling more accurate simulation outcomes. This section provides a detailed overview of the application of HyperLynx software in EMC simulation, along with the principle, specific implementation method, and operational steps for conducting EMC co-simulations analysis with HFSS.

HyperLynx software allows for the creation of simulation models for circuit boards, IC packages, and entire electronic systems. It facilitates the analysis of electromagnetic compatibility issues and the prediction of their impact on system performance and reliability.

In EMC co-simulations, the integration between HyperLynx software and HFSS software is achieved through an interface. The underlying principle is as follows:

1. Building a HyperLynx simulation model

Begin by constructing a simulation model of the circuit board or IC package in HyperLynx software. Ensure that the required components and wiring are accurately incorporated into the model.

2. Exporting the HyperLynx Simulations Model

Export the HyperLynx simulation model in a file format accepted by HFSS, such as GDSII or STEP.

3. Establishing an HFSS simulation model

In HFSS software, recreate the same structure as the HyperLynx simulation model and configure the simulation parameters and material properties accordingly.

4. Importing the HyperLynx Simulations Model

Import the HyperLynx simulation model into HFSS software as part of the HFSS simulation model.

5. Completing the simulation analysis

Conduct the simulation analysis using the co-simulation results to evaluate the electromagnetic compatibility. The following are the specific implementation method and operational steps for EMC co-simulation analysis using HyperLynx software and HFSS software:

1. Building a HyperLynx simulation model

Construct a simulation model of the circuit board or IC package in HyperLynx software. Add components, wires, power supplies, etc., and configure simulation parameters and material properties as required.

2. Exporting the HyperLynx Simulations Model

Export the HyperLynx simulation model to a file format compatible with HFSS, such as GDSII or STEP.

3. Establishing an HFSS simulation model

Create a new project in HFSS software and build a simulation model with the same structure as the HyperLynx model. Set simulation parameters and material properties during the model creation.

4. Importing the HyperLynx Simulations Model

In HFSS software, select the option to import the HyperLynx simulation model. Choose the appropriate file format, such as GDSII or STEP, when importing the model.

5. Setting simulation parameters

After importing the HyperLynx simulation model, configure the simulation parameters. Specify parameters like simulation frequency, simulation time, and grid precision as per requirements.

6. Completing the simulation analysis

Once the simulation parameters are set, perform the simulation analysis and evaluate the electromagnetic compatibility issues using the co-simulations results. During the analysis process, leverage the visual analysis tools provided by HFSS software for result analysis and optimization.

When integrating HyperLynx software and HFSS software for co-simulations, various approaches can be employed, including file import, API interface, and EDA tool chain. Among these, file import is the simplest and most commonly used method.

5.7 Optimization Methods in EMC Analyses

5.7.1 Parameter Optimization

Parameter optimization aims to enhance system performance by adjusting the parameters within the system. Common methods for parameter optimization include sensitivity analysis, response surface method, and genetic algorithm. Sensitivity analysis calculates the impact of parameter changes on the objective function, allowing for the identification of important parameters and their optimization. The response surface method establishes a mathematical model linking the objective function and parameters, enabling the determination of optimal parameter values and reducing the need for extensive experimentation. Genetic algorithms, based on natural evolution principles, utilize gene mutation, crossover, and selection operations to obtain the optimal parameter combination. Sensitivity analysis is suitable for complex systems and objective functions, providing quick determination of important parameters. The response surface method predicts optimal parameter combinations but requires expertise and skill. Genetic algorithms are adept at handling parameter interactions and global optimization but necessitate extensive calculations, experiments, and expertise.

5.7.2 Structure Optimization

Structure optimization techniques involve adjusting the structure of an electronic system to improve its electromagnetic compatibility. This includes material optimization, geometric shape optimization, and layout optimization. Material optimization entails selecting appropriate materials through theoretical calculations and experimental testing to enhance electromagnetic compatibility. Geometric shape optimization focuses on reducing the size of structures and adjusting their shapes to minimize electromagnetic radiation and sensitivity, thereby improving system interference and radiation resistance. Finite element methods, optimization algorithms, and parametric modeling can be utilized for these optimization techniques. For example, in microwave filter design, optimizing performance can be achieved by reducing transmission line size and adjusting structure shape. Similarly, in antenna design, adjusting the shape and size can alter radiation direction and frequency response, leading to performance optimization.

5.7.3 Layout Optimization

Layout optimization involves adjusting the position and arrangement of components within a system to improve its EMC performance. The objective is to minimize mutual interference between components and decrease electromagnetic radiation and sensitivity. Commonly employed methods for layout optimization include finite element methods, optimization algorithms, and parametric modeling. In circuit board design, adjusting component position and arrangement can mitigate electromagnetic radiation and sensitivity, leading to layout optimization and improved performance. Microwave filter design utilizes structural optimization techniques such as size reduction and shape adjustment to enhance EMC. It is crucial to consider actual operating conditions, environmental factors, electromagnetic field coupling, nonlinear effects, convergence, and robustness of optimization algorithms during layout optimization. However, layout optimization techniques still face challenges in EMC analysis, such as multi-scale, nonlinear, and multi-physical field problems, which require innovative methods and techniques for resolution. In summary, structure optimization techniques are widely utilized and essential in EMC analysis, enabling the enhancement of system EMC performance.

5.7.4 Challenges in Optimization Processes

Before commencing the optimization process in EMC analyses, it is essential to define the objective function and constraints [45–47, 49]. The objective function represents the performance index to be optimized, while the constraints represent the conditions that must be satisfied within the system, such as power consumption, size limitations, and others. Defining the objective function and constraints ensures the effectiveness of the optimization results. The objective function typically comprises measured or simulated performance indicators like signal-to-noise ratio, transmission rate, anti-jamming capability, and more. Constraints can be categorized as hard or soft constraints. In complex systems, there may be multiple objective functions and constraints, requiring the consideration of multi-objective optimization algorithms and problem-solving methods.

The selection of an appropriate optimization algorithm depends on the problem's characteristics. Commonly used optimization algorithms include sensitivity analysis, response surface method, genetic algorithm, particle swarm optimization, differential evolution algorithm, simulated annealing algorithm, and others. Sensitivity analysis determines the impact of parameters on system performance, while the response surface method quickly identifies optimal parameter values. Genetic algorithms handle complex multi-parameter and nonlinear optimization problems, particle swarm algorithms are suitable for continuous and global optimization problems, and the differential evolution algorithm is effective for multi-parameter and nonlinear optimization problems. The simulated annealing algorithm is useful for complex multi-parameter and global optimization problems. The choice of optimization algorithm should consider the problem's characteristics, algorithm advantages and disadvantages, as well as convergence, search efficiency, and stability.

Uncertain factors should be considered during the optimization process. Uncertainty can be addressed by increasing the sample size, introducing probability distributions, and utilizing randomization. Increasing the sample size improves the statistical significance and reliability of the model, probability distributions better reflect the randomness of uncertainty factors, and randomization reduces their influence on the optimization results. Monte Carlo simulation is a method that considers uncertainty factors by obtaining a large number of samples through random sampling to accurately reflect their impact.

For multi-objective optimization problems, it is crucial to consider the relationship between objective functions and the feasibility of the optimization results. Solving multi-objective optimization problems involves combining multiple objective functions into a comprehensive index for optimization, searching for feasible solution sets, and determining Pareto optimal solutions. A Pareto optimal solution is one in which improving the values of some objective functions is impossible without degrading others. Multi-objective optimization problems can be addressed using algorithms like NSGA-II (Non-dominated Sorting Genetic Algorithm II) and MOGA (Multi-objective Genetic Algorithm). These algorithms identify Pareto optimal solutions or feasible solution sets, providing engineers with various solutions to choose from. When dealing with large-scale optimization problems, it is important to select efficient optimization algorithms and computational methods, as well as conduct reasonable parameter sampling and pruning of the optimization process. Methods for solving large-scale optimization problems include incremental optimization, parallel computing, and approximate computing. Incremental optimization involves optimizing a subset of parameters in the parameter space and gradually adjusting the parameter space based on optimization results until reaching the final solution. Parallel computing accelerates the optimization process by utilizing multiple computers or processors simultaneously.

5.8 Development and Prospect

In EMC analyses, artificial intelligence (AI) technology holds significant potential for widespread application [49, 50]. AI can automate and optimize EMC testing, leading to reduced testing time and costs, improved testing accuracy and efficiency. Additionally, AI can optimize the configuration of test environments and equipment. In EMC design, AI can enhance electromagnetic compatibility by leveraging deep learning algorithms for design and circuit optimization. Moreover, AI can expedite electromagnetic field calculations, improve accuracy, and predict and fit the behavior and effects of electromagnetic fields. Despite the extensive applications of AI in EMC analysis, challenges such as data acquisition and processing, algorithm interpretability, robustness, and universality need to be addressed through future research and the development of new algorithms and methods.

The application of new materials and technologies is increasingly crucial in EMC analysis. Materials like electromagnetic wave absorption materials, impedance matching materials, and flexible conductive materials show promise in suppressing electromagnetic radiation, enhancing absorption, and reducing electromagnetic interference. Technologies such as wireless charging, 5G communication, and optical communication also have potential in EMC analysis but face challenges related to cost, performance stability, and interference limitations. Continued research and development of new materials and technologies will expand their application potential in EMC analysis, offering more efficient and accurate methods to improve the electromagnetic compatibility of electronic products.

The coupling of multiphysics fields is a prominent research topic in the EMC field. Multiphysics field coupling refers to the interaction and interference caused by different physical fields, including electromagnetic, thermal, mechanical, and fluid fields. Researchers are focused on establishing multiphysics coupling models and simulation software, as well as developing simulation technologies. Integrating experiments and simulations and advancing new multiphysics coupling theories and methods are also key research directions. Investigating multiphysics coupling problems holds theoretical significance and practical value. Emerging technology fields such as 5G communication, autonomous driving, Internet of Things, and new energy pose new challenges for electromagnetic compatibility. Meeting these challenges requires addressing higher design requirements, stricter interference limits, and ensuring reliable operation in complex electromagnetic environments. Thus, researchers need to develop new EMC analysis methods and techniques to enhance the performance and reliability of electronic devices in these complex environments. EMC analysis technology is continuously evolving, necessitating ongoing exploration and development of new methods and technologies. Researchers must strive to meet these challenges, improve the electromagnetic compatibility of electronic equipment, and ensure their normal operation in complex electromagnetic environments.

5.9 Simulations of RF Links Based on CST and ADS

5.9.1 Overall Framework of Simulations

In the design of modern electronic devices, RF link design is a crucial task and its performance impacts the performance of the entire system. With the increase of frequency, the electromagnetic compatibility issue of the circuit is becoming more severe, therefore, accurate simulation analysis of RF links is a necessary means to ensure its performance. This article mainly introduces how to perform PCB board-level RF link simulation based on CST and ADS software. The following content mainly includes the confirmation and export of PCB files, the import and simulation of CST, the simulation of ADS, and the confirmation of actual measurement steps (Fig. 5.3).

5.9.2 PCB Preparation

Firstly, the platform selection: we opt to export the ODB++ format from PCB software due to its universality, which can be used in platforms such as AD, PADS, and Cadence. Of course, if preferred, one can independently study its exclusive format based on CST's better support for PADS and Cadence.

In the step of confirming RF trace routing, the Si9000 software is utilized for the simulation of coplanar waveguide routing. According to the manufacturer's provided information, appropriate modifications will be made to the RF trace width in the PCB. Regarding the PCB export, in AD software, you can directly export ODB++ files.

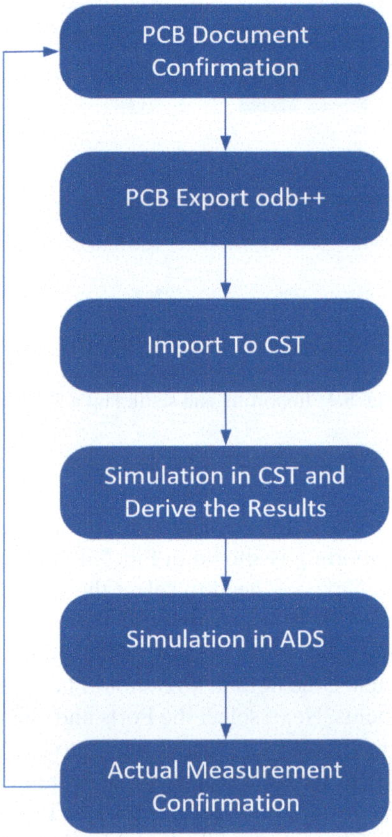

Fig. 5.3 Overall framework of simulations based on CST and ADS

After the export, two compressed files will be added to the project file directory, and we can choose to import the .zip file. In PADS, one can choose to export in .asc or ODB++ format. It's important to note that high-version PADS might not support .asc format.

5.9.3 CST Imports and Simulations

When importing into CST software, the specific requirements of the simulation need to be considered. If you want to simulate stray full-band problems, you will choose the time domain; if you want to simulate matching problems within a specific frequency band alone, you will choose the frequency domain. Generally speaking, as long as the grid is set properly, the results of time-domain calculations are basically the same as those in the frequency domain, and the simulation speed is also very fast. Now, you can start importing the previously exported odb++ files. After importing,

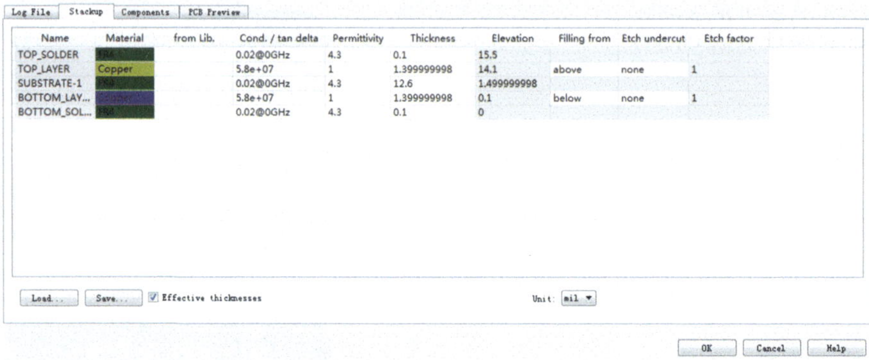

Fig. 5.4 Importing and setting up a RF wiring in CST

you can set the frequency in the CST software. The narrower the frequency band of frequency-domain simulation, the faster the simulation speed. The effect after importing is shown in Fig. 5.4, and uncheck the option at the Import components.

Next, we need to select the desired network. If there are multiple networks, you can locate them in the schematic and then copy them for filtering in the Filter section. Here, we select the size of the circuit board to be cut. If there are no specific requirements, the default size should be sufficient to meet the simulation accuracy requirements. Next, select the Ports and check the "Only show pins of selected nets" option. Import all the selected pins into the box on the right. Additionally, we also need to select the reference network. Again, import all the selected reference networks into the right box. Finally, click OK to start the simulation (Fig. 5.5).

The contents displayed in CST after importing are as follows. The software has already set up the simulation ports. The frequency of the simulation can be modified in the Frequency section. To speed up the simulation, you can adjust the size of the air box in the background. On the side closer to the ground, you can change "10" to "0." Select "setup solver" and then choose "start" to begin the simulation. The simulation time depends on the size of the board and the computational performance (memory and CPU frequency). After the simulation is complete, the Touchstone will be highlighted. Click on it and select the default configuration to obtain an SnP file. "N" represents the number of ports in the simulation (Fig. 5.6).

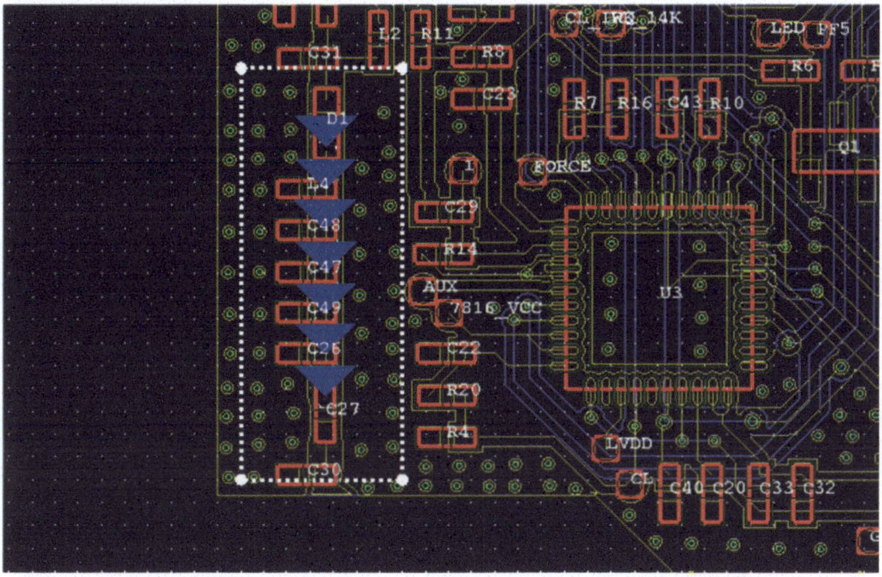

Fig. 5.5 Size selection of circuit board

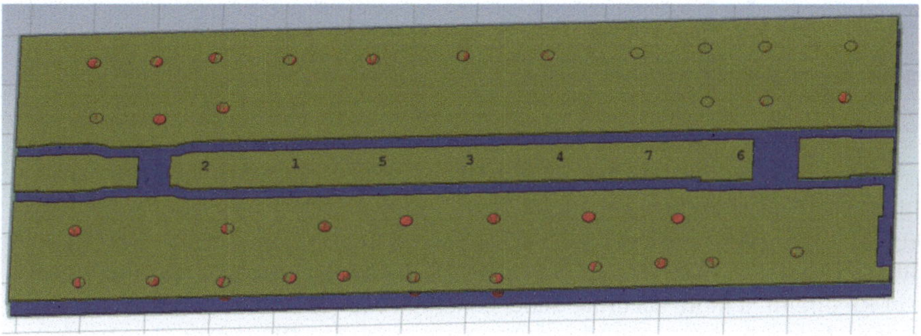

Fig. 5.6 Establishment of CST model

5.9.4 ADS Simulations

Installation steps for ADS and how to install Murata's library will not be described in detail here. In ADS, two ports are connected to ports 2 and 6, respectively. The file called s7p in the folder is invoked in SnP (which can be searched directly in the search bar in the top left corner). Ports 4 and 5 are connected to the actual test results. By selecting the simulation button, the results can be obtained. Here, we can directly select templates in the insert menu or manually add them. The accurate simulation of the S_{21} and S_{12} parts in the simulation process has always been challenging, and it is greatly influenced by the properties of the material (Fig. 5.7).

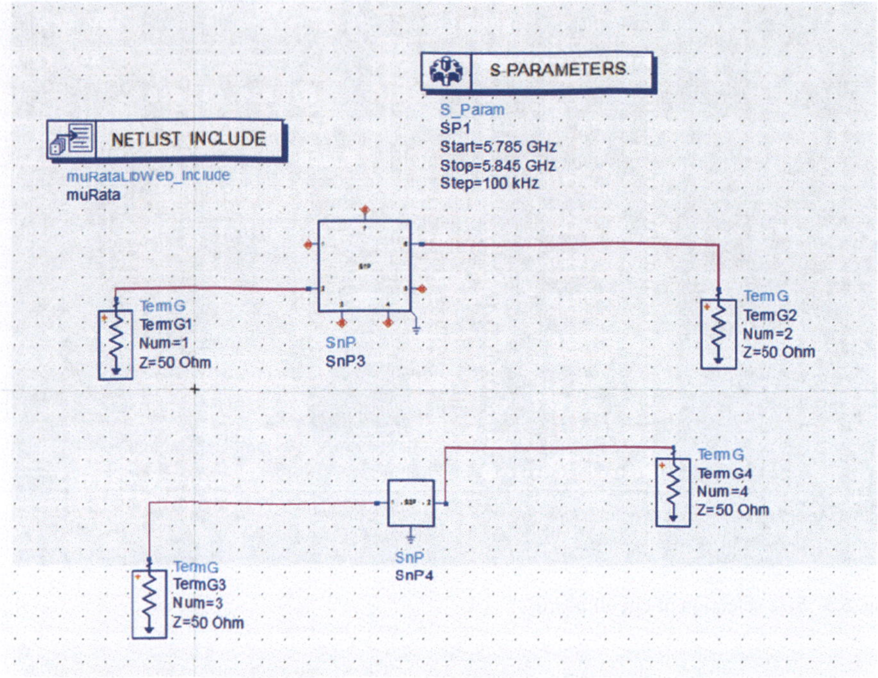

Fig. 5.7 Circuit configuration in ADS

If there is a need for matching in the circuit, please add the SnP file provided by Murata or other component manufacturers to the required ports. In the frequency range below 3.0 GHz, it is generally possible to achieve simulation results consistent with test results. By conducting CST and ADS simulation analysis on the PCB board-level RF circuit, a more accurate understanding of its performance can be obtained, enabling better decision-making in the design process. RF circuit simulation is a complex process that requires a deep understanding of circuit design, simulation software, and electromagnetic theory (Fig. 5.8).

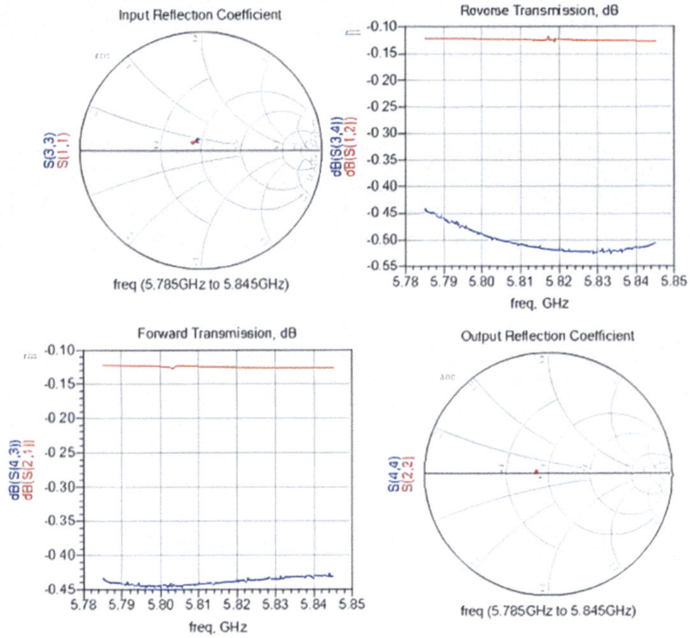

Fig. 5.8 ADS simulation result

Problems

P5.1 Co-simulation of ADS with Simbeor

When it comes to PCB layer stackup and Signal Integrity (SI) impedance modeling, obtaining accurate dielectric material properties from reliable sources is crucial. One important parameter for precise impedance modeling is the relative permittivity (ϵ_r), also known as the dielectric constant (Dk). The effective dielectric constant (Dkeff) can be approximated using the following equation [51]:

$$Dk_{\text{eff}} \approx \frac{H_{\text{smooth}}}{(H_{\text{smooth}} - 2R_z)} \times Dk$$

- H_{smooth} represents the dielectric core thickness obtained from the laminate supplier's *Dk*/*Df* table data sheet or the pressed prepreg thickness from the PCB stackup drawing.
- R_z denotes the 10-point mean roughness measured by a mechanical profilometer.
- D_k refers to the dielectric constant obtained from the laminate supplier's *Dk*/*Df* table data sheet.

In this exercise, the variation of Dk_{eff} with frequency is simulated using the causal Huray-Bracken roughness model. Simbeor is utilized for modeling, and Keysight ADS is used for simulation purposes [52].

P5.2 Dynamic linking of SIwave as a radiation source to HFSS for shielding analysis

In this exercise, the Data Link feature is utilized to establish collaborative analysis between HFSS and SIwave, Ansoft's signal integrity analysis tool. SIwave is used to analyze the internal printed circuit board (PCB) within the aircraft body, and the resulting analysis outcomes are employed as a radiation source for electromagnetic field analysis in HFSS.

References

1. Ott HW (2011) Electromagnetic compatibility engineering. John Wiley & Sons
2. Paul CR, Scully RC, Steffka MA (2022) Introduction to electromagnetic compatibility. John Wiley & Sons
3. Paul CR, Scully RC, Steffka MA (2022) Introduction to electromagnetic compatibility. John Wiley & Sons
4. Boyer A, Dhia BS, Sicard E (2007) Characterization of the electromagnetic susceptibility of integrated circuits using a near field scan. Electron Lett 43(1):15
5. Chung DDL (2001) Electromagnetic interference shielding effectiveness of carbon materials. Carbon 39(2):279–285
6. Kaur M, Kakar S, Mandal D (2011) Electromagnetic interference. In: 2011 3rd international conference on electronics computer technology, vol 4. IEEE
7. Kaur M, Kakar S, Mandal D (2011) Electromagnetic interference. In: 2011 3rd international conference on electronics computer technology, vol 4. IEEE
8. Savage E, Radasky W (2012) Overview of the threat of IEMI (intentional electromagnetic interference). In: 2012 IEEE international symposium on electromagnetic compatibility. IEEE
9. Eliardsson P et al (2015) Military HF communications considering unintentional platform-generated electromagnetic interference. In: 2015 international conference on military communications and information systems (ICMCIS). IEEE
10. Jakubowski K, Paś J, Rosiński A (2021) The issue of operating security systems in terms of the impact of electromagnetic interference generated unintentionally. Energies 14(24):8591
11. Boyer A, Dhia BS, Sicard E (2007) Characterization of the electromagnetic susceptibility of integrated circuits using a near field scan. Electron Lett 43(1):15
12. Xie H et al (2010) Spice models for radiated and conducted susceptibility analyses of multi-conductor shielded cables. Prog Electromagn Res 103:241–257
13. Leone M (2005) Radiated susceptibility on the printed-circuit-board level: simulations and measurement. IEEE Trans Electromagn Compat 47(3):471–478
14. Javor ER, Anderson T (1998) Design of a Helmholtz coil for low frequency magnetic field susceptibility testing. In: 1998 IEEE EMC symposium. international symposium on electromagnetic compatibility. Symposium record (Cat. No. 98CH36253), vol 2. IEEE
15. Herrmann GF (1963) Resonance and high frequency susceptibility in canted antiferromagnetic substances. J Phys Chem Solids 24(5):597–606
16. Crovetti PS, Fiori F (2011) Distributed conversion of common-mode into differential-mode interference. IEEE Trans Microw Theory Tech 59(8):2140–2150
17. Jayasree PVY et al (2012) EMI filter design for reducing common-mode and differential-mode noise in conducted interference. Int J Electron Commun Eng 5(3):319–329

18. Miloudi M, Bendaoud A, Miloudi H (2017) Common and differential modes of conducted electromagnetic interference in switching power converters. Revue Roumaine Science Technique-Électrotechnique et Énergétique 62(3):246–251
19. Celozzi S et al (2023) Electromagnetic shielding: theory and applications. John Wiley & Sons
20. Grcev L, Dawalibi F (1990) An electromagnetic model for transients in grounding systems. IEEE Trans Power Deliv 5(4):1773–1781
21. Kaiser KL (2005) Electromagnetic Shielding. CRC Press
22. Cangellaris AC (1996) Frequency-domain finite element methods for electromagnetic field simulation: fundamentals, state of the art, and applications to EMI/EMC analysis. In: Proceedings of symposium on electromagnetic compatibility. IEEE
23. Wu TL, Buesink F, Canavero F (2013) Overview of signal integrity and EMC design technologies on PCB: fundamentals and latest progress. IEEE Trans Electromagn Compat 55(4):624–638
24. Perez R (ed) (2013) Handbook of electromagnetic compatibility. Academic Press
25. Guru BS, Hiziroglu HR (2009) Electromagnetic field theory fundamentals. Cambridge University Press
26. Thidé B (2009) Electromagnetic field theory
27. Sato T, Watanabe R, Niita K (2006) Development of a calculation method for estimating specific energy distribution in complex radiation fields. Radiat Prot Dosim 122(1–4):41–45
28. Prokhorets MI et al (2007) Point-Kernel method for radiation fields simulation. Probl Atom Sci Tech
29. Carobbi CFM (2016) Bayesian inference in action in EMC-fundamentals and applications. IEEE Trans Electromagn Compat 59(4):1114–1124
30. Sevgi L (2017) A practical guide to EMC engineering. Artech House
31. Tesche FM, Ianoz M, Karlsson T (1996) EMC analysis methods and computational models. John Wiley & Sons
32. Weston DA (2016) Electromagnetic compatibility: methods, analysis, circuits, and measurement. CRC Press
33. Bruns HD, Schuster C, Singer H (2007) Numerical electromagnetic field analysis for EMC problems. IEEE Trans Electromagn Compat 49(2):253–262
34. Leferink F, Keyer C, Melentjev A (2016) Static energy meter errors caused by conducted electromagnetic interference. IEEE Electromagn Compat Mag 5(4):49–55
35. Keiser BE (1979) Principles of electromagnetic compatibility. Dedham
36. Li E-P et al (2010) Progress review of electromagnetic compatibility analysis technologies for packages, printed circuit boards, and novel interconnects. IEEE Trans Electromagn Compat 52(2):248–265
37. Tellini B et al (2001) Conducted and radiated interference measurements in the line-pantograph system. IEEE Trans Instrum Meas 50(6):1661–1664
38. Weston DA (2016) Electromagnetic compatibility: methods, analysis, circuits, and measurement. CRC Press
39. Rajasekaran K, Jayalakshmi J, Jayasankar T (2013) Design and analysis of stepped impedance microstrip low pass filter using ADS simulation tool for wireless applications. Int J Sci Res Publ 3(8):1–5
40. Rondon-Pinilla E et al (2013) Modeling of a buck converter with a SiC JFET to predict EMC conducted emissions. IEEE Trans Power Electron 29(5):2246–2260
41. Antonini G, Cristina S, Orlandi A (1996) EMC characterization of SMPS devices: circuit and radiated emissions model. IEEE Trans Electromagn Compat 38(3):300–309
42. Bouesse GF et al (2007) Asynchronous logic Vs Synchronous logic: concrete results on electromagnetic emissions and conducted susceptibility. In: 6th international workshop on electromagnetic compatibility of integrated circuits (EMC Compo'07)
43. Badini L et al (2017) A rationale for statistical correlation of conducted and radiated susceptibility testing in aerospace EMC. IEEE Trans Electromagn Compat 59(5):1576–1585
44. Omid M, Kami Y, Hayakawa M (1997) Field coupling to nonuniform and uniform transmission lines. IEEE Trans Electromagn Compat 39(3):201–211

45. Tesche FM, Ianoz M, Karlsson T (1996) EMC analysis methods and computational models. John Wiley & Sons
46. Deschrijver D et al (2011) Automated near-field scanning algorithm for the EMC analysis of electronic devices. IEEE Trans Electromagn Compat 54(3):502–510
47. Capraro GT (1981) An intrasystem EMC analysis program. In: AGARD Electromagnetic Compatibility 22 p (SEE N82-10306 01-32)
48. Slama JBH et al (2009) Relevant parameters of SPICE3 MOSFET model for EMC analysis. In: 2009 IEEE international symposium on electromagnetic compatibility
49. Choo V et al (1994) Implementation of intelligent EMC analysis and design techniques. In: Proceedings of IEEE symposium on electromagnetic compatibility
50. Surekha M et al (1990) Frame-based knowledge systems for EMC analysis. In: IEEE international symposium on electromagnetic compatibility
51. https://www.signalintegrityjournal.com/articles/2573-a-tale-of-two-data-sheets-part-1-what-you-need-to-know-about-dielectric-constant
52. Simonovich B PCB fabrication: what SI/PI engineers need to know for first time modeling success. In: DesignCon 2021 spring break webinar series, April 12–16, 2021

Chapter 6
Co-simulations of Thermal Properties

6.1 Introduction

Thermal simulations refer to the numerical simulations aiming to the thermodynamic properties of devices or systems, usually including the temperature distribution and thermal transfer process based on thermal fields. The thermal simulations play a crucial role in predicting and evaluating various thermal performance parameters, such as temperature, thermal stress, and thermal loss, under different working conditions, so that the thermodynamic characteristics of analyzed devices or systems can be made clear. This is achieved through the analysis for the fundamental modes of thermal transfer, i.e., conduction, convection, and radiation based on the results of thermal simulations [1–3].

Thermal simulations have been widely used in modern engineering designs and scientific research since they can offer valuable insights into the thermal performances of various devices or systems under different operating conditions. The typical applications include the cooling design of electronic components, thermal design of mechanical equipment, thermal-insulation design of buildings, thermal dissipation design of aerospace devices, and temperature-control design of medical instruments and equipment [2, 4–7]. One of the key benefits from the thermal simulations is that the designers can quickly know the effects of thermal designs so that they can take an action immediately. Based on the results of thermal simulations, engineers or researchers can predict the change of thermal parameters under diverse working conditions, facilitating the design of thermal systems or investigation of thermal mechanisms. Also, the thermal simulations can significantly enhance the efficiency and reduce the costs. Traditional methods often rely on numerous experiments to investigate the design schemes and validate design effects, which will consume much time and resource. In contrast, the simulation methods can save much time and resource while keeping good results. The thermal simulations can help engineers or researchers identify potential problems that might otherwise be overlooked. Since the modern software tools used in the thermal simulations have

M. S. Tong and X. Y. Li, *Co-simulations of Microwave Circuits and High-Frequency Electromagnetic Fields*, https://doi.org/10.1007/978-981-99-8307-0_6

become more and more intelligent and powerful, they can automatically identify potential problems in designs or investigations in advance so that unfavorable results can be avoided early. Furthermore, the thermal simulations can also provide more accurate data. The conducted experiments in the traditional methods often have many uncontrollable factors during the implementation processes so that the experimental results usually include many unpredictable errors. In contrast, the thermal simulations can give rise to much more precise and reliable results since their errors can be predicted and controlled, and many optimization methods can be used to improve the accuracy [8–10].

The thermal simulations can provide an indispensable support for engineering designs and scientific research by enabling engineers and researchers to quickly predict and assess the thermal performances of objects under various complicated conditions. Such a support has the characteristics of high efficiency, high accuracy, low costs, ability of identifying potential problems, and ability of optimizing designs [11, 12]. This chapter will introduce the methods of thermal simulations based on different kinds of advanced software. It also includes the co-simulations for the interaction between thermal simulations and electromagnetic simulations, which is needed to predict the impact of thermal performances on electromagnetic performances.

6.2 Definitions and Applications of Thermal Parameters

6.2.1 Thermal Transfer

Thermal conduction is the process of thermal transfer within a substance, occurring between regions with different temperatures [13, 14]. It always flows from a higher temperature region to a lower temperature region. Among the three fundamental modes of thermal transfer, thermal conduction is the most common and significant [15]. According to Fourier's law of thermal conduction, the rate of thermal conduction is directly proportional to the temperature gradient and inversely proportional to the thermal conductivity of the medium through which the thermal is transferred. This relationship can be expressed as follows:

$$q = -k\frac{dT}{dx} \tag{6.1}$$

where q represents the thermal flux, k is the thermal conductivity of the material, T denotes the temperature, and x represents the direction of thermal transfer. Thermal conductivity k is a physical property that characterizes a substance's ability to conduct thermal, and its magnitude is determined by the material's intrinsic thermal conductive properties.

6.2.2 Thermal Convection

Thermal convection refers to the process of thermal transfer resulting from fluid flow, where thermal is transferred through the movement of the fluid [15, 16]. Thermal convection is typically categorized into natural convection and forced convection. Natural convection occurs due to changes in fluid density and volume caused by temperature variations, while forced convection arises from fluid flow driven by external forces and temperature differences. The mathematical representation of thermal convection can be described using Newton's law of cooling, given by

$$q = hA(T - T_{\infty}) \tag{6.2}$$

where q represents the thermal flux, h is the thermal transfer coefficient, A denotes the thermal transfer area, T is the surface temperature, and T_{∞} is the fluid temperature. The thermal transfer coefficient h is a parameter that characterizes the thermal transfer performance of a fluid, and its magnitude is influenced by factors such as fluid's thermal conductivity, density, and flow rate.

6.2.3 Thermal Radiation

Thermal radiation refers to the process in which the electromagnetic waves, emitted from the surface of an object due to its temperature, propagate through air or vacuum, and are absorbed and re-emitted by other objects [17, 18]. Unlike conduction and convection, thermal radiation can transfer thermal directly without the need for a medium, allowing thermal transfer even in a vacuum. According to Stefan-Boltzmann law, the thermal flux associated with thermal radiation is proportional to the fourth power of the surface temperature of an object:

$$q = \epsilon\sigma A(T^4 - T_{\infty}^4) \tag{6.3}$$

where q represents the thermal flux, ϵ is the emissivity of the surface, σ is the Stefan-Boltzmann constant, A is the surface area, T is the surface temperature, and T_{∞} is the ambient temperature.

Thermal simulation finds wide-ranging applications, including thermal dissipation design for electronic components, thermal design of mechanical equipment, building thermal-insulation design, thermal dissipation design for aerospace devices, and temperature control of medical equipment. In electronic component thermal dissipation design, thermal simulation enables the prediction of temperature distribution and thermal loss under various working conditions, assisting engineers in making informed design decisions. In mechanical equipment design, thermal simulation helps analyze temperature distribution and thermal stress under different operating conditions, ensuring equipment remains within safe temperature limits for extended periods. For architectural design, thermal simulation allows for prediction of temperature distribution and thermal flow through building exterior walls in different environments, facilitating effective thermal-insulation design. Thermal simulation is also useful in analyzing thermal dissipation design for aerospace devices, predicting temperature distribution and thermal loss under various operating conditions, and aiding in the development of appropriate designs. In the case of medical devices, thermal simulation assists in predicting temperature distribution and thermal loss under different working conditions, enabling engineers to design effective temperature-control systems that ensure safe operation without harm to the human body [19–22].

6.3 Types and Technologies of Thermal Simulations

Thermal simulations can be classified into different types based on different criteria, such as the objects of simulations, the scopes of simulations, and the purposes of simulations. The following sections give the detailed types and technologies of thermal simulations according to these criteria [23].

6.3.1 Classification by Simulated Objects

Thermal simulations can be divided into the following two categories based on the object of simulation:

1. Independent object thermal simulation: This type of simulation focuses on the thermal analysis of a single object or device. It requires consideration of the object's three-dimensional geometry, material properties, thermal source location, boundary conditions, and other factors.

2. System-level thermal simulation: System-level thermal simulation aims to analyze the thermal behavior of a system consisting of multiple independent objects, such as the thermal dissipation system of an automobile engine or the thermal flow distribution inside an electronic equipment cabinet. System-level thermal simulation takes into account factors like thermal transfer and exchange between multiple objects, as well as thermal exchange between objects and the environment.

6.3.2 Classification by Scopes of Simulations

Thermal simulations can be divided into the following two categories based on the scope of simulation:

1. Local thermal simulations: This type of simulation focuses on analyzing the thermal behavior of a specific part of an object, such as an electronic component chip

or a particular area of a radiator. Local thermal simulation considers the local geometry of the object, material properties, thermal source location, boundary conditions, and other relevant factors.

2. Global thermal simulations: Global thermal simulation aims to analyze the thermal behavior of the entire object or system, such as the cooling system of an automobile engine or the thermal-insulation system of building exterior walls. Global thermal simulation takes into account the overall geometry of the object, material properties, thermal source location, boundary conditions, and other relevant factors.

6.3.3 Classification by Purposes of Simulations

Thermal simulation can be divided into the following two categories based on the purpose of simulation:

1. Design optimization simulations: Design optimization simulation involves evaluating and optimizing the performance of products or systems using thermal simulation during the design phase. It helps engineers predict the temperature distribution, thermal stress distribution, and thermal flow distribution under different conditions, enabling them to optimize design schemes, reduce costs, and improve performance. Popular thermal simulation software for design optimization includes ANSYS, COMSOL, STAR-CCM+, and others.

2. Fault diagnosis simulations: Fault diagnosis simulation aims to identify the cause of a failure in a product or system using thermal simulation. It helps engineers determine the causes and solutions for faults by analyzing the temperature distribution, thermal stress distribution, thermal flow distribution, and other relevant information about the faulty location during the troubleshooting process. Commonly used thermal simulation software for fault diagnosis includes FLUENT, FloEFD, and others.

6.3.4 Technologies of Thermal Simulations

Thermal simulation refers to a series of techniques used to simulate and analyze the thermal transfer process of objects or systems. Currently, commonly used thermal simulation includes the following [24–27]:

1. Solution of the heat conduction equation

The solution of the heat conduction equation involves thermal simulation analysis based on the heat conduction equation for objects or systems. Heat conduction equations can describe the internal thermal transfer process of objects or systems. Commonly used heat conduction equations include the heat conduction equation, convection heat conduction equation, thermal radiation equation, and others. The solution of the heat conduction equation requires consideration of the object or system's geometry, material properties, thermal source location, boundary conditions, and other factors.

2. Finite element method

The finite element method is a numerical analysis-based thermal simulation. It divides a complex object or system into smaller elements, solves the heat conduction equation for each element, and then conducts thermal simulation analysis of the entire object or system. The finite element method considers the object or system's geometry, material properties, thermal source location, boundary conditions, and other factors.

3. Computational fluid dynamics

Computational fluid dynamics (CFD) is a thermal simulation based on numerical analysis. It treats the fluid and thermal transfer processes inside the object or system as a whole and conducts thermal simulation analysis using numerical simulation methods. The computational fluid dynamics method considers the dynamic characteristics of the fluid, thermal transfer characteristics, the object or system's geometry, material properties, thermal source location, boundary conditions, and other factors.

4. Monte Carlo method

The Monte Carlo method is a thermal simulation based on stochastic mathematics. It simulates the internal thermal transfer process of the object or system using random sampling, enabling thermal simulation analysis of the entire object or system. The Monte Carlo method takes into account the object or system's geometry, material properties, thermal source location, boundary conditions, and other factors. Popular thermal simulation software for this method includes MCNP, Geant4, and others.

6.4 Principle of Thermal Simulations

6.4.1 Heat Conduction Equation

Thermal conduction is a fundamental mechanism of thermal transfer in matter and a core aspect of thermal simulation. The heat conduction equation is a mathematical model that describes the process of thermal conduction and serves as a theoretical foundation for thermal simulation [28–32]. The heat conduction equation is used to describe the change in temperature distribution within a body [33–38]. Its basic form is given by

$$\frac{\partial T}{\partial t} = \alpha \nabla^2 T \tag{6.4}$$

where T represents the temperature field, t denotes time, α is the thermal diffusivity, and $\nabla^2 T$ represents the Laplacian of the temperature field. Thermal conduction arises from molecular interactions, where thermal is transferred between molecules due to temperature differences. Assuming the presence of a temperature gradient within an object, molecules in the high-temperature region possess higher energy compared to those in the low-temperature region. This energy difference drives the transfer of energy from the high-temperature region to the low-temperature region until equilibrium is reached. Consequently, thermal conduction is the process by which thermal flows from regions of higher temperature to regions of lower temperature.

Consider a small region inside an object. Let the temperature within the region be T, and suppose there is a small piece of matter with a length of ΔX inside the region, with a temperature of $T + \Delta T$. The amount of thermal change within this small piece of matter can be expressed as follows:

$$Q = \rho c_p \Delta x \Delta T \tag{6.5}$$

where ρ represents the density of the substance, and c_p denotes the specific thermal capacity of the substance. Applying the first law of thermodynamics, we have

$$\frac{\partial Q}{\partial t} = \frac{\partial(\rho c_p \Delta x \Delta T)}{\partial t} \tag{6.6}$$

Dividing both sides of the equation by ΔX and taking the limit, we obtain

$$\frac{\partial Q}{\partial t} = \rho c_p \frac{\partial T}{\partial t} \Delta x \tag{6.7}$$

As ΔX approaches 0, the equation becomes

$$\frac{\partial Q}{\partial t} = \rho c_p \frac{\partial T}{\partial t} dx \tag{6.8}$$

Through further transformations, we arrive at

$$\frac{\partial T}{\partial t} = \frac{1}{\rho c_p} \frac{\partial Q}{\partial t} \tag{6.9}$$

According to the fundamental law of thermal conduction, the rate of thermal transfer from a region of higher temperature to a region of lower temperature is proportional to the temperature gradient. This relationship is described by Fourier's law:

$$\mathbf{q} = -k \nabla T \tag{6.10}$$

where $\mathbf{q}$ represents the thermal flux passing through a unit area per unit time, k is the thermal conductivity coefficient, and ∇T denotes the temperature gradient. As per the principle of thermal conduction, the direction of $\mathbf{q}$ is opposite to the temperature gradient, indicating that thermal flows from regions of higher temperature to regions of lower temperature. Consider a small region V with a volume of ΔV and a surface area of ΔS. The thermal flow through this region can be expressed as follows:

$$Q = -k \Delta S \nabla T \Delta t \tag{6.11}$$

By rearranging the equation, we obtain

$$\frac{Q}{\Delta V \Delta t} = -k\frac{\Delta S}{\Delta V}\frac{\nabla T}{\Delta x} \tag{6.12}$$

As ΔV approaches 0, the equation becomes

$$\frac{\partial Q}{\partial V} = -k\nabla^2 T \tag{6.13}$$

where $\nabla^2 T$ represents the Laplacian of the temperature field and signifies the second spatial derivative of the temperature. Substituting (6.13) into (6.9), we obtain

$$\frac{\partial T}{\partial t} = \alpha\nabla^2 T \tag{6.14}$$

where $\alpha = \frac{k}{\rho c_p}$ represents the thermal diffusivity.

To sum up, the basic form of the heat conduction equation is (6.14), which describes the temporal variation of the internal temperature distribution within an object.

The heat conduction equation possesses several important properties that significantly contribute to the understanding and application of thermal simulation. The heat conduction equation is a partial differential equation that describes the temporal evolution of the object's internal temperature distribution. Solving the heat conduction equation analytically is challenging, and numerical methods are required for its solution. The equation exhibits the principle of linear superposition, meaning that by applying different boundary conditions and initial conditions, the total temperature distribution of the system can be obtained through superposition. Furthermore, the heat conduction equation remains invariant with respect to time and space. This implies that for identical boundary and initial conditions, solutions to the heat conduction equation exhibit similarities across different time and spatial scales.

Due to the heat conduction equation being a partial differential equation, exact analytical solutions are typically unattainable through analytical methods, necessitating the use of numerical techniques. Commonly employed numerical methods include the finite difference method, finite element method, and boundary element method. The finite difference method transforms the continuous partial differential equation into discrete difference equations for solution. It discretizes the computational domain into a grid of discrete points and computes the temperature changes at each grid point. By employing difference approximations in both time and space, a set of difference equations is derived. The numerical solution of the temperature distribution is obtained through iterative solution of these difference equations. The finite element method decomposes the computational domain into a series of small finite elements. It establishes a set of basis functions that represent the temperature variations within each finite element, and solves the resulting algebraic equations to obtain the numerical solution of the temperature distribution. The finite element method offers high accuracy and wide applicability, making it suitable for solving thermal simulation models involving complex geometric shapes and nonlinear materials. The boundary element method decomposes the computational domain into a series of small boundary elements. It formulates the integral equation for temperature distribution on each boundary element and solves these integral equations to obtain the numerical solution of the temperature distribution. The boundary element method is characterized by its high efficiency, accuracy, and low computational complexity, making it suitable for solving thermal simulation models with special geometries.

The heat conduction equation represents a core problem in thermal simulation. By employing appropriate numerical methods to solve it, we can obtain the temporal variation of temperature distribution within an object, providing an essential theoretical basis for practical engineering applications.

6.4.2 Boundary and Initial Conditions

Boundary conditions and initial conditions are vital aspects of thermal simulation models that greatly impact the simulation and calculation results of the thermal conduction process. In this section, we will provide a detailed introduction to boundary conditions and initial conditions in thermal simulation models. This includes their definitions, classifications, specific implementation methods, and factors that influence them. Boundary conditions pertain to the thermal input and output within the thermal simulation model. They encompass temperature boundary conditions, thermal flow boundary conditions, and convection boundary conditions. The specification of boundary conditions directly affects the solution of the heat conduction equation and the accuracy of the calculation results. Therefore, the selection and setting of boundary conditions hold immense importance in thermal simulation models.

Temperature boundary conditions involve fixing the temperature of a surface to a constant value within the thermal simulation model. This is commonly used to describe the temperature at the boundaries. Temperature boundary conditions generally take two forms. First, there is the temperature fixed boundary condition, where specific surfaces or points in the model have their temperatures fixed. Second, there is the periodic boundary condition, where the temperature of two adjacent surfaces in the model is equal. This can be used to simulate objects with a periodic structure. The mathematical expression for temperature boundary conditions is given by

$$T(x, y, z, t) = T_0 \tag{6.15}$$

where T_0 represents the fixed temperature value, and x, y, z denote spatial positions, while t represents time. Temperature boundary conditions are typically employed to describe the boundary temperature of solid objects. For example, in the design of thermal dissipation for electronic components, temperature boundary conditions can be used to represent the surface temperature of the radiator.

Thermal flow boundary conditions involve fixing the thermal flow of a surface to a constant value within the thermal simulation model. This is typically used to describe the thermal flow at the boundaries. Thermal flow boundary conditions also take two forms. First, there is the thermal flow fixed boundary condition, where the thermal flow is fixed by selecting specific surfaces or points in the model. Second, there is the periodic thermal flow boundary condition, where the thermal flow of two adjacent surfaces in the model is equal. This can be used to simulate objects with a periodic structure. The mathematical expression for thermal flow boundary conditions is given by

$$\mathbf{q}(x, y, z, t) = \mathbf{q}_0 \tag{6.16}$$

where $\mathbf{q}_0$ represents the fixed value of thermal flow and $\mathbf{q}(x, y, z, t)$ denotes the thermal flow vector. The variables x, y, z represent spatial positions, while t represents time. Thermal flow boundary conditions are commonly used to describe the surface thermal flow of liquid or gas objects. For example, in the design of a cooling fan, the thermal flow of the air blown by the fan can be described using thermal flow boundary conditions.

Convective boundary conditions describe the thermal flow of a surface in a thermal simulation model as convective thermal flow flux with the surrounding medium. They are commonly used to represent convective thermal transfer of fluids at the surface. Convective boundary conditions can be categorized into two forms. The first is the natural convective boundary condition, which describes the natural convective thermal transfer of the fluid without external forces. The second is the forced convection boundary condition, which describes the convective thermal transfer under the influence of external forces such as fans or pumps in the fluid. The mathematical expression for convective boundary conditions is given by

$$\mathbf{q}(x, y, z, t) = h\,(T_s - T_\infty) \tag{6.17}$$

where h represents the surface thermal transfer coefficient, T_s is the surface temperature, and T_∞ is the ambient medium temperature. Convective boundary conditions are typically used to describe thermal transfer on the surface of fluid objects. For instance, in the design of thermal dissipation for electronic components, convective boundary conditions can be utilized to represent the thermal transfer on the surface of a radiator due to airflow generated by a fan.

Initial conditions in thermal simulation models depict the internal temperature distribution of an object at the beginning of the simulation. They serve as the starting point for solving the heat conduction equation. The selection of initial conditions significantly influences the accuracy of simulation results and solution efficiency. Initial conditions can be classified into two forms. The first is the uniform initial condition, where the temperature distribution inside the object is uniform. This form is suitable for describing situations where the object is in thermal equilibrium. The second is the non-uniform initial condition, where the temperature distribution inside the object is non-uniform. This form is suitable for describing scenarios where the object starts to thermal or cool at a specific temperature. The mathematical expression for the initial condition is given by

$$T(x, y, z, t = 0) = T_0 \tag{6.18}$$

where T_0 represents the initial value of the internal temperature distribution of the object, while x, y, and z denote spatial positions. The setting of initial conditions has a significant impact on the accuracy of thermal simulation results, necessitating a reasonable selection and setting based on the specific circumstances.

The setting of boundary conditions and initial conditions is a crucial aspect of thermal simulation models, as it greatly influences the accuracy of calculation results and solution efficiency. Several factors affect boundary conditions and initial conditions, including:

1. Material and geometry of the object: The material and geometry of the object are important factors in determining the selection of boundary and initial conditions. Different materials exhibit varying thermal conductivities, necessitating appropriate choices of boundary conditions and initial conditions based on these properties. For example, metals and plastics possess distinct thermal conduction properties, demanding different boundary conditions and initial conditions. Furthermore, the geometry of the object also influences the selection of boundary and initial conditions. When simulating the thermal transfer process of an automobile engine, for instance, appropriate boundary conditions and initial conditions must be chosen to ensure accurate and reliable calculation results.

2. Numerical setting of boundary and initial conditions: The numerical settings of boundary conditions and initial conditions also impact the calculation results of thermal simulation models. Setting values that are too large or too small can affect the calculation of the thermal conduction process. Thus, it is essential to select appropriate numerical ranges when setting boundary conditions and initial conditions, ensuring accuracy and reliability in the calculation results.

3. Ambient temperature and thermal flow: The temperature and thermal flow of the surrounding environment are important factors affecting boundary conditions and initial conditions. Different temperatures and thermal flows in the surrounding environment require the selection of different boundary conditions and initial conditions. For instance, when simulating the thermal dissipation of electronic components, considerations must be given to the temperature and thermal flow of the air surrounding the radiator, necessitating appropriate boundary conditions and initial conditions for accurate and reliable calculation results.

6.4.3 Thermal Source Modeling

In thermal simulation, a thermal source refers to an object or medium that can generate or absorb thermal. This includes electronic devices, circuit boards, radiators, mechanical equipment, buildings, and more. Accurate modeling of thermal sources helps bring simulation results closer to real-world scenarios, providing reliable data and information for optimization design. This section outlines the principles and methods of thermal source modeling.

In thermal source modeling, common types of thermal sources include constant temperature thermal sources, constant power thermal sources, and nonlinear thermal sources [39–42].

A constant temperature thermal source refers to an object or medium that continuously generates or absorbs thermal at a specific temperature. Examples include constant temperature boxes and constant temperature water tanks.

A constant power thermal source refers to an object or medium that continuously generates or absorbs a certain amount of power. Examples include electronic devices, circuit boards, and radiators.

A nonlinear thermal source refers to a thermal source whose thermal characteristics change with temperature or power. Examples include thermistors and thermocouples.

When modeling a thermal source, it is important to consider its thermal characteristics, such as thermal capacity, thermal conductivity, thermal resistance, and thermal diffusivity. Thermal capacity represents the thermal absorbed or released per unit mass of material for a unit temperature change. Thermal conductivity represents the rate of thermal conduction per unit length, unit time, and unit temperature difference in the material. Thermal resistance represents the conduction resistance of thermal per unit area, unit thickness, and unit temperature difference. Thermal diffusivity represents the rate of thermal diffusion per unit area, unit time, and unit temperature difference. These parameters significantly influence the temperature distribution and thermal transfer process of the thermal source. When modeling a thermal source, various methods can be used [39, 43, 44]:

1. Theoretical calculation method: This method involves using mathematical models based on the physical characteristics and shape of the thermal source to calculate its temperature distribution and thermal transfer process. Examples include solving the heat conduction equation using finite element methods or analyzing the thermal transfer process through thermal resistance networks.

2. Experimental test method: This method involves obtaining temperature distribution and thermal transfer data of the thermal source through experimental tests. Techniques such as infrared thermal imaging and thermocouples can be used to measure the temperature of the thermal source. The obtained data can be used in the thermal simulation model.

3. Empirical method: This method involves modeling the thermal source based on empirical formulas or existing thermal simulation models. It may involve using standard thermal source models or utilizing pre-existing thermal simulation models for the simulation.

The choice of method for thermal source modeling depends on the specific situation. Simple thermal sources may be adequately modeled using empirical methods, while complex thermal sources require the application of theoretical calculation methods or experimental testing. Additionally, attention should be given to selecting appropriate mesh generation methods and solution algorithms to ensure accuracy and computational efficiency.

The following steps outline the process of thermal source modeling using electronic devices as an example:

1. Determine the type and characteristics of the thermal source: Electronic devices are typically considered constant power thermal sources. Their thermal characteristics include thermal capacity, thermal conductivity, thermal resistance, and thermal diffusivity.

2. Establish the thermal source model: Create a thermal source model based on the shape and material characteristics of the electronic device. This can be achieved using three-dimensional modeling software or by employing standard thermal source models.

3. Set boundary conditions and initial conditions: Define the boundary conditions and initial conditions based on the specific situation. This includes specifying the power of the electronic device, the thermal conduction properties of the materials, the temperature of the surrounding environment, and other relevant factors.

4. Perform mesh division and solve: Divide the model into a mesh and apply a suitable solution algorithm to solve the equations. This step involves selecting an appropriate mesh generation method and solution algorithm.

5. Analyze the simulation results: Evaluate the temperature distribution and thermal transfer characteristics of the electronic device based on the simulation results. Use this analysis to optimize the design.

Thermal source modeling is a critical aspect of thermal simulation models and plays a significant role in simulating complex thermal transfer processes. Through proper thermal source modeling, the accuracy and reliability of the thermal simulation model can be improved, providing valuable data and information for optimization design.

6.4.4 Mesh Generation and Solution Algorithms

Mesh generation and solution algorithms are crucial components of thermal simulation models as they significantly impact the accuracy and efficiency of the simulation results. This section focuses on the principles and common methods used in meshing and solving algorithms [45, 46].

Meshing is a fundamental step in thermal simulation models that involves dividing the model into small grid cells to enable the discretization solution of each cell [47]. To ensure accuracy and efficiency of numerical solutions, the appropriate grid size and number should be selected based on the specific situation. Common types of meshes include structured and unstructured meshes. A structured grid consists of regular rectangular or cubic cells, which are easy to construct and solve. On the other hand, an unstructured grid consists of cells with various shapes, providing more flexibility but posing greater challenges in solving. The accuracy and efficiency of mesh generation significantly influence the precision and computational time of thermal simulation results. Generally, a finer mesh leads to more accurate results but also increases the computational time. Thus, there is a tradeoff between accuracy and efficiency in meshing.

The solution algorithm is another critical aspect of thermal simulation models as it determines the solution accuracy and computational efficiency. Commonly used algorithms include implicit, explicit, and iterative algorithms.

1. Implicit algorithm: The implicit algorithm is a stable method suitable for most cases in heat conduction equations. It involves advancing the unknown temperature field value from time step t_n to time step t_{n+1} by solving a system of equations. The unknown quantity in the equation set is the temperature field value at time step t_{n+1}, while the right side of the equation set includes the temperature field value at time t_n and the thermal source term. The implicit algorithm avoids the influence of time step on calculation accuracy but requires longer computation time and substantial memory to store the matrix.

2. Explicit algorithm: The explicit algorithm is a fast solution algorithm but lacks stability. It computes the temperature field value at time t_{n+1} based on the known temperature field value at time t_n and the thermal source term. The explicit algorithm is computationally efficient but requires adherence to time-step restrictions to prevent numerical instability.

3. Iterative algorithm: Iterative algorithms are used for solving nonlinear problems. Nonlinear terms in heat conduction equations often involve temperature-dependent parameters like thermal conductivity and thermal capacity, which can be effectively addressed using iterative algorithms. Common iterative algorithms include Jacobi iteration, Gauss-Seidel iteration, and Gauss-Jordan iteration with all pivoting. Numerical methods are employed to solve the heat conduction equation through numerical calculations. Commonly used numerical methods include the finite element method, finite difference method, and finite volume method.

6.5 Co-simulations of Thermal Properties

Co-simulations involve the integration of simulation models from multiple physical domains, such as electromagnetic, thermal, and structural, to co-simulate the behavior and performance of a system [48]. It enables designers to gain a comprehensive understanding of the system's working conditions and optimize its design and performance. Thermal simulation plays a crucial role in co-simulations by facilitating a better understanding of thermal transfer processes in the system and optimizing thermal management and cooling designs. This section introduces the basic concepts of thermal co-simulations, including the process of co-simulations, the establishment of co-simulations models, and solution methods for such models.

The process of thermal co-simulations includes model establishment, boundary condition setting, solution, and result analysis. The specific steps are as follows:

1. Model establishment: The first step is to establish a complete thermal co-simulations model, which includes simulation models from electromagnetic, thermal,

structural, and other physical domains. When building the model, it is essential to consider the interaction and influence between these physical domains to ensure model accuracy and reliability.

2. Setting of boundary conditions: After establishing the complete thermal co-simulations model, it is necessary to set the boundary conditions for each physical domain, describing the inputs and outputs of the system. In co-simulations, the boundary conditions of electromagnetic, thermal, structural, and other physical domains need to be coupled to ensure the accuracy and reliability of the model.

3. Solution: Once the boundary conditions are set, solve the thermal co-simulations model using an appropriate solution method. Commonly used solution methods include the finite element method, boundary element method, finite volume method, and others. During the solution process, it is crucial to consider the interaction and influence between different physical domains to ensure the accuracy and reliability of the model.

4. Result analysis: After completing the solution of the thermal co-simulations model, analyze and evaluate the results. Various tools can be utilized to visualize and analyze the results, such as temperature contour plots and thermal flux maps. Through result analysis, a comprehensive understanding of the system's thermal transfer can be obtained, enabling optimization of thermal management and cooling designs.

The establishment of a thermal co-simulations model requires considering the interaction and influence between multiple physical fields to ensure model accuracy and reliability. The following aspects should be considered when building the model:

1. Establishment of electromagnetic and thermal coupling model:

In combined thermal simulation, it is necessary to couple the electromagnetic and thermal fields and establish an electromagnetic and thermal coupling model. This coupling model can be achieved by formulating a set of equations that describe electromagnetism and thermal and by considering the interaction between thermal sources and the electromagnetic field. For example, in the thermal dissipation design of electronic products, it is necessary to model the circuit board, radiator, and fan as a whole, considering the interaction between thermal sources and the electromagnetic field of electronic components. This approach enables obtaining more accurate temperature distribution and thermal flow distribution.

2. Establishment of thermal and structural coupling model:

In thermal co-simulations, it is necessary to couple the thermal and structural fields and establish a thermal and structural coupling model. This coupling model can be achieved by formulating the heat conduction equation and the structural equations while considering the interaction of thermal expansion and structural deformation. For example, in the thermal dissipation design of automobile engines, it is necessary to model the engine structure and thermal dissipation system, considering the interaction of thermal expansion and structural deformation. This approach enables obtaining more accurate temperature distribution and stress distribution.

3. Establishment of coupled model of thermal and fluid:

In combined thermal simulation, it is necessary to couple the thermal and fluid fields and establish a coupled model of thermal and fluid. This coupling model can be achieved by formulating equations of fluid dynamics and thermal conduction, taking into account the temperature distribution and thermal transport of the fluid. For example, in the design of air conditioning systems, it is necessary to couple air flow and thermal transfer to obtain more accurate temperature distribution and thermal flow distribution.

The solution of a combined thermal simulation model needs to consider the interaction and influence of multiple physical fields. Commonly used solution methods include the finite element method, boundary element method, finite volume method, and others. The following aspects need to be considered in the solution process:

1. Mesh: The thermal co-simulations model needs to be meshed in the solving process. This involves determining the grid size and the number of grids to ensure the accuracy and efficiency of the numerical solution. In thermal co-simulations, multigrid partitioning is often performed to account for the interactions and effects between multiple physical domains and obtain more accurate results.

2. Setting of boundary conditions: During the solution process, the boundary conditions of the thermal co-simulations model need to be set to describe the input and output of the system. In thermal co-simulations, the boundary conditions of multiple physical fields need to be coupled to ensure the accuracy and reliability of the model.

3. Numerical solution method: The appropriate numerical solution method needs to be selected during the solving process, such as implicit algorithm, explicit algorithm, or iterative algorithm. In thermal co-simulations, it is important to consider the interaction and influence between multiple physical fields. Therefore, a solution method suitable for multigrid should be chosen to ensure the efficiency and accuracy of the solution.

4. Result analysis: After solving the thermal co-simulations model, the results need to be analyzed and evaluated. Various tools can be used to visualize and analyze the results, such as temperature contour plots, and thermal flux maps. Through result analysis, the thermal transfer within the system can be understood, and thermal management and cooling design of the system can be optimized.

6.5.1 Interaction between Thermal Simulations and Electromagnetic Simulations

Thermal simulation and electromagnetic simulation are two common engineering design techniques used to analyze the temperature distribution and electromagnetic field distribution of an object, respectively. In certain application scenarios, the thermal conduction process and electromagnetic field distribution of the object are coupled, necessitating the coupling of thermal simulation and electromagnetic simulation to obtain more accurate results. The interaction between the thermal field and electromagnetic field primarily involves the conversion and transfer of electromagnetic energy and thermal energy. In the electromagnetic field, electromagnetic waves generate thermal energy, while in the thermal field, thermal energy produces electromagnetic waves. Therefore, it is essential to consider the interaction between thermal conduction and electromagnetic waves in the coupling analysis of thermal simulation and electromagnetic simulation.

The effects of thermal conduction and electromagnetic wave interaction can be described by thermal resistance and electromagnetic impedance. Thermal resistance characterizes the resistance in the thermal conduction process, while electromagnetic impedance describes the resistance in electromagnetic wave propagation. During the interaction between thermal conduction and electromagnetic waves, changes in thermal resistance and electromagnetic impedance can result in variations in the propagation speed of thermal and electromagnetic waves, thus affecting the temperature distribution and electromagnetic field distribution of the object.

In the coupling analysis of thermal simulation and electromagnetic simulation, it is necessary to combine the heat conduction equation and Maxwell's equations to solve the temperature distribution and electromagnetic field distribution of the object. Maxwell's equations describe the distribution and variation of the electromagnetic field, encompassing the electric field and magnetic field distributions and variations. By combining Maxwell's equations with the heat conduction equation, coupled equations for the thermal and electromagnetic fields can be obtained to solve the temperature distribution and electromagnetic field distribution of the object.

To combine the heat conduction equation with Maxwell's equations, we need to consider the impact of the electromagnetic field on the thermal field and vice versa. A common method of coupling these two fields is through the current density (**J**) and the Joule heat from which they originate, respectively, and the relationship can be described by

$$\frac{\partial T}{\partial t} = \alpha \nabla^2 T + Q_e \tag{6.19}$$

$$\nabla \times \boldsymbol{E} = -\frac{\partial \boldsymbol{B}}{\partial t} \tag{6.20}$$

$$\nabla \times \boldsymbol{H} = \boldsymbol{J} + \frac{\partial \boldsymbol{D}}{\partial t} \tag{6.21}$$

$$\nabla \cdot \boldsymbol{D} = \rho_e \tag{6.22}$$

$$\nabla \cdot \boldsymbol{B} = 0 \tag{6.23}$$

$$\nabla \cdot \boldsymbol{J} = 0 \tag{6.24}$$

$$\nabla \cdot Q_e = 0 \tag{6.25}$$

where T is the temperature field, Q_e is the thermal source generated by the electromagnetic field, α is the thermal diffusion coefficient, $\boldsymbol{E}$ and $\boldsymbol{H}$ are the electric field and magnetic field, respectively, $\boldsymbol{D}$ and $\boldsymbol{B}$ are the electric displacement and magnetic induction, respectively, $\boldsymbol{J}$ is the current density, and ρ_e is the charge density. The first equation in the set represents the thermal conduction process inside the object and the influence of the thermal source generated by the electromagnetic field on the temperature field. The second to sixth equations describe the distribution and variation of the electromagnetic field, and the fifth and sixth equations are part of Maxwell's equations. The last equation in the system represents the passive nature of the thermal source generated by the electromagnetic field.

After discretizing the equations, the finite element method can be used to solve them. In the solving process, the object needs to be meshed, and the equations on the mesh need to be solved. The influence of thermal conduction and electromagnetic wave interaction should be considered, and the necessary corrections should be applied.

For example, at the point (x, y, z) in the thermal medium, the thermal transfer at time t satisfies the partial differential equation of thermal conduction, known as the Heat Conduction Equation (HTE), given by

$$\rho_m c_m \frac{\partial T(x, y, z)}{\partial t} = k_t \nabla^2 T(x, y, z) - V_S \left(T(x, y, z) - T_a\right) + P_d(x, y, z, t) \tag{6.26}$$

where T [°C] represents the transient temperature of the object, t [s] is the time of the process, k_t [W/(m· °C)] is the thermal conductivity of the material, ρ_m [kg/m^3] is the density of the material, c_m [J/(kg· °C)] is the specific thermal at constant pressure of the material, V_S [W/(m^3· °C)] is the thermal capacity flow product of the cooling flow, T_a [°C] is the temperature of the cooling flow, and P_d [W/m^3] is the power density of the internal thermal source.

An electromagnetic and thermal integrated analysis method for microwave passive circuits based on the spectral element method can be employed. The method involves the following steps:

1. Solve Maxwell's equations using a time-domain spectral element method to obtain the instantaneous distribution of electric and magnetic fields within the circuit under the influence of a high-power pulse. Additionally, determine the electromagnetic losses at the current moment.

2. Assuming that all electromagnetic losses within the model are converted into thermal energy, incorporate the obtained thermal energy into the heat conduction equation to derive the temperature distribution at each point during the current time.

3. Utilize the relationship between the dielectric parameters and temperature to determine the electrical parameters of the material for the next moment. Recalculate the electromagnetic field equation to obtain the electromagnetic losses.

4. Repeat the above steps until the predetermined thermaling time is reached.

By employing this integrated electromagnetic and thermal analysis, the distribution of internal temperature within the model over time under the influence of different pulses can be accurately obtained. The modeling approach is flexible and enables convenient subdivision. The resulting matrix exhibits good sparsity, leading to high computational efficiency.

6.5.2 *Effect of Thermal Simulations on Electromagnetic Performances*

Electromagnetic interference (EMI) and electromagnetic compatibility (EMC) are crucial considerations in electronic equipment design. Thermal simulation plays a significant role in analyzing and optimizing the thermal dissipation design of electronic equipment, thereby impacting its electromagnetic performance. This section focuses on describing the effects of thermal simulation on electromagnetic performance [49, 50].

When components experience thermaling and expansion, their relative positions can change, thereby influencing the electromagnetic performance of electronic equipment. For instance, thermal expansion of wires can alter the distance between them, affecting the distribution and transmission performance of the electromagnetic field. By leveraging thermal simulation, designers can analyze and optimize component layout and thermal dissipation design, minimizing the impact of thermal expansion on electromagnetic performance.

Temperature typically affects the performance of electronic components. Parameters such as current gain and cutoff frequency of a transistor, for example, often vary with temperature. Hence, it is necessary to consider the impact of temperature on component performance and mitigate its effects.

During the operation, the electronic equipment will generate much heat. Failure to dissipate the heat can result in excessive temperatures, thereby impacting the electromagnetic performance of the equipment. For instance, elevated temperatures of wires can lead to reduced transmission performance of the electromagnetic field.

The thermal conductivity of materials generally affects their electromagnetic properties. In microwave circuit design, for example, the thermal conductivity of materials influences dissipated power and temperature distribution within the circuit, consequently affecting its transmission performance. Therefore, when designing electronic equipment, it is vital to consider the influence of material thermal conductivity on electromagnetic performance. Thermal simulation assists designers in analyzing and optimizing material selection and thermal dissipation design, thereby reducing the impact of thermal conductivity on electromagnetic performance.

The significant influence of thermal simulation on electromagnetic performance necessitates the integration of thermal simulation and electromagnetic simulation through co-simulations in electronic equipment design. Co-simulations enables comprehensive consideration of the effects of thermal, temperature, thermal conductivity, and material thermal expansion on electromagnetic performance, facilitating more accurate analysis and optimization of electronic device design.

The co-simulations of thermal simulation and electromagnetic simulation can be achieved through the following two methods:

1. Direct coupling method: This approach directly couples the solving processes of thermal simulation and electromagnetic simulation, allowing for mutual interaction between thermal simulation and electromagnetic simulation. The direct coupling method offers the advantage of more accurate analysis and optimization of

thermal and electromagnetic performance. However, it requires the support of high-performance computers due to its computational intensity.

2. Indirect coupling method: The indirect coupling method involves separate solving of thermal simulation and electromagnetic simulation, followed by matching and transferring of the simulation results. The indirect coupling method offers the advantage of relatively lower computational requirements. However, it necessitates data conversion and transfer, which introduces potential error transfer issues.

6.6 Co-simulations of Thermal Properties with ADS

6.6.1 Co-simulations of FloTHERM with ADS

FloTHERM is a kind of professional thermal simulation software that supports thermal co-simulations with ADS. It allows the import of circuit boards or chips designed in ADS for thermal simulation analysis, enabling the evaluation of thermal characteristics and thermal dissipation effects of the circuit. This information can then be used to optimize circuit design and thermal dissipation schemes.

FloTHERM is a kind of three-dimensional thermal simulation software based on the finite element method, specifically designed for analyzing the thermal characteristics of circuit boards, chips, thermal sinks, and other components. The thermal simulation principles in FloTHERM encompass the following aspects:

1. Heat conduction equation: The heat conduction equation serves as the fundamental equation for solving the thermal field in FloTHERM. It describes the thermal transfer law inside the object. By utilizing the finite element method, FloTHERM can solve the heat conduction equation and obtain the temperature distribution of the circuit board or chip.

2. Boundary conditions and initial conditions: Boundary conditions and initial conditions are vital parameters for thermal simulation analysis. They include the temperature of the radiator, wind speed and temperature of the vents, power distribution of the circuit board or chip, and more. In FloTHERM, these boundary conditions and initial conditions can be set based on the actual situation to perform thermal simulation analysis.

3. Thermal source modeling: Accurate modeling of the power distribution within the circuit board or chip is crucial for thermal simulation analysis. FloTHERM allows the input of power distribution data from ADS, which can be incorporated into the thermal simulation model as a thermal source for analysis.

4. Mesh generation and solution algorithm: Mesh generation and solution algorithm are key technologies in thermal simulation analysis. FloTHERM supports various meshing methods and solution algorithms, including fluid grid, thermal-electric coupled grid, and solid grid. By selecting the appropriate mesh generation and solution algorithm according to the specific situation, the accuracy and efficiency of the numerical solution can be ensured.

The interface between FloTHERM and ADS encompasses the following aspects:

1. Importing the geometric model of the circuit board or chip: Before conducting thermal simulation analysis, the geometric model of the circuit board or chip needs to be imported into FloTHERM. Once the circuit design is completed in ADS, it can be exported to the STEP file format and then opened in FloTHERM to import the geometric model of the board or chip.

2. Importing circuit power distribution data: After completing the circuit design in ADS, the circuit's power distribution data can be exported to an ASCII format file. This file can then be opened in FloTHERM to import the power distribution data into the thermal simulation model.

3. Adjusting the boundary conditions and initial conditions: In FloTHERM, the boundary conditions and initial conditions, such as the temperature of the radiator, the wind speed and temperature of the vents, and the power distribution of the circuit board or chip, can be set based on the actual situation for thermal simulation analysis.

4. Thermal design optimization: After conducting thermal simulation analysis, the temperature distribution of the circuit board or chip, as well as the temperature distribution of the radiator, can be obtained. These results enable the evaluation of the radiator's thermal dissipation capacity and the thermal characteristics of the circuit. Based on the analysis results, the circuit design can be optimized. This may involve increasing the thermal dissipation area of the radiator and the size of the vents, or adjusting the layout of the radiator. The objective is to ensure the circuit's performance and reliability.

6.6.2 Co-simulations of Icepak with ADS

Icepak is a kind of computational fluid dynamics (CFD) software designed for the thermal design and analysis of electronic devices. It enables engineers to simulate and analyze the internal airflow and thermal dissipation performance of electronic equipment, providing crucial insights for design and optimization. By creating a three-dimensional model and conducting simulations, Icepak can generate data such as temperature distribution and airflow patterns under different operating conditions. When combined with ADS, Icepak facilitates the comprehensive analysis of both electromagnetic and thermal performance of electronic devices. This approach allows for a more accurate representation of the actual thermal behavior, thereby enhancing product reliability and stability.

In the combined thermal simulation of Icepak and ADS, the process involves completing electromagnetic simulation in ADS to obtain power consumption and temperature rise data of electronic components. These data are then imported into Icepak as input conditions for thermal simulation. In Icepak, a three-dimensional model of the electronic device is established, and relevant boundary conditions such as radiators and fans are set.

There are primarily two methods for data exchange between Icepak and ADS: file import/export and API interface calls. File import/export involves saving electromagnetic simulation results as a specific file format in ADS and then importing that file into Icepak. Similarly, Icepak's thermal simulation results can be saved as a file for further analysis in ADS. On the other hand, API interface calls offer a more advanced approach to data exchange. Users can write scripts or programs that utilize the API interfaces of Icepak and ADS to achieve real-time data exchange and control. This method enables tighter software integration, leading to improved simulation efficiency and accuracy.

Specific operation steps for the combined thermal simulation of Icepak and ADS include:

1. Complete electromagnetic simulation in ADS:

(a) Establish the electromagnetic model of the electronic equipment, including components, wiring, and boundary conditions.

(b) Set simulation parameters such as frequency range and port parameters.

(c) Run the electromagnetic simulation to obtain data such as power consumption and temperature rise for each component.

(d) Export the electromagnetic simulation results to a specific file format, such as an .s2p file.

2. Perform thermal modeling of electronic devices in Icepak:

(a) Create a new Icepak project and import the CAD model.

(b) Add various thermal components to the project, such as radiators and fans.

(c) Set the thermal power of each element based on the electromagnetic simulation results.

(d) Define the boundary conditions for thermal simulation, including ambient temperature and flow velocity.

3. Run Icepak thermal simulation:

(a) Set the thermal simulation parameters in Icepak, such as mesh generation and solver settings.

(b) Run the thermal simulation and monitor convergence in real time during the process.

(c) Upon completion of the thermal simulation, obtain results such as temperature distribution and flow conditions within the electronic equipment.

4. Analyze thermal simulation results and optimize the design:

(a) Analyze the thermal performance of the electronic equipment based on the electromagnetic simulation results.

(b) Identify areas and components with thermal issues, such as overthermaling or excessive thermal resistance.

(c) Optimize the design of the electronic equipment to address thermal problems, such as improving radiator structure or adjusting fan placement.

(d) Update the electromagnetic simulation model and rerun the simulation to verify the effectiveness of the optimization.

5. Apply the optimized design to the actual product:

(a) Update the CAD model and manufacturing process documents based on the optimized design.

(b) Produce and test actual products to verify the reliability and stability of the design.

(c) Based on the test results, further refine the design to enhance performance and product quality.

6.7 Co-simulations of Thermal Properties with HFSS

There are two main interfaces for HFSS: ANSYS Workbench and HFSS Script. ANSYS Workbench is an integrated development environment of ANSYS, which can integrate multiple ANSYS modules in one platform. As a ANSYS Workbench module, HFSS can be integrated with other modules of ANSYS to realize thermal co-simulations. Through the ANSYS Workbench, the power distribution data of HFSS can be imported and the thermal simulation analysis of ANSYS can be completed conveniently. HFSS Script is a scripting language that can be used to control the operation and parameter settings of HFSS. Through HFSS Script, the power distribution data of HFSS can be exported as a txt file as the input of ANSYS thermal simulation.

6.7.1 Co-simulations of FloTHERM with HFSS

FloTHERM is a kind of widely used thermal simulation software that finds application in various fields such as electronics, automotive, and aerospace. Similarly, HFSS is a kind of widely utilized electromagnetic simulation software that serves as a critical tool in electronics, automotive, and aerospace industries. In electronic design, the thermal and electromagnetic characteristics are closely intertwined, making the application of thermal simulation and electromagnetic simulation vital for the design and optimization of electronic products.

FloTHERM is a kind of specialized thermal simulation software primarily used for the thermal analysis and design of electronic products and circuit boards. It offers support for various mesh generation methods such as fluid mesh, thermal-electrical-coupled mesh, and solid mesh. Furthermore, FloTHERM takes into account factors like the material and thickness of the circuit board, the material and shape of the thermal sink, and the size and location of the vent during thermal simulation analysis. It also supports multiphysics field coupling simulations, including electromagnetic-thermal and structure-thermal coupling simulations.

When co-simulating FloTHERM and HFSS, the first step involves establishing the geometric and physical models of the circuit. In HFSS, electromagnetic simulation is conducted to obtain information such as electromagnetic field distribution and power distribution. The power distribution information is then exported to FloTHERM to generate a thermal simulation model for subsequent analysis. In FloTHERM, thermal simulation analysis takes place by considering parameters such as the material and thickness of the circuit board, the material and shape of the thermal sink, and the size and location of the vent. The appropriate meshing method and solution algorithm are selected to ensure accurate and efficient solutions. Ultimately, the temperature distribution and thermal flow distribution of the circuit can be obtained.

The co-simulations of FloTHERM and HFSS follows these steps:

1. Establish the geometric and physical model of the circuit:

In HFSS, create the geometric and physical models of the circuit for electromagnetic simulation, considering the material, geometry, and boundary conditions. Export the power distribution information to FloTHERM at the end of the electromagnetic simulation.

2.Generate the thermal simulation model:

In FloTHERM, generate the thermal simulation model based on the geometric model and power distribution information of the circuit. Set parameters such as thermal conductivity, thermal capacity, density, material properties, and radiator layout.

3.Perform thermal simulation analysis:

Conduct thermal simulation analysis in FloTHERM, setting parameters such as the solution algorithm, mesh generation method, and convergence criterion for accurate and efficient solutions. Obtain the temperature distribution and thermal flow distribution of the circuit.

4.Optimize the circuit design:

(a) Evaluate the thermal characteristics of the circuit and thermal dissipation capacity of the thermal sink based on the temperature distribution and thermal flow distribution.

(b) Optimize the layout and thermal dissipation design of the circuit to enhance its performance and reliability. Consider measures like increasing the thermal dissipation area of the radiator, adjusting the size of the vent, and modifying the radiator layout. During the electromagnetic simulation in HFSS, factors such as the material, geometry, boundary conditions, and electromagnetic performance of the circuit need to be considered.

6.7.2 Co-simulations of Icepak with HFSS

Icepak is a type of ANSYS thermal simulation software designed to evaluate the thermal performance of electronic products and optimize thermal designs. By combining thermal simulation with HFSS, designers can gain a better understanding of the circuit's thermal characteristics, optimize thermal designs, and enhance circuit performance and reliability. Icepak offers various thermal conduction models, meshing methods, and solving algorithms, enabling quick and accurate analysis of electronic product thermal characteristics. It supports thermal co-simulations with HFSS and provides interfaces and data formats for direct import of models, meshes, and boundary conditions from HFSS to generate thermal simulation models for analysis.

In the thermal simulation process using HFSS and Icepak, the circuit model from HFSS needs to be imported into Icepak for thermal simulation analysis. Initially, the geometric model and material properties of the circuit are defined in HFSS, and electromagnetic field distribution and power distribution are calculated. Subsequently, the derived electromagnetic field distribution and power distribution are imported into Icepak to generate the corresponding thermal simulation model. Icepak uses meshing and solving algorithms to calculate the temperature distribution and thermal flow distribution of the circuit, enabling evaluation of thermal designs and thermal characteristics.

The steps for implementing combined thermal simulation of HFSS and Icepak are as follows:

1. Establish HFSS model: Define the geometric model and material properties of the circuit in HFSS and specify the electromagnetic wave source and power distribution.

2. Export HFSS model: Export the HFSS model in a format readable by Icepak, such as STL or SAT. Specify the desired meshing accuracy and mesh size during export.

3. Import Icepak model: Import the HFSS model into Icepak and perform meshing and boundary condition settings. Icepak provides an automatic meshing tool, but manual mesh refinement is also possible.

4. Define the thermal sink and boundary conditions: Specify the material and size of the thermal sink, as well as the size and location of vents in Icepak. Icepak offers a standard radiator library for selection or customization based on specific requirements.

5. Perform thermal simulation analysis: Conduct thermal simulation analysis in Icepak to calculate the temperature distribution and thermal flow distribution of the circuit. Icepak provides analysis tools, such as temperature contour plots and thermal flow maps, for visualizing and analyzing simulation results.

The operation steps for thermal co-simulations of HFSS and Icepak are as follows:

1. Establish the geometric model and material properties of the circuit in HFSS, and define the electromagnetic wave source and power distribution.

2. Export the HFSS model in a format readable by Icepak, such as STL or SAT.

3. Import the HFSS model into Icepak and set up meshing and boundary conditions.

4. Define the thermal sink and boundary conditions in Icepak, including the material and size of the thermal sink, as well as the size and location of vents.

5. Conduct thermal simulation analysis in Icepak to calculate the temperature distribution and thermal flow distribution of the circuit.

6. Based on the simulation results, carry out optimization design such as adjusting the radiator layout or increasing the size and location of vents to improve circuit performance and reliability.

In the latest version of ANSYS SIwave, the SIwave-Icepak electro-thermal co-simulations function is integrated. Designers can utilize the SIwave software interface environment to simultaneously invoke the SIwave DC simulator and Icepak 3D thermal radiation simulator for electro-thermal coupling analysis. This allows obtaining results such as current density distribution and temperature distribution during PCB operation. Such analysis aids in evaluating the impact of temperature changes on PCB performance, predicting hot spots in temperature distribution for thermal design, and enabling proactive measures.

6.8 Practices and Skills for Thermal Simulations

6.8.1 Model Preparation

Before conducting thermal simulation analysis, it is essential to prepare the model, which includes geometric modeling, material modeling, and setting boundary and initial conditions. The quality and accuracy of model preparation directly impact the reliability and accuracy of thermal simulation analysis. Therefore, special attention should be given to the following aspects during model preparation.

Geometric modeling serves as the foundation for thermal simulation analysis. An accurate geometric model enables a better representation of the actual circuit, leading to improved accuracy and reliability of the thermal simulation analysis. When performing geometric modeling, the following points should be considered:

1. The geometric model should closely resemble the actual circuit's shape and size, including the circuit board and chip geometry, thermal sink shape and layout, among others.

2. The geometric model should be simplified whenever possible to reduce model complexity and computation. However, it is crucial to ensure that the simplified model accurately captures the circuit's actual characteristics.

3. The geometric model should be standardized to facilitate subsequent tasks such as material assignment and mesh generation.

Material modeling is a crucial aspect of thermal simulation analysis. A precise material model enables a better representation of the circuit's thermal characteristics, enhancing the accuracy and reliability of the thermal simulation analysis. When conducting material modeling, the following considerations should be taken into account:

1. Material properties should be as accurate as possible, encompassing thermal conductivity, specific thermal capacity, density, and other relevant parameters.

2. Standardization of material properties is recommended to facilitate subsequent tasks such as thermal source modeling and solution algorithms.

3. For complex structures like multilayer circuit boards and chips, it is essential to model the interfaces between layers based on the actual configuration.

The proper setting of boundary and initial conditions is crucial for thermal simulation analysis. Accurate boundary and initial conditions provide a better representation of the circuit's actual operating conditions, leading to improved accuracy and reliability of the thermal simulation analysis. When defining boundary and initial conditions, the following points should be considered:

1. Boundary conditions and initial conditions should strive for accuracy, encompassing parameters such as ambient temperature, wind speed, and radiator temperature.

2. The normalization of boundary and initial conditions is recommended to facilitate subsequent tasks such as solution algorithms and result analysis.

3. The boundary conditions and initial conditions should closely resemble the actual working conditions to ensure the reliability of the thermal simulation analysis.

After completing the model preparation, it is crucial to perform a thorough check and correction of the model to ensure its accuracy and integrity. The inspection and correction process involves the following aspects:

1. Geometric model inspection and correction: Carefully examine and correct the geometric model to ensure its completeness and accuracy.

2. Material model check and correction: Verify and rectify the material model to ensure its accuracy and integrity.

3. Boundary conditions and initial conditions check and correction: Verify and rectify the boundary conditions and initial conditions to ensure their accuracy and completeness.

4. Thermal source modeling check and correction: Verify and rectify the thermal source modeling to ensure its accuracy and completeness.

During the model checking and modification process, special attention should be given to the accuracy and integrity of the parameters. This helps to avoid errors in the thermal simulation analysis caused by inaccurate or missing parameters.

Mesh generation and solution algorithm selection are critical steps in thermal simulation analysis, as they significantly impact the accuracy and efficiency of the analysis. The following considerations should be taken into account when performing meshing and selecting the solution algorithm:

1. Meshing: The mesh should be as refined as possible to enhance the accuracy of the thermal simulation analysis. However, excessively fine grid division increases computation and solution time. Therefore, a balance should be struck based on actual requirements.

2. Solution algorithm: The chosen solution algorithm should aim to achieve both accuracy and efficiency in the thermal simulation analysis. Commonly used solution algorithms include finite element method, finite difference method, and finite volume method, among others. The selection should be based on specific needs.

3. Computing resources and time constraints: When selecting meshing and solving algorithms, it is crucial to consider the constraints posed by computing resources and time. This ensures the efficiency and reliability of the thermal simulation analysis.

By carefully considering these factors, an appropriate balance can be achieved, leading to accurate and efficient thermal simulation analysis while taking into account the available computing resources and time constraints.

After completing the thermal simulation analysis, it is important to analyze and verify the results in order to assess the accuracy and reliability of the analysis. The analysis and verification of the results involve the following aspects:

1. Temperature distribution: Analysis and verification: The temperature distribution is analyzed and verified to assess the accuracy and reliability of the thermal sim-

ulation analysis. If the observed temperature distribution deviates from the expected results, further analysis is required to check the accuracy and completeness of the model.

2. Thermal flow distribution analysis and verification: The thermal flow distribution is analyzed and verified to evaluate the accuracy and reliability of the thermal simulation analysis. If the observed thermal flow distribution differs from the expected results, further analysis is necessary to assess the accuracy and completeness of the model.

3. Radiator performance evaluation: The performance of the radiator is evaluated to determine its effectiveness and reliability. If the radiator's performance does not meet expectations, further analysis and optimization of the radiator design are required.

4. Results verification and comparison: The results of the thermal simulation analysis are verified and compared to evaluate their accuracy and reliability. Common verification methods include experimental verification and theoretical calculations.

By conducting thorough analysis and verification of the results, any discrepancies or deviations can be identified and addressed. This ensures the accuracy and reliability of the thermal simulation analysis, providing valuable insights for further improvements or optimizations in the design process.

6.8.2 Parameter Setups and Optimizations

Parameter setting and optimization are critical aspects of thermal simulation practice as they directly impact the accuracy and reliability of simulation results. This section provides an overview of thermal simulation parameter settings and optimization techniques to assist designers in conducting effective thermal simulation analysis.

1. Material parameter settings:

Setting the material parameters is a crucial step in thermal simulation analysis. Different materials possess distinct parameters such as thermal conductivity, specific thermal capacity, and density, which significantly influence the solution of the heat conduction equation. Thus, it is essential to establish appropriate material parameters based on the actual scenario before conducting thermal simulation analysis.

2. Mesh parameter settings:

When selecting the meshing method, it is necessary to consider the geometry and material distribution of the thermal simulation model and choose the appropriate meshing method and parameters accordingly. Commonly used mesh generation methods include structured mesh, unstructured mesh, quadrilateral mesh, and triangular mesh. To ensure both accuracy and efficiency in mesh generation, it is crucial to select the appropriate method based on the actual situation.

When setting the grid parameters, it is important to consider the following:

1. Grid density: Grid density is a crucial parameter that directly impacts the accuracy of the numerical solution. Therefore, it is essential to select an appropriate grid density based on the specific circumstances. Typically, enhancing the grid density

in computational models improves the precision of the results, albeit at the expense of increased computational effort and time.

2. Mesh size: Mesh size is a key parameter that affects the efficiency of the numerical solution. It is necessary to choose an appropriate mesh size based on the specific requirements. Smaller mesh sizes typically enhance the accuracy of solutions but also increase computational time requirements.

3. Mesh shape: The choice of mesh shape can also affect the accuracy and efficiency of the numerical solution. It is important to select the appropriate mesh shape based on the complexity of the object being analyzed. Square and rectangular grids are typically more accurate and efficient, but for objects with intricate shapes, triangular or quadrilateral grids may be more suitable.

After completing the thermal simulation analysis, it is crucial to analyze and verify the simulation results to assess their reliability and accuracy. The temperature distribution is a key output of thermal simulation analysis, providing insights into the thermal characteristics and effectiveness of the circuit's design. When analyzing the temperature distribution, the following factors should be taken into account:

1. Material and thickness of the circuit board: The material and thickness of the circuit board play a significant role in the temperature distribution of the circuit. During thermal simulation analysis, it is important to select appropriate materials and thickness for the circuit board based on the specific situation to ensure the reliability of the simulation results.

2. Shape and material of the thermal sink: The shape and material of the thermal sink also have a significant impact on the temperature distribution of the circuit. In thermal simulation analysis, it is essential to choose the appropriate shape and material for the thermal sink, considering the specific circumstances, to ensure the accuracy of the simulation results.

3. Size and position of the vents: The size and position of vents can influence the temperature distribution of the circuit. When conducting thermal simulation analysis, it is necessary to consider the size and placement of vents to ensure the reliability of the simulation results.

Thermal flow distribution is another crucial output of thermal simulation analysis, providing insights into the thermal dissipation capacity of the thermal sink and the thermal characteristics of the circuit. When analyzing the thermal flow distribution, the following factors should be taken into consideration:

1. Power distribution of the circuit: The power distribution within the circuit is a significant factor influencing the thermal flow distribution. During thermal simulation analysis, it is essential to input accurate power distribution information for the circuit based on the specific situation to ensure the accuracy of the simulation results.

2. Impact of the thermal flow distribution: In thermal simulation analysis, it is important to select the appropriate shape and material for the thermal sink, considering the specific circumstances, to ensure the reliability of the simulation results.

3. Size and position of the vents: The size and position of vents can affect the thermal flow distribution. When conducting thermal simulation analysis, it is necessary to consider the size and placement of vents to ensure the accuracy of the simulation results.

6.9 Co-simulations of Thermal Properties with HFSS and Icepak

With the rapid advancement of modern electronic product integration, the thermal density of products is increasing. Consequently, the demand for thermal analysis of electronic products is also growing. However, traditional electromagnetic simulations cannot consider the influence of temperature, and traditional thermal simulations cannot consider the effects of distributed Joule thermaling caused by electromagnetic losses. This leads to significant errors when using separate electromagnetic and thermal simulations. Ansys simulation offers a significant advantage in multiphysics co-simulations. Ansys has integrated electromagnetic field tools such as HFSS, Maxwell, Q3D, and thermal simulation tool Icepak in the electronic desktop. Users can easily transfer models, material parameters, and electromagnetic loss simulation results from the electromagnetic field tools to Icepak for thermal modeling and simulation. They can also return the results to the electromagnetic field tools to evaluate the impact of temperature rise on electromagnetic performance, achieving bidirectional electro-thermal coupling analysis. This section will demonstrate how to use Ansys HFSS and Icepak to achieve electro-thermal bidirectional coupling simulation through an example simulation of a Hybrid Ring (Fig. 6.1).

Step 1: Simulating Electromagnetic Loss of a Hybrid Ring in HFSS

First, we need to open ANSYS Electronics Desktop and create an HFSS project. We model the Hybrid Ring based on its geometric parameters. During the modeling process, since we need to perform multiphysics analysis involving electrical and thermal coupling, we must select the thermal-related material property settings in HFSS when defining the material properties. If necessary, we can manually set the temperature-dependent electro-thermal parameters of the materials. Then, in the HFSS menu, we click on "Set Object Temperature" and select the options "Include Temperature Dependence" and "Enable Feedback" in the pop-up menu. We set the initial temperature to the ambient temperature of 22 degrees Celsius, but this value

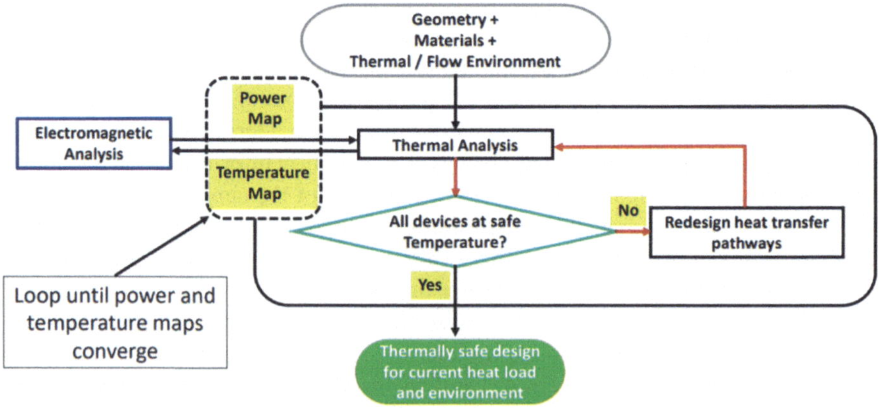

Fig. 6.1 Process of thermal modeling and simulation analysis

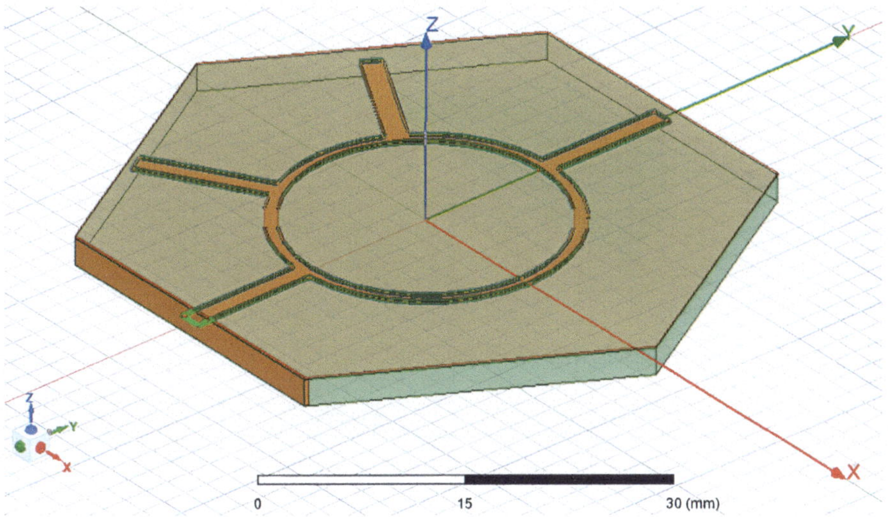

Fig. 6.2 Modeling of hybrid ring in HFSS

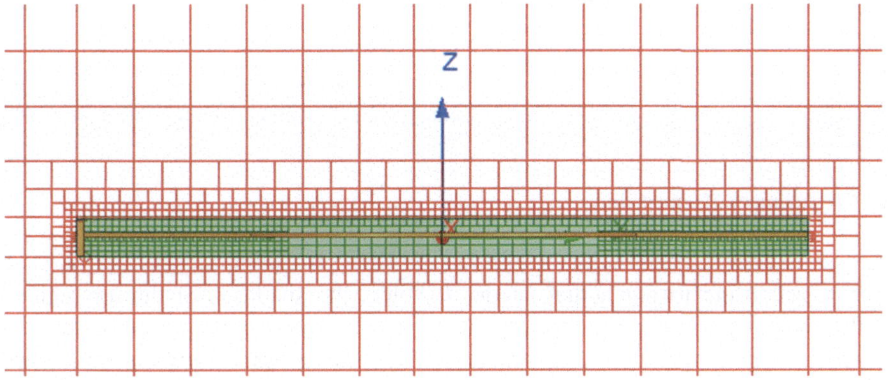

Fig. 6.3 Grid partitioning of the hybrid ring model

can be manually adjusted as needed. Once we have configured the electromagnetic simulation settings, including excitations and boundary conditions, we can proceed with and complete the electromagnetic field simulation analysis (Fig. 6.2).

Step 2: Implementing Electro-thermal Coupled Simulations in Icepak
1. After completing the HFSS electromagnetic field analysis, we create an Icepak project within the same engineering project. 2. We return to the HFSS project, select the Hybrid Ring model, and then copy and paste it into the Icepak project. This transfers the corresponding geometry model and material properties to Icepak. At this point, we can manually check if the material properties have the thermal parameter option selected. 3. We can adjust the size of the simulation solution domain, the

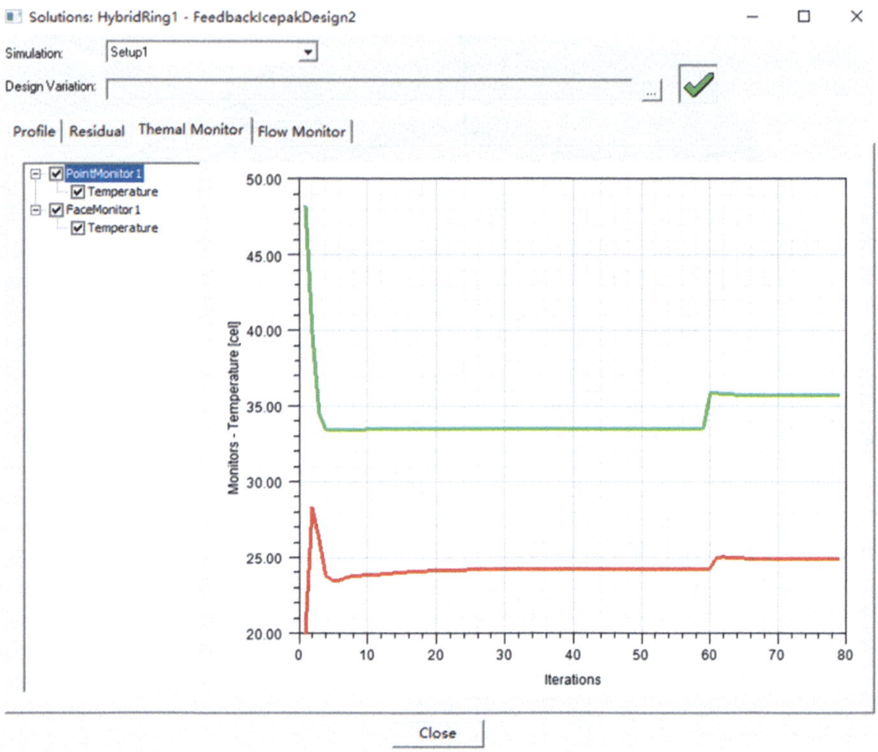

Fig. 6.4 Iterations of temperature

Region, according to our actual needs. 4. Select the six faces of the air region, set the boundary condition to Opening, and leave the Opening properties at default, allowing the air temperature to pass through the boundary without pressure. 5. Select all the geometry models of the Hybrid Ring, right-click, and choose EM loss. In the pop-up menu, select "Use This Project" to link the HFSS electromagnetic loss to Icepak as a thermal source. 6. Right-click on Mesh in the left Project Manager panel, choose Edit Global Region, and select the appropriate mesh settings based on the requirements of simulation accuracy and speed. You can also manually set the mesh on the Advanced page. 7. The meshing process in thermal simulation is performed separately. Right-click on Mesh, choose Generate Mesh to generate the grid. 8. Once the mesh generation is complete, the Mesh Viewer will automatically pop up. Select the corresponding planes or geometries to view the meshing results (Fig. 6.3).

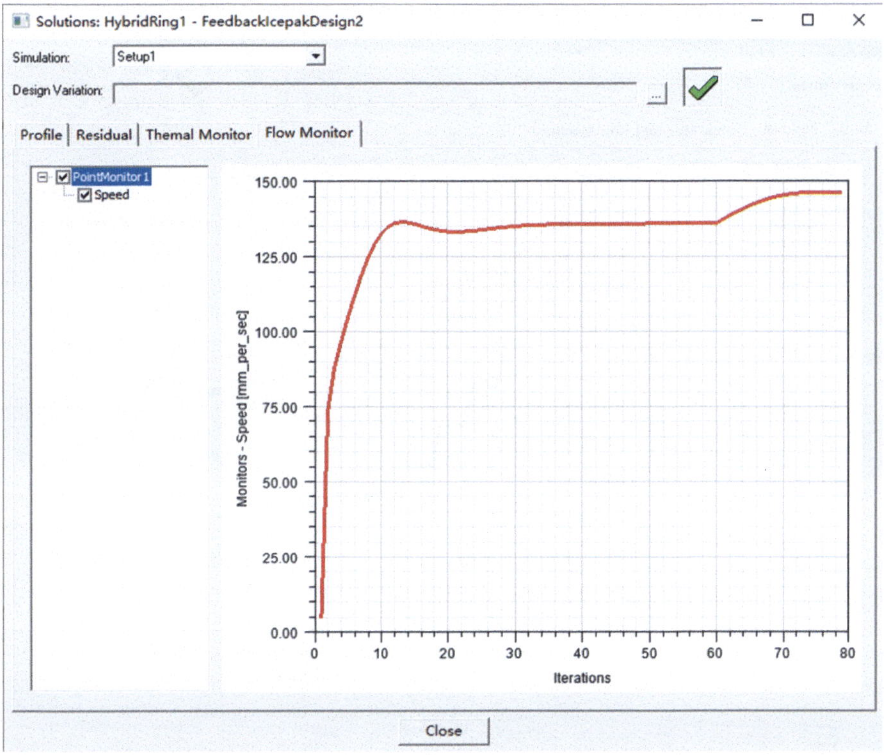

Fig. 6.5 Iterations of speed

9. Select "Steady State" and "Temperature and Flow" in the Solution Type. 10. In the Design Settings, you can set the simulation initial state, which is by default at room temperature. 11. Now you can add the Solution Setup. In the popped-up menu, you can set the maximum calculation iterations, the type of thermal analysis solution, and set the convergence criteria on the Convergence page. 12. Set monitoring points to assist convergence evaluation. Choose the center point on the upper surface of the air region and the upper surface of the ring as monitoring points. 13. Right-click on setup to enable bidirectional coupling. You can set the number of couplings and the maximum iterations per coupling. 14. After completing the setup, you can start the simulation. Icepak and HFSS will automatically perform the bidirectional coupling analysis without the need for manual intervention. After the simulation is completed, you can right-click on setup, click on "Thermal Monitor" or "Flow Monitor" to check the convergence status (Figs. 6.4, 6.5, and 6.6).

15. Select Ring, right-click and choose Plot Fields, select Temperature, and check "Plot on surface only" in the popped-up menu to output the distribution of temperature field results in the bidirectional coupled simulation (Fig. 6.7).

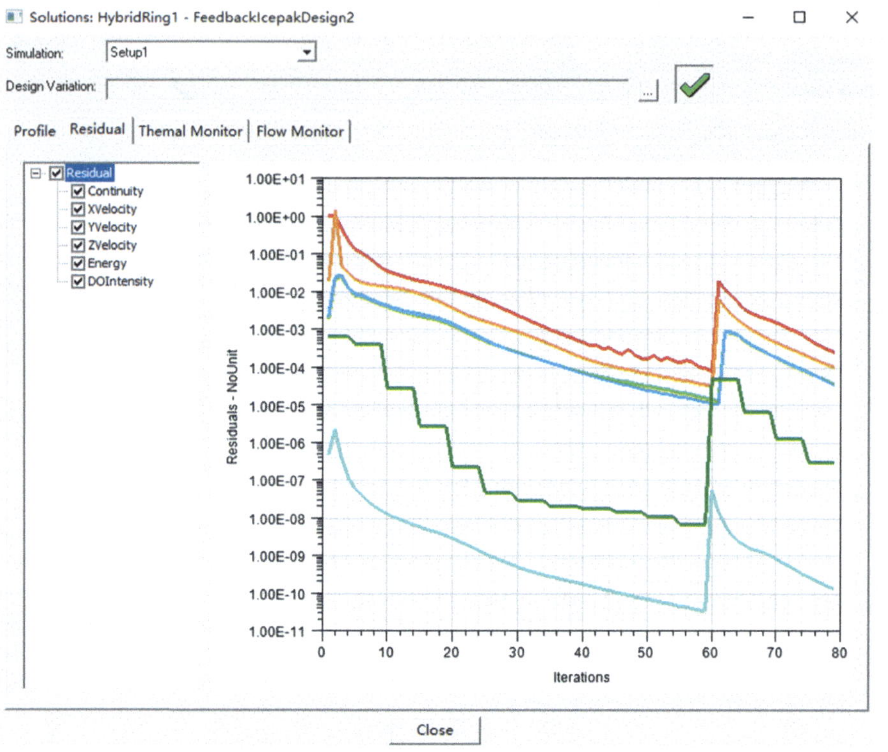

Fig. 6.6 Iterations of residual

Using simulation methods for digital design evaluation has become a norm in modern product development. The Ansys electromagnetic-thermal simulation platform possesses powerful multiphysics analysis capabilities, enabling us to quickly optimize electromagnetic fields and corresponding Joule thermaling through simulation design. This assists businesses in efficiently accomplishing more challenging product designs. During the process of combined thermal simulation with HFSS and Icepak, we can accurately predict and analyze the electromagnetic and thermal behavior of the product, while iteratively optimizing the design to ensure performance and reliability.

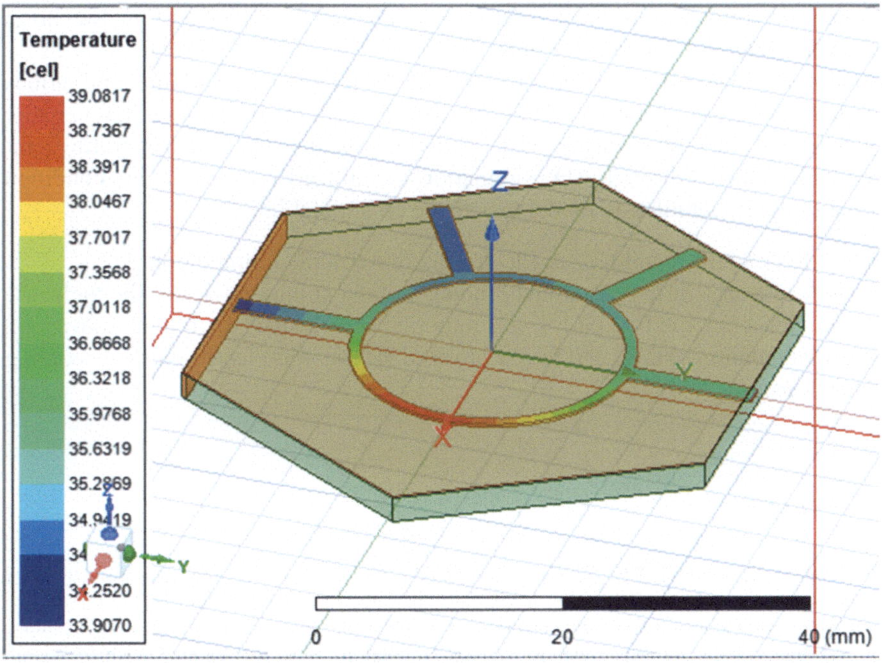

Fig. 6.7 Distribution of temperature field

Problems

P6.1 Synergistic Optimization Design of Low-Noise Amplifier (LNA) Chip
To address the design challenges of the LNA-MMIC chip, circuit and electromagnetic simulation models of the LNA-MMIC chip are separately established using ADS. Additionally, a thermal simulation model of the chip's packaging is created using Flotherm. The MDIDS software is utilized to accomplish multidisciplinary collaborative optimization of the LNA-MMIC chip.

The optimization objectives, employing the built-in DemoKit-Non-Linear standard library in ADS, are as follows:

- Frequency range: 8–12 GHz.
- Small-signal gain: greater than 11 dB.
- Noise figure: less than 2.5 dB.
- Stability factor: greater than 1.
- 1 dB compression point output power: greater than 10 dBm.

The ADS circuit and electromagnetic co-simulation design include small-signal S-parameter circuit-electromagnetic co-simulation design and harmonic balance circuit-electromagnetic co-simulation design.

Flotherm software is utilized to package and model the bare die of the low-noise amplifier, obtaining chip temperature indicators. The thermal packaging optimization of the LNA-MMIC chip aims to simulate the temperature variations of various components during operation and ensure that the maximum operating temperature is below 61 °C, thereby ensuring proper chip functionality. The chip's operating temperature is obtained through ADS simulation and utilized as the excitation condition for the Flotherm simulation model to achieve co-simulation.

P6.2 ANSYS HFSS and Icepak Electro-thermal Coupled Simulation
The surface losses of the metal layer model and the volume losses of the dielectric layer model, computed in HFSS, are imported into Icepak for forced convection cooling calculations. HFSS and Icepak are used for one-way electro-thermal coupling calculations.

References

1. Bechtold T, Rudnyi EB, Korvink JG (2005) Dynamic electro-thermal simulation of microsystems-a review. Journal of Micromechanics and Microengineering 15(11):R17
2. L'Eplattenier P et al (2009) Introduction of an electromagnetism module in LS-DYNA for coupled mechanical-thermal-electromagnetic simulations. Steel Res Int 80(5):351–358
3. Rajapakse AD, Gole AM, Wilson PL (2005) Electromagnetic transients simulation models for accurate representation of switching losses and thermal performance in power electronic systems. IEEE Trans Power Deliv. 20(1):319–327
4. Peuportier B, Thiers S, Guiavarch A (2013) Eco-design of buildings using thermal simulation and life cycle assessment. J Clean Prod 39:73–78
5. Sarno C, Tantolin C (2010) Integration, cooling and packaging issues for aerospace equipments. In: 2010 design, automation & test in europe conference & exhibition (DATE 2010). IEEE
6. Lee J, Gerlach D. W, Joshi Y. K (2008) Parametric thermal modeling of thermal transfer in handheld electronic devices. In: 2008 11th intersociety conference on thermal and thermomechanical phenomena in electronic systems. IEEE
7. Sloos T et al (2022) A multi-criteria decision analysis of implanted biomedical device antenna: electro-thermal simulation, design, and data analysis. Simulation 98(10):863–873
8. Welle B, Haymaker J, Rogers Z (2011) ThermalOpt: A methodology for automated BIM-based multidisciplinary thermal simulation for use in optimization environments. Building simulation 4:Tsinghua Press
9. Yigit S, Ozorhon B (2018) A simulation-based optimization method for designing energy efficient buildings. Energy Build. 178:216–227
10. Zhang L et al (2022) Optimization method of the thermal control index of space remote sensors based on force-thermal coupling algorithm. Appl. Thermal Eng. 216:119058
11. Mara TA, Tarantola S (2008) Application of global sensitivity analysis of model output to building thermal simulations. Build. Simul. 1. Springer, Berlin, Heidelberg
12. Székely V et al (1997) Electro-thermal and logi-thermal simulation of VLSI designs. IEEE Trans Very Large Scale Integr (VLSI) Syst 5(3):258–269
13. Chung DDL (2001) Materials for thermal conduction. Appl Therm Eng 21(16):1593–1605
14. Lepri S, Livi R, Politi A (2003) Thermal conduction in classical low-dimensional lattices. Phys Rep 377(1):1–80
15. Jiji LM, Jiji LM (2006) Thermal convection. Springer, New York
16. Busse FH (1978) Non-linear properties of thermal convection. Rep Prog Phys 41(12):1929
17. Siegel R (2001) Thermal radiation thermal transfer. CRC Press

18. Idso SB, Jackson RD (1969) Thermal radiation from the atmosphere. J Geophys Res 74(23):5397–5403
19. Remsburg R (2017) Thermal design of electronic equipment. CRC Press
20. Remsburg R (1998) Advanced thermal design of electronic equipment. Springer Science & Business Media
21. Al-Homoud MS (2005) Performance characteristics and practical applications of common building thermal insulation materials. Build Environ 40(3):353–366
22. Wei T et al (2017) Optimization design of heat dissipation structure of integrated motor drives for application in aerospace. In: 2017 20th international conference on electrical machines and systems (ICEMS). IEEE
23. Horváth G, Poppe A (2005) The Sissy electro-thermal simulation system-Based on modern software technologies. THERMINIC 2005. TIMA Editions
24. Howell JR, Perlmutter M (1964) Monte Carlo solution of thermal transfer through radiant media between gray walls. pp 116–122
25. Lewis RW et al (1996) The finite element method in heat transfer analysis. Wiley
26. Sun Z et al (2010) Efficient finite element analysis/computational fluid dynamics thermal coupling for engineering applications. J Turbomach 132:3
27. Haas BL et al (1994) Rates of thermal relaxation in direct simulation Monte Carlo methods. Phys Fluids 6(6):2191–2201
28. Kakaç S, Yener Y, Naveira-Cotta CP (2018) Heat conduction. CRC Press
29. Narasimhan TN (1999) Fourier's heat conduction equation: history, influence, and connections. Rev Geophys 37(1):151–172
30. Thomas PH (1958) On the thermal conduction equation for self-heating materials with surface cooling. Trans Faraday Soc 54:60–65
31. Brysk H, Campbell PM, Hammerling P (1975) Thermal conduction in laser fusion. Plasma Phys 17(6):473
32. Nettleton RE (1960) Relaxation theory of thermal conduction in liquids. Phys Fluids 3(2):216–225
33. Narasimhan TN (1999) Fourier's thermal conduction equation: History, influence, and connections. Rev Geophys 37(1):151–172
34. Özisik MN, Özışık MN, Özışık MN (1993) Heat conduction. Wiley
35. Hahn DW, Özisik MN (2012) Heat conduction. Wiley
36. Grigull U, Sandner H (1984) Heat conduction, vol 7. Springer, Berlin
37. Cheng PING, Hsu CT (1998) Heat conduction. Transport phenomena in porous media. Pergamon Press, Elsevier Science, Oxford, pp 57–76
38. Beck JV et al (1992) Heat conduction using Green's function. Taylor & Francis
39. Wang T, Tseng KJ, Zhao J (2015) Development of efficient air-cooling strategies for lithium-ion battery module based on empirical thermal source model. Appl Therm Eng 90:521–529
40. Kumar R, Abbas IA (2013) Deformation due to thermal source in micropolar thermoelastic media with thermal and conductive temperatures. J Comput Theor Nanosci 10(9):2241–2247
41. Vican JGBF et al (2002) Development of a microreactor as a thermal source for microelectromechanical systems power generation. Proc Combust Inst 29(1):909–916
42. Marin M, Hobiny A, Abbas I (2021) Finite element analysis of nonlinear bioheat model in skin tissue due to external thermal sources. Mathematics 9(13):1459
43. Mirkoohi E et al (2019) Thermal source modeling in selective laser melting. Materials 12(13):2052
44. Teixeira PRF, Araújo DB, da Cunda LAB (2014) Study of the Gaussian distribution thermal source model applied to numerical thermal simulations of TIG welding processes. Journal/Conference Name
45. Srinivasan K, et al (2004) Vehicle thermal management simulation using a rapid omni-tree based adaptive Cartesian mesh generation methodology. In: Heat transfer summer conference, vol 46911
46. Yang S et al (2015) Volume element model mesh generation strategy and its application in ship thermal analysis. Adv Eng Softw 90:107–118

47. Zeng K et al (2015) Comparison of 3DSIM thermal modeling of selective laser melting using new dynamic meshing method to ANSYS. Mater Sci Technol 31(8):945–956
48. Lu T, Jin JM (2016) Electrical-thermal Co-simulations for analysis of high-power RF/microwave components. IEEE Trans Electromagn Compat 59(1):93–102
49. Rajapakse AD, Gole AM, Wilson PL (2005) Electromagnetic transients simulation models for accurate representation of switching losses and thermal performance in power electronic systems. IEEE Trans Power Deliv 20(1):319–327
50. Wang CS et al (2010) Coupled structural-electromagnetic-thermal modelling and analysis of active phased array antennas. IET Microwaves, Anten Propagat 4(2):247–257

Chapter 7
Co-simulations of Structures

7.1 Introduction

Structural simulations refer to using various sorts of computer software to simulate and analyze various structural problems in engineering so that the optimal designs of structures can be achieved. These analyses may include static analysis, dynamic analysis, thermodynamic analysis, and fluid-mechanic analysis for structures [1–6]. The structural simulations have become indispensable for optimized structural designs and have been widely used in various areas. The primary application areas include (a) Mechanical engineering where the simulations can enable a quick analysis of strength, stiffness, stability, and other characteristics of mechanical components, as well as the evaluation of mechanical system performances [7–10]. The structural simulations are also utilized for design optimizations and fault diagnoses. For instance, in engine designs, the simulation software can simulate the vibration and noise of engines under different operating conditions, predict their fatigue life, and improve their designs to reduce failure rates. (b) Aerospace engineering where the simulations are very essential in the analyses and designs of aerospace vehicles such as aircrafts, satellites, and rockets [11–14]. The simulation software can simulate the material stress and deformation under various load conditions and thus help optimize the structural designs with optimal performances and reliability for aerospace vehicles. For example, in aircraft designs, the simulations are used to evaluate the stiffness and strength of fuselages, optimize wing aerodynamics, and improve landing gear designs for aircrafts. (c) Automotive engineering where the simulations are extensively employed for the structural designs and safety evaluation of vehicles [15–18]. The simulations can predict the change of stress and deformation of the body under various collision conditions, facilitating the assessment of safety performances for vehicles. (d) Constructional engineering where the structural simulations are extensively applied for the analyses and designs of building structures [19–22]. By simulating the stress change and deformation of buildings under an earthquake,

M. S. Tong and X. Y. Li, *Co-simulations of Microwave Circuits and High-Frequency Electromagnetic Fields*, https://doi.org/10.1007/978-981-99-8307-0_7

wind, snow, and other loads, one can predict the structural parameters of buildings so that their seismic resistance and safety can be enhanced by optimizing their structural designs.

With the continuous development of computer technologies and numerical methods, the simulation software tools have become more and more powerful and their applications have also been greatly expanded and deepened. In the future, the simulation technologies for structures will develop toward the functions of multiphysics simulations and high-performance computing. Multiphysics simulations involve the co-simulations of multiple physical fields such as mechanical, thermal, electrical, and magnetic [23–27]. Such simulations can more accurately predict the behaviors and performances of simulated structures so as to achieve an optimization for their structural designs.

Due to the complexity and large scale of multiphysics problems, the high-performance computing techniques are needed for multiphysics simulations. The high-performance computing techniques employ the parallel computing and distributed computing schemes to significantly enhance the computational speed and efficiency in simulations. The techniques can provide strong support for the fast and accurate simulations of more complex structures. In addition, the artificial intelligence (AI) technology can also be incorporated into structural simulations. Through machine learning and deep learning techniques, the simulation software tools can become much more intelligent and stronger so that large simulation data can be quickly and smartly analyzed and handled, further enhancing the capability of structural designs and optimizations. This chapter will discuss the structural simulations and co-simulations in detail, including basic principles, simulation methods, result analyses, optimization methods, and applications.

7.2 Goals and Challenges of Structural Simulations

In structural simulations, the primary goal is to accurately predict the behavior of objects under different working conditions. This requires the construction of a suitable mathematical model that can be processed by a computer. The model should encompass important parameters such as geometry, material properties, and boundary conditions to facilitate accurate calculations during the simulation process.

Another important goal is design optimization. Structural simulation enables the comparison of different design schemes, determination of the optimal design, and reduction of testing and manufacturing costs. Design parameters can also be optimized, such as weight reduction or material strength enhancement, to improve overall performance.

The ultimate goal is design validation. Simulations' results can be compared with experimental results to verify the feasibility of the design and provide guidance for improvement. This helps identify design issues, minimizing potential losses and waste during actual manufacturing.

However, structural simulation faces several challenges and difficulties, which include the following aspects:
Modeling and Meshing of Complex Geometry:
Simulating the mechanical behavior of objects requires accurate modeling and discretization into finite element models. Complex geometries, such as buildings and cars, often necessitate intricate geometric processing and meshing. This process consumes time and resources, and requires modeling and meshing stability and accuracy.
1. Modeling and Parameter Determination of Mechanical Properties:
Mechanical properties of materials are crucial in structural simulation, directly impacting the accuracy of results. However, material properties are diverse, encompassing elasticity, plasticity, creep, fracture, etc., requiring complex modeling and parameter determination. Material properties are also influenced by factors like temperature and stress history, necessitating extensive testing and analysis.
2. Treatment of Complex Boundary Conditions:
Structural simulation deals with complex external forces on objects, such as gravity, wind load, and seismic forces. These forces cause deformation and stress distribution, necessitating the treatment of complex boundary conditions. Determining boundary conditions often involves considering factors like load size, direction, and location, requiring thorough testing and analysis.
3. Calculation and Solution of Large-Scale Simulations:
Advancements in computer technology allow large-scale parallel calculations and solutions in structural simulation. However, practical simulation models often contain numerous nodes and elements, requiring complex calculations and solutions. Nonlinear and dynamic problems often necessitate iterative solutions, calling for efficient algorithms and optimized calculation methods.
4. Guarantee of Accuracy and Reliability:
Ensuring accuracy and reliability in structural simulation is crucial for obtaining correct results. Due to modeling uncertainties, parameter settings, and boundary condition processing, simulation results may have errors and deviations. Analyzing and evaluating accuracy and reliability are necessary. Comparing and verifying multiple sets of simulations improves the reliability of the results.

Structural simulation faces various challenges and difficulties that require comprehensive technical research and optimization. The subsequent sections will introduce and analyze key technologies in structural simulation.

7.3 Categories and Methods of Structural Simulations

Structural simulation refers to the process of predicting and analyzing the performance and behavior of a physical structure under various conditions by simulating its working state and behavior through computers. Structural simulation technology is widely used in fields such as aerospace, automotive, construction, electronics, and mechanical manufacturing. Through structural simulation, the cost and time of product development can be effectively reduced, while improving the quality and reliability of the product. Structural simulation can be divided into statics simulation, dynamics simulation, thermodynamics simulation, and fluid-mechanic simulations [28–31]. Among them, statics simulation is mainly used to study the static mechanical properties of structures, dynamics simulation is primarily used to study the dynamic response and vibration characteristics of structures, thermodynamics simulation is mainly used to study the deformation and thermal stress of structures under thermal loads, and fluid-mechanic simulations mainly simulates the motion and interaction of fluids through mathematical models and efficient computational algorithms. This section will introduce the methods and applications of these three types of simulation, respectively.

7.3.1 Static Simulations

Static simulation is a computational process that predicts and analyzes the deformation and failure behavior of structures under static loads by simulating the mechanical properties and behavior of the physical structure. In engineering design, static simulation technology is widely employed in analyzing and designing mechanical components, building structures, bridges, towers, and other structures. The methods commonly used for static simulation include the finite element method, boundary element method, and finite-difference method, with the finite element method being one of the most prevalent [32–34].

The fundamental principle of static simulation is based on the elasticity principle, which relates stress (σ) to strain (ϵ) through the equation $\sigma = E\epsilon$, where E represents Young's modulus. By creating a finite element model of the structure, it is discretized into numerous small units. The strain and stress of each unit are calculated based on the boundary and load conditions, allowing for the comprehensive determination of strain and stress distribution throughout the entire structure. This information enables the prediction of the deformation and failure state of the structure.

When an elastic body with a volume V and Young's modulus E is subjected to an external force F, it will deform to a certain extent. To study this deformation, the elastic material can be discretized into N small units, each with varying shapes and sizes. The stress, strain, and stiffness matrices of each element are calculated, and the strain, stress, and stiffness matrices of the entire structure are comprehensively determined to predict its deformation and failure state. The finite element method, which is widely used for static simulation, divides the structure into multiple elements, each with unique shapes and sizes. By calculating the stress, strain, and stiffness matrices of these elements, the overall strain, stress, and stiffness matrices of the structure are obtained, enabling the prediction of its deformation and failure state. The finite element method consists of two stages: element analysis and system analysis.

In the element analysis stage, the structure is divided into small finite elements, and the stress, strain, and stiffness matrices of each element are calculated to determine their deformations and establish element relationships. In the system analysis stage, the entire structure is analyzed as a whole. By solving the equations considering boundary and load conditions, the strain, stress, and stiffness matrices of the complete structure are calculated, leading to a comprehensive assessment of its deformation and failure state. The finite element method employs various element types, such as triangles, quadrilaterals, hexahedra, and tetrahedra, and the selection of appropriate element shapes and sizes can enhance the accuracy and efficiency of calculations.

In the finite element method, it is necessary to establish the stiffness matrix of the structure, which is constructed by combining the stiffness matrices of each element. The stiffness matrix of each element is calculated based on the material properties and element geometry. The resulting stiffness matrix is a symmetric matrix where each element represents the stiffness value in a particular direction. Additionally, boundary conditions (e.g., fixed support, free support, clamping support) and load conditions (e.g., concentrated load, uniformly distributed load, pressure load) must be considered. Different boundary and load conditions can be incorporated into the simulation process based on the requirements of the practical problem.

7.3.2 Dynamic Simulations

Dynamic simulation is a fundamental technique in structural analysis used to investigate the behavior and vibration characteristics of structures subjected to dynamic loads. It finds wide applications in mechanical engineering, automotive industry, aerospace sector, electronics, and other related fields.

The dynamic simulation process is based on Newton's second law, which states that the acceleration of an object is directly proportional to the force applied to it and inversely proportional to its mass. In dynamic simulation of structures, it is assumed that the structure consists of multiple discrete units, with each unit experiencing vibrations when subjected to external loads. These vibrations are then transmitted to adjacent units, eventually defining the vibration characteristics of the entire structure. Thus, the objective of dynamic simulation is to calculate the acceleration, velocity, and displacement of each unit to determine the overall response and vibration characteristics of the structure under various loads.

The fundamental equations for dynamic simulation are expressed as follows:

$$m \cdot a + c \cdot v + k \cdot x = f(t) \tag{7.1}$$

where m represents the mass matrix of the structure, a denotes the acceleration matrix, c is the damping matrix, v represents the velocity matrix, k denotes the stiffness matrix, x represents the displacement matrix, and $f(t)$ is the external load matrix applied to the structure.

According to Newton's second law, the equation can be rewritten as

$$m \cdot a = f(t) - c \cdot v - k \cdot x \tag{7.2}$$

By discretizing the structure into numerous small elements and determining the mass, stiffness, and damping matrices for each element, the dynamic simulation problem can be transformed into solving a large-scale linear equation system:

$$M \cdot a + C \cdot v + K \cdot x = F(t) \tag{7.3}$$

where M, C, and K represent the mass, damping, and stiffness matrices of the structure, respectively. a, v, and x denote the acceleration, velocity, and displacement matrices of the structure, while $F(t)$ is the external load matrix applied to the structure. Commonly used methods for dynamic simulation include the finite element method, boundary element method, and multibody dynamics method. Among these methods, the finite element method is the most widely employed approach. It involves dividing the structure into small elements, which can vary in shape and size. By calculating the acceleration, velocity, and displacement of each element, the overall vibration characteristics and response of the structure are determined, allowing for predictions of structural stability and fatigue life under vibration and impact loads.

1. Finite element method

The finite element method is a widely used structural simulation method that involves discretizing the structure into numerous small elements. By calculating the vibration characteristics and response of each element, the method determines the overall vibration characteristics and response of the entire structure.

2. Boundary element method

The boundary element method (BEM) is a technique that discretizes the structure's surface, focusing solely on the vibration and response characteristics of the surface. It decomposes the structure's surface into small elements, analyzes the vibration and response of each element, and then combines the results to obtain the overall vibration characteristics and response of the structure. Unlike the finite element method, the boundary element method does not require discretization of the entire structure, making it advantageous for handling large and complex structures. Additionally, the boundary element method can be applied to vibration analysis across different frequency ranges, providing specific advantages in certain applications.

3. Multibody dynamics method

The multibody dynamics method is employed to address structural vibration problems, particularly those related to multibody systems. This method decomposes the structure into multiple rigid or flexible bodies, analyzes the vibration of each body, and then combines the analysis results to determine the overall vibration characteristics and response of the structure.

The solving process of the multibody dynamics method involves three steps: model establishment, system equation determination, and system equation solving. During model establishment, the structure is decomposed into multiple rigid or flexible bodies, with appropriate material and physical parameters assigned to each body. When determining the system equations, the interaction between rigid or flexible bodies and the external load is considered to derive the equations of motion for the system. Various methods, including analytical and numerical techniques, can be utilized to solve the system equations.

During dynamic simulation, the following considerations should be taken into account:

(1) Selection of appropriate load and boundary conditions: Dynamic simulation requires careful consideration of the effects of load and boundary conditions on the structure. Therefore, it is crucial to choose the appropriate load and boundary conditions to obtain accurate simulation results.

(2) Selection of suitable materials and physical parameters: Materials and physical parameters significantly influence the vibration characteristics and response of the structure. It is important to choose appropriate materials and physical parameters, and conduct material testing and parameter calibration to ensure accurate representation.

(3) Validation of simulation results: The accuracy of dynamic simulation results should be verified and validated. This can be achieved by comparing the simulation results with experimental data, ensuring the reliability and accuracy of the simulated behavior.

(4) Attention to numerical stability: Dynamic simulation often involves complex and large-scale calculations. It is essential to prioritize numerical stability and computational accuracy to ensure reliable simulation results.

7.3.3 Thermodynamic Simulations

Thermodynamic simulation is a crucial method in structural simulation, primarily employed to analyze and predict the deformation and failure state of structures under high-temperature environments. Practical applications of thermodynamic simulation often involve complex physical fields and various coupling effects, necessitating the use of sophisticated numerical methods and software tools for analysis.

Thermal conduction refers to the transfer of thermal energy within a substance, and its fundamental principle can be described by the thermal conduction equation. In the one-dimensional case, the thermal conduction equation can be expressed as

$$\frac{\partial T}{\partial t} = \frac{\alpha}{\rho c} \frac{\partial^2 T}{\partial x^2} + \frac{Q}{\rho c} \tag{7.4}$$

where T represents temperature, t denotes time, x is the spatial coordinate, α signifies the thermal conductivity, ρ represents the density, c denotes the specific thermal capacity, and Q denotes the thermal source intensity. The first term in the equation describes the thermal conduction process, while the second term accounts for the thermal source. A similar form of the thermal conduction equation can be written for the three-dimensional case. The thermal conduction equation is a fundamental equation in thermodynamic simulation, enabling the calculation of temperature distribution and thermal stress distribution in structures subjected to thermal loads.

Thermal stress refers to the stress changes in a structure resulting from temperature variations. Its magnitude and distribution are influenced by the geometric shape, material properties, and temperature change rate of the structure. Thermal stress analysis is a vital aspect of thermodynamic simulation, enabling the prediction of deformation and failure states of structures under high-temperature environments.

In thermal stress analysis, the linear thermoelastic theory is commonly employed to describe the relationship between thermal stress and temperature. According to this theory, there exists a linear relationship between the stress change $\Delta\sigma$ and the temperature change ΔT in a structure:

$$\Delta\sigma = \alpha\Delta T \tag{7.5}$$

where α represents the linear thermal expansion coefficient, which can be determined through experimental measurements or referenced from material literature.

According to the formula mentioned above, the stress distribution and thermal stress distribution of the structure in a high-temperature environment can be calculated. When the thermal stress reaches the yield strength of the material, the structure will fail.

Thermodynamic simulation is an important method in structural simulation, primarily used for analyzing and predicting the deformation and failure state of structures in high-temperature environments. In practical applications, thermodynamic simulation often involves complex physical fields and various coupling effects, necessitating the use of sophisticated numerical methods and software tools for analysis. thermal conduction refers to the transfer process of thermal energy within a substance, and its basic principle can be described by the thermal conduction equation. In the one-dimensional case, the thermal conduction equation can be written as

$$\frac{\partial T}{\partial t} = \frac{\alpha}{\rho c}\frac{\partial^2 T}{\partial x^2} + \frac{Q}{\rho c} \tag{7.6}$$

where T represents temperature, t represents time, x represents spatial coordinates, α represents thermal conductivity, ρ represents density, c represents specific thermal capacity, and Q represents thermal source intensity. In the above equation, the first term describes the thermal conduction process, and the second term describes the effect of the thermal source. In the three-dimensional case, the thermal conduction equation can be written in a similar form. The thermal conduction equation is one of the fundamental equations in thermodynamic simulation, enabling the calculation of temperature distribution and thermal stress distribution in the structure under thermal loads.

Thermal stress refers to the change of stress in the structure caused by temperature change, and its magnitude and distribution are related to the geometric shape, material properties, and temperature change rate of the structure. Thermal stress analysis is one of the important contents of thermodynamic simulation, which can be used to predict the deformation and failure state of structures in a high-temperature environment.

In thermal stress analysis, the linear thermoelastic theory is usually used to describe the relationship between thermal stress and temperature. According to this theory, when a structure is subjected to a change in temperature, there is a linear relationship between the change in stress $\Delta\sigma$ and the change in temperature ΔT, that is

$$\Delta\sigma = \alpha\Delta T \tag{7.7}$$

where α represents the coefficient of linear thermal expansion, which can be obtained by experimental measurement or material literature review. According to the above formula, the stress distribution and thermal stress distribution of the structure in a high-temperature environment can be calculated. When the thermal stress reaches the yield strength of the material, the structure will fail.

When performing a thermal stress analysis, the linear elastic properties of the material may also be considered. According to the theory of linear elasticity, there is a linear relationship between stress and strain, which can be expressed by Hooke's law. In one dimension, Hooke's law can be written as

$$\sigma = E\varepsilon \tag{7.8}$$

where σ is stress, E is the elastic modulus, and ε is strain. Strain can be determined by the stress-strain curve of stress and material. Combined with the linear thermal expansion coefficient and Hooke's law, the total stress variation of the structure at high temperature can be obtained

$$\Delta\sigma = \alpha E\Delta T \tag{7.9}$$

This equation takes into account the thermal stress due to temperature changes and the linear elastic properties of the material. Through thermodynamic simulation and thermal stress analysis, the stress distribution, temperature distribution, and prediction of deformation and failure state of the structure in a high-temperature environment can be obtained. This is very important for the design and optimization of structures and materials working under high-temperature conditions, which can help improve the safety and reliability of structures.

It is important to note that the above discussion covers only basic concepts and equations in thermodynamic simulation and thermal stress analysis. In practical applications, additional factors and more complex models may be involved. The specific simulation and analysis methods should be determined based on the specific problems and research fields, and actual calculations and analysis should be conducted using professional simulation software and tools.

Thermodynamic simulation is a complex numerical analysis method that commonly employs numerical methods such as the finite element method, boundary element method, finite-difference method, and others. Due to the typically large-scale calculations involved in thermodynamic simulation, high-performance computers, and parallel computing technology are often required to accelerate the calculations. Additionally, since thermodynamic simulation encompasses the coupling effects of various physical fields, multi-field coupling simulation analysis is necessary to obtain accurate simulation results.

Currently, there are several thermodynamic simulation software packages available on the market. Popular options include ANSYS, ABAQUS, COMSOL, among others. These software packages offer powerful simulation capabilities and find wide application in areas such as static simulation, dynamic simulation, thermodynamic simulation, fluid simulation, and other types of simulation analysis.

Taking ANSYS as an example, its thermodynamic simulation module comprises three main components: thermal conduction analysis, thermal stress analysis, and coupled field analysis. Users can select the appropriate analysis module and calculation method based on their specific needs to carry out thermodynamic simulation analysis. ANSYS features an intuitive user interface and comprehensive post-processing functions that enable visualization and analysis of simulation results, facilitating engineers in making informed decisions and optimizing designs.

Similarly, ABAQUS is a kind of powerful finite element analysis software that supports static simulation, dynamic simulation, thermodynamic simulation, coupled field simulation, and other simulation analyses. ABAQUS boasts high scalability and a modular design, enabling the simulation analysis of complex structures and multiple physical fields. It also provides an intuitive user interface and flexible post-processing functions, facilitating users in analyzing and optimizing simulation results.

Apart from ANSYS and ABAQUS, there are specialized thermodynamic simulation software options available, such as TAS (Thermal Analysis System), THERMICA, and others. These software packages offer unique simulation capabilities and are tailored for specific application fields. Therefore, the choice of simulation software should align with the specific requirements of the application at hand.

7.3.4 Fluid-Mechanic Simulations

In the field of engineering, structural simulation is a technique that employs computer models to simulate and analyze the behavior of actual physical systems. A crucial branch within structural simulation is Computational Fluid Dynamics (CFD), which is the science of studying and analyzing fluid flow through numerical methods on a computer. This section will focus on introducing the fundamental principles and equations of fluid dynamics simulation [35–37]. The foundation of fluid dynamics simulation is the Navier-Stokes equations, which are a set of nonlinear partial differential equations that describe how the velocity field of a fluid changes over time. The general form of the Navier-Stokes equations is

$$\rho \left(\frac{\partial \mathbf{v}}{\partial t} + \mathbf{v} \cdot \nabla \mathbf{v} \right) = -\nabla p + \mu \nabla^2 \mathbf{v} + \mathbf{f} \tag{7.10}$$

where ρ is the density of the fluid, $\mathbf{v}$ is the velocity field, t is time, p is pressure, μ is dynamic viscosity, and $\mathbf{f}$ is the body force (such as gravity).

In fluid dynamics simulation, the continuity equation is also often used to ensure the conservation of mass within the flow. For incompressible fluids, the continuity equation is

$$\nabla \cdot \mathbf{v} = 0 \tag{7.11}$$

For compressible fluids, the continuity equation becomes

$$\frac{\partial \rho}{\partial t} + \nabla \cdot (\rho \mathbf{v}) = 0 \tag{7.12}$$

The Navier-Stokes equations are typically challenging to solve directly, especially in the presence of complex geometries and boundary conditions. Therefore, fluid dynamics simulation usually employs discrete methods such as the Finite Element Method (FEM), Finite Volume Method (FVM), and Finite-Difference Method (FDM).

FEM is a technique that transforms continuous physical equations into discrete algebraic equations. In FEM, the solution space is decomposed into a set of basis functions, usually polynomials. For fluid flow problems, the Galerkin method can be used to write the Navier-Stokes equations in a weak form:

$$\int_{\Omega} \rho \left(\frac{\partial \mathbf{v}}{\partial t} + \mathbf{v} \cdot \nabla \mathbf{v} \right) \cdot \delta \mathbf{v} d\Omega = \int_{\Omega} (-\nabla p + \mu \nabla^2 \mathbf{v} + \mathbf{f}) \cdot \delta \mathbf{v} d\Omega \tag{7.13}$$

where Ω is the fluid domain, and $\delta \mathbf{v}$ is the test function. In the Finite Volume Method, control volumes are used to discretize the fluid domain. The Navier-Stokes equations and the continuity equation are approximated in integral form over each control volume:

$$\int_V \rho \left(\frac{\partial \mathbf{v}}{\partial t} + \nabla \cdot (\mathbf{v}\mathbf{v}) \right) dV = \int_V \Big(- \nabla p + \mu \nabla^2 \mathbf{v} + \mathbf{f} \Big) dV \tag{7.14}$$

where V is the control volume. Fluid dynamics simulation also involves the handling of boundary conditions, such as wall conditions, inlet and outlet conditions. At the wall, it is commonly assumed that there is no relative slip between the fluid and the wall (no-slip boundary condition), that is

$$\mathbf{v} = \mathbf{0} \quad \text{on} \quad \partial \Omega_{\text{wall}} \tag{7.15}$$

Inlet and outlet boundary conditions typically need to be set based on the nature of the problem. The applications of fluid dynamics simulation are diverse and include meteorology, aerospace, automotive industry, biomedical engineering, and more. Through simulation, engineers and scientists can gain a deeper understanding of the complexities of fluid flow, providing valuable information for design and optimization. When performing fluid dynamics simulation, various factors need to be considered, including grid discretization, time step selection, iterative solvers, and convergence criteria. For transient problems, time-dependent equations need to be solved, and an appropriate time integration method should be chosen. Fluid dynamics simulation is an extremely important and powerful tool in structural simulation. By solving the Navier-Stokes equations and the associated continuity equation, it can be used to simulate and analyze complex fluid flow problems. However, due to the nonlinearity and complexity of these equations, fluid dynamics simulation often requires significant computational resources and expertise to set up and interpret results correctly.

7.4 Principles of Structural Simulations

The principle of structural simulation is based on numerical simulation techniques such as finite element analysis (FEA) and computational fluid dynamics (CFD). Finite element analysis involves discretizing the actual object into a finite number of small elements, establishing a local coordinate system in each element, and solving for displacement and strain to obtain information such as deformation and stress distribution. Computational fluid dynamics, on the other hand, utilizes mathematical models to describe fluid movement and various physical phenomena within the fluid, including thermal transfer and mass transfer. By solving the equations, characteristics such as flow velocity, pressure, and temperature of the fluid can be determined.

In the structural simulation, the object is discretized into either a finite element model or a computational fluid dynamics model. For the finite element model, the object is modeled using CAD software and then imported into finite element analysis software for discretization. In the case of a computational fluid dynamics model, the fluid domain is discretized into a finite number of small grids. Boundary conditions and physical parameters are set, and the computational fluid dynamics software is employed to solve the equations.

When dealing with structural simulation problems, it is crucial to define the appropriate constraint conditions and load conditions based on the object's boundary conditions, loading methods, and material parameters. By selecting the suitable solution algorithm, deformation, stress, vibration, and other characteristics of the object under force can be obtained [1, 2, 38–40]. Optimization and fault diagnosis can be performed by modifying parameters such as the object's geometric shape, material parameters, and loading modes.

7.4.1 Stress and Strain Analyses

Stress, denoted as σ, represents the force per unit area. Assuming an object is subjected to a force F and has an area A, the stress can be expressed as [41]

$$\sigma = \frac{F}{A} \tag{7.16}$$

where σ is measured in N/m^2 or Pa (Pascal). Stress acts perpendicular to the object's surface in the same direction as the force applied. In the mechanics of materials, stress is typically categorized as tensile stress, compressive stress, or shear stress. Tensile stress elongates an object when deformed by a force, usually represented as a positive value. Compressive stress compresses an object when deformed by a force, generally represented as a negative value. Shear stress induces tangential forces within an object when deformed by a force.

Stress and strain analyses are the fundamental tools extensively used in mechanical engineering, civil engineering, materials science, and related fields. It is employed to examine the deformation behavior of a material under external loads and determine the induced stress distribution within the material [41, 42]. Understanding stress and strain concepts is crucial as they provide deep insights into material properties and structural behavior. Strain, represented as ε, quantifies the amount of deformation per unit length. Assuming an object has an initial length L and a deformed length L', the strain of the object can be expressed as

$$\varepsilon = \frac{L' - L}{L} \tag{7.17}$$

Strain is dimensionless. It can be categorized as longitudinal strain, transverse strain, or shear strain. Longitudinal strain is the ratio of elongation to the initial length of an object in the direction of tension. Transverse strain is the ratio of compression to the initial length of an object perpendicular to the direction of tension. Shear strain is the ratio of the amount of deformation in the shear direction to the object's length.

A relationship exists between stress and strain. In the elastic region, stress is proportional to strain. This ratio, often referred to as the modulus of elasticity, describes the material's stiffness. The modulus of elasticity, denoted as E, represents the ratio of tensile stress per unit area to the corresponding longitudinal strain and can be expressed as

$$E = \frac{\sigma}{\varepsilon} \tag{7.18}$$

where E is measured in Pa. In the plastic deformation region of a material, the stress-strain relationship becomes more complex. When the material starts to deform, the stress-strain curve is no longer linear but exhibits a curved shape. This behavior is a consequence of microstructural changes within the material, such as lattice slip or proliferation, influencing the deformation mechanism. During this stage, an increase in strain leads to an increase in stress, but the rate of stress increase gradually diminishes until it reaches a maximum value, known as the material's yield strength. Once the stress exceeds the yield strength, plastic deformation occurs. At this point, an increase in strain no longer causes an increase in stress but rather maintains a constant stress level or causes stress reduction. This phenomenon arises due to the plastic deformation inside the material, which results in a decrease in strength and stiffness. In the plastic deformation region, the stress-strain relationship is no longer a single function but a highly nonlinear and complex relationship [43, 44].

In engineering practice, stress and strain analysis methods typically involve theoretical analysis, numerical simulation, and experimental measurement. These methods can be used individually or in combination to obtain more accurate and comprehensive results.

Theoretical analysis is a method that calculates the stress and strain of materials based on theoretical formulas and mathematical models. This approach is suitable for simple geometries and boundary conditions, such as bending and torsion of straight or curved bars, beams, or plates. In theoretical analysis, it is important to determine the mechanical properties of materials, such as the elastic modulus, yield strength, and fracture strength.

Numerical simulation is a computer-based method that simulates the stress and strain of materials under different conditions. This method is suitable for complex geometries and unconventional boundary conditions, including three-dimensional structures, complex loading scenarios, and multi-physical field interactions. Techniques like finite element analysis (FEA) or computational fluid dynamics (CFD) are commonly used in numerical simulations. The process involves discretization, numerical solution, and post-processing. When conducting numerical simulations, it is crucial to select appropriate material models and boundary conditions, and to perform reasonable verification and validation.

Experimental measurement is a method of directly measuring the stress and strain of materials using laboratory equipment. This approach allows for direct observation of material behavior and provides real data to validate theoretical models and numerical simulation results. Experimental measurements can be performed using mechanical testers, tensile testers, compression testers, bending testers, and other specialized equipment. It is important to select suitable test methods and measurement techniques, and to comply with relevant test standards and specifications when conducting experimental measurements.

To measure the stress and strain of materials, special instruments and equipment are typically required. Stress can be measured using a stress sensor, also known as a strain gauge. A strain gauge is a type of sensor that measures the deformation of an object when subjected to stress. When an object is stressed, the strain gauge generates an electrical signal. This signal can be amplified by an amplifier, converted into a digital signal, and then analyzed and processed by a computer. Strain gauges have an accuracy typically ranging from 0.1% to 1% and can be used to measure various types of stress, including tensile, compressive, and shear stress [45].

Measurement of strain can be accomplished using strain gauges or optical strain gauges. Strain gauges calculate strain by measuring changes in electrical resistance. On the other hand, optical strain gauges utilize optical principles to calculate strain by analyzing the reflection or transmission of light on the surface of an object. Optical strain gauges typically offer an accuracy between 0.01% and 0.1% and can be used to measure very small strains.

Stress and strain analysis plays a vital role in various engineering applications. It is used to design and optimize mechanical and structural components, predict the performance and lifespan of materials and structures, and assess their safety and reliability. Below are some key applications of stress and strain analysis:

1. Mechanical design: Stress and strain analysis helps determine the appropriate size, shape, and materials of mechanical components, ensuring they can withstand expected loads and maintain stability.
2. Structural analysis: Stress and strain analysis is employed to evaluate the stability, safety, and reliability of structural components.
3. Material evaluation: Stress and strain analysis assists in evaluating the mechanical properties, fracture behavior, and fatigue life of materials.
4. Fatigue analysis: Stress and strain analysis aids in assessing the behavior of materials and structures subjected to repeated stress loading, predicting their lifespan and safety.
5. Plastic analysis: Stress and strain analysis is used to evaluate the behavior of materials and structures in regions of plastic deformation.
6. Multi-physical field analysis: Stress and strain analysis can be combined with other physical field analysis methods, such as thermodynamic analysis, hydrodynamic analysis, and electromagnetic analysis, to evaluate the behavior of materials and structures under various conditions.

7.4.2 Material Models and Mechanical Properties

Material models are mathematical representations that characterize the physical behavior of materials. They play a crucial role in comprehending and forecasting the mechanical properties and behavior of materials. Material models encompass a range of types, such as linear elasticity, plasticity, viscoelasticity, creep, fracture, and more. In this section, we will delve into these material models and elucidate their significance in engineering applications.

The linear elastic model is one of the most fundamental models used to describe materials. It assumes that the material behaves linearly elastically when subjected to external forces, meaning that the stress is directly proportional to the strain. This proportionality constant is commonly known as the elastic modulus and is indicative of a material's stiffness [46–48]. The relationship between stress and strain in the linear elastic model can be expressed by Hooke's law. This model is applicable within the small strain range, such as slight deformations in elastic bodies or minor deformations in metals. In practical applications, the validity of the linear elastic model can be assessed by observing the magnitude of strain and elastic modulus changes.

Plastic deformation occurs when a material is subjected to stress that surpasses its yield strength. The plastic model describes the irreversible plastic behavior of materials, assuming that materials can undergo permanent deformation when exposed to external forces [49]. Plastic models often employ rheological methods to depict the stress-strain relationship of materials. A commonly used plasticity model is the von Mises plasticity model, which is based on rheological theory and characterizes the plastic behavior of materials as a flow phenomenon. The von Mises plasticity model can be expressed by the following equation:

$$\sigma = K\epsilon^n \tag{7.19}$$

where K represents the rheological constant, and n denotes the plasticity index of the material, typically ranging from 0.1 to 0.5. This model is suitable for describing the plastic behavior of metals and other materials, with its parameters determined through experimental investigations in practical applications. Plastic models find wide applications, such as in mechanical design, where they aid in determining the ultimate load-carrying capacity and fatigue life of materials, thus enabling the design of more robust and reliable components.

The viscoelastic model describes the behavior of a material when subjected to stress, exhibiting both elastic and viscous characteristics. This model combines an elastic model and a viscous model to capture the stress-strain relationship of the material [50]. The elastic model represents the material's elastic behavior, which can be either linear or nonlinear. On the other hand, the viscous models capture the viscous behavior of materials and are commonly represented by the Dirichlet model (also known as the standard viscous model) or the Maclaurin model. These models introduce a damping force to account for the internal energy dissipation within the material. Viscoelastic models find wide-ranging applications in fields such as seismology, construction engineering, automotive engineering, aerospace engineering, and more. In these applications, the viscoelastic model is used to describe the dynamic response and energy dissipation behavior of structures. It enables the analysis of the structure's response under time-varying loads, providing insights into its stability and performance.

Creep is a special deformation behavior of materials, which is usually caused by long-term exposure of materials under high temperature or high pressure. Creep model is a model describing the creep behavior of materials, which can be used to predict the long-term performance of materials under high temperature or high pressure environment [51]. Creep models usually use relaxation models or constitutive models to describe the stress-strain relationship of materials. In the relaxation model, the strain of the material is a function of time, while in the constitutive model, the strain of the material is a function of stress. These models usually use polynomial or exponential functions to describe the stress-strain relationship of the material. The creep model is mainly used in engineering applications under high temperature or high pressure environment, such as nuclear power plants, aero-engines, automotive engines, and so on. In these applications, creep models can be used to predict the long-term performance of materials in high temperature or high pressure environments to help design more durable and reliable components.

A fracture model is utilized to describe the behavior of materials under fracture conditions, allowing for the prediction of fracture strength and fracture toughness. Generally, fracture models can be classified into two categories: linear and nonlinear fracture mechanics. In linear fracture mechanics, fracture strength and fracture toughness are considered inherent properties of materials, providing a basis for predicting their fracture behavior [52]. Within this model, fracture toughness is commonly described as the relationship between crack growth rate and stress intensity factor. Experimental techniques such as single ring tests and double ring tests are employed in practical applications to determine the fracture parameters of materials. On the other hand, nonlinear fracture mechanics is a more intricate model that takes into account various physical and mechanical behaviors of materials prior to fracture. Such models often employ numerical simulation techniques like finite element analysis and computational fluid dynamics to accurately describe the fracture behavior. While nonlinear fracture mechanics introduces additional complexity in practical applications, it can offer more precise and comprehensive results.

Fracture models find extensive applications in diverse fields, including materials science, engineering structural design, aerospace engineering, automotive engineering, and more. These models are used to evaluate the strength and toughness of materials, assisting in the design of safer and more reliable structures and components. By enabling accurate prediction of fracture behavior, fracture models contribute to enhancing the structural integrity and performance of engineering systems in various applications.

7.4.3 Boundary Conditions and Loads

In the analysis of structural mechanics and mechanics of materials, determining the appropriate boundary conditions and loads is of paramount importance. These conditions significantly influence the distribution of stress and strain in materials and structures, ultimately affecting their stability and performance. This section introduces the fundamental concepts of boundary conditions and loads, as well as practical approaches to determining them in engineering applications.

Boundary conditions encompass the constraints imposed on a material or structure at its external boundaries, exerting a substantial impact on the distribution of stress and strain. In practical engineering, boundary conditions are typically classified into three categories: displacement boundary conditions, force boundary conditions, and mixed boundary conditions.

Displacement boundary conditions involve applying specific displacements or deformations at the external boundaries of materials or structures. Examples include fixations, slippage, or rotations. This type of boundary condition allows for the simulation of practical scenarios such as fixing or imposing constraints on both ends of a bar, or restricting the movement of a structure in a particular direction.

Force boundary conditions pertain to the application of specific forces or pressures at the external boundaries of materials or structures. This may involve imposing forced displacements or loads, or applying external loads or constraints to a specific side of the material or structure.

Mixed boundary conditions refer to a combination of displacements and forces. They involve applying external loads while also limiting displacements on one side of a material or structure. Such boundary conditions are commonly employed to simulate complex practical situations, including restricted movement or intricate constraints.

Loads are external forces or pressures applied to a material or structure, resulting in deformation or failure. In practical engineering, loads are typically categorized as static loads or dynamic loads.

Static loads refer to stable forces or pressures applied to a material or structure, such as gravity, support forces, or pressure. This type of load is commonly associated with static stress analysis of materials or structures, providing insights into their stability and load-carrying capacity.

Dynamic loads, on the other hand, involve unsteady forces or pressures exerted on materials or structures. Examples include vibration, explosions, or wind forces. Dynamic loads necessitate dynamic stress analysis to predict the response and failure characteristics of materials or structures. Parameters such as frequency, amplitude, and duration of the load must be considered when conducting dynamic stress analysis.

Boundary conditions and loads play a crucial role in determining the stress and strain distributions of materials and structures. In static stress analysis, they govern the stability and load-carrying capacity of materials and structures, while also influencing the magnitude and spatial distribution of stress and strain. In dynamic stress analysis, boundary conditions and loads impact the response and failure characteristics of materials and structures, including factors such as vibration frequency, amplitude, and mode.

When conducting stress and strain analysis of materials and structures, it is essential to determine the type and magnitude of boundary conditions and loads. This process involves careful analysis and calculations based on the specific circumstances, along with compliance with relevant standards and specifications.

Displacement boundary conditions are determined by considering the constraints and limitations of the structure. For instance, in reinforced concrete structures, fixed boundary conditions are often imposed on supporting walls or columns to ensure structural stability and strength. When calculating displacement boundary conditions, factors such as the geometric shape of the structure, type of constraints, material properties, and working conditions must be taken into account.

Force boundary conditions are determined by considering external loads and constraints acting on the structure. In bridge structures, for example, factors like vehicle loads, wind loads, and snow loads need to be considered when determining the external loads on the structure. Similar to displacement boundary conditions, the calculation of force boundary conditions requires consideration of the structure's geometry, type of constraints, material properties, and working conditions.

Mixed boundary conditions are determined based on the specific circumstances. In some complex structures, a combination of displacement and force needs to be considered to simulate the actual constraints and working conditions. Calculating mixed boundary conditions necessitates an understanding of the structure's geometry, type of constraints, material properties, and working conditions.

When determining boundary conditions and loads, it is important to refer to relevant standards and codes, such as the bridge design specifications of the American Society of Civil Engineers (ASCE) or the oil and gas pipeline design specifications of the American Petroleum Institute (API). These standards and codes provide detailed guidance and recommendations on boundary conditions and loads, aiding engineers and scientists in conducting more accurate and reliable stress and strain analyses.

7.4.4 Mesh Generation and Solution Algorithms

Mesh generation and solution algorithms are fundamental steps in structural simulation. Mesh generation involves dividing the continuous domain into discrete finite elements, while the solution algorithm calculates physical quantities such as displacement, strain, and stress for each element. In this section, we will discuss the basic principles of meshing and solution algorithms, as well as the commonly used methods and tools in practice.

Meshing is a crucial initial step in structural simulation, aiming to divide the domain into discrete finite elements. These elements are designed and optimized based on application requirements and commonly include shapes such as triangles, quadrilaterals, hexagons, and tetrahedrons.

During meshing, several factors should be considered, including the model's geometry, material properties, boundary conditions, and numerical stability. To achieve accurate simulation results, it is essential to choose an appropriate mesh generation method, determine the mesh density, and optimize and adjust the mesh accordingly.

Commonly used meshing methods include:

1. Delaunay triangulation: This method, based on a point set, generates a triangular mesh using the minimum circumscribed circle between connected points. It is suitable for models of arbitrary shape, offering good numerical stability and mesh quality.

2. Unstructured grid partitioning: This method can handle complex three-dimensional structures of any shape by segmenting the domain into a series of unstructured smaller units. It provides higher mesh quality and better adaptability.

3. Structured grid partitioning: This method is suitable for simple geometric shapes like planar or axisymmetric structures. It divides the domain into a series of regular smaller units, offering better computational efficiency and numerical stability.

When meshing, the following points should be noted:

1. The mesh density should be sufficient to capture detailed features in the structure accurately.

2. The mesh quality should be high to ensure the accuracy and stability of the calculation results.

3. Meshing should be tailored to different material properties and boundary conditions to obtain realistic simulation results.

4. Meshing should be feasible in terms of computational resources and time to ensure computational efficiency and practicality.

The solution algorithm is the central component of structural simulation, responsible for calculating physical quantities such as displacement, strain, and stress on each finite element. Typically, the solution algorithm involves solving a large system of algebraic equations, which can contain millions of unknowns and equations. The accuracy, efficiency, and stability of the solution algorithm are crucial for obtaining high-quality and reliable simulation results.

In the solution algorithms, the following three methods are commonly used:

1. Direct method: This approach directly solves the system of linear algebraic equations using techniques like Gaussian elimination or LU decomposition. Direct methods offer high accuracy and stability. However, they can be computationally complex for large and dense equation systems, requiring significant computational resources and time.

2. Iterative method: This method approximates the solution of a system of equations through iterative calculations. Iterative methods provide fast computation and require less memory. However, it is important to choose an appropriate iterative algorithm and convergence criteria, as they may encounter issues of numerical instability.

3. Hybrid method: This approach combines the advantages of both direct and iterative methods to solve large systems of linear algebraic equations. Hybrid methods often incorporate special techniques like preconditioning, acceleration techniques, and solver reuse to enhance computational speed and improve numerical stability.

In the solution algorithm, the following points should be considered:

1. Select an appropriate solution algorithm and solver, adjusting and optimizing them based on the model's characteristics and available computing resources.

2. Choose suitable numerical schemes and ensure the fulfillment of numerical stability conditions.

3. Select and process appropriate boundary conditions and initial conditions, making necessary adjustments as required.

4. Choose the appropriate convergence criteria and the number of iterations, monitoring and controlling them effectively.

Commonly used meshing and solving algorithm tools in practice include finite element analysis software, computational fluid dynamics software, and multiphysics simulation software. These tools offer engineers and scientists efficient, accurate, and reliable capabilities for structural simulation, aiding in the design of safer, more reliable, and efficient products and structures.

Here are some commonly used meshing tools:

1. Gmsh: A kind of open-source 3D finite element meshing software that handles complex geometries and unstructured meshes. It supports various file formats (e.g., STEP, IGES, STL, OBJ) and offers rich post-processing functions.

2. ANSYS Meshing: A type of commercial finite element meshing software capable of handling diverse geometric shapes and complex structures. It features automatic mesh generation and optimization, supports multiple mesh types, and provides mesh quality indicators.

3. Abaqus CAE: A type of commercial finite element simulation software that includes meshing, solution, and post-processing functions. It supports various mesh types and material models, boasts a powerful solver, and offers multiphysics simulation capabilities.

Commonly used algorithm tools include PETSc, Trilinos, and FEniCS.

7.4.5 Result Analysis and Verification

After performing structural simulation analysis, it is essential to analyze and verify the simulation results. This analysis and verification process helps determine the accuracy and reliability of the results, as well as evaluate the safety and reliability of the structure. In this section, we will introduce the fundamental principles and methods of result analysis and verification, along with common techniques used in structural simulation result analysis and verification.

Result analysis involves data processing, chart creation, statistical analysis, and comparative analysis of simulation results. Its fundamental principle is to generate simulation results based on the simulation model and parameters and present the results in a visual manner. Result analysis typically involves the following steps:

1. Data processing: Import the simulation results into data processing software and perform necessary cleaning and format conversion operations.

2. Chart creation: Use graphing software to create charts of simulation results, such as stress-strain curves, displacement-time curves, stress contour plots, and deformation contour plots.

3. Statistical analysis: Conduct statistical analysis on the simulation results, including calculations of mean, variance, standard deviation, and confidence intervals.

4. Comparison and analysis: Compare and analyze different simulation results, such as comparing results from different models, different parameters, and different loading conditions.

The purpose of result analysis is to assist engineers in understanding the simulation results and drawing appropriate conclusions and decisions based on the analysis. Result analysis also provides valuable information for subsequent optimization and improvement efforts.

Result verification involves comparing and validating simulation results against actual measurement data. The fundamental principle is to use experimental measurements or reliable reference data to verify the accuracy and reliability of simulation results. Result verification typically includes the following steps:

1. Measurement data acquisition: Use measuring instruments and equipment to measure the actual structure and record the measurement data.

2. Data processing: Import the measurement data into data processing software and perform necessary cleaning and format conversion operations.

3. Simulations' results generation: Generate simulation results using simulation software and compare them with the measured data.

4. Results analysis: Analyze and compare the simulation results with the measurement results to evaluate the accuracy and reliability of the simulation results.

The purpose of result verification is to assess the accuracy and reliability of the simulation results and provide a reference for subsequent simulation analyses. Result verification can also help assess the accuracy and reliability of simulation models and parameters, and provide engineers with valuable information and guidance to optimize and improve the design and performance of structures.

In structural simulations, there are several commonly used techniques for result analysis and verification:

1. Stress and strain analysis: This technique involves analyzing and comparing the stress and strain of the structure to evaluate its strength and stability. It includes generating stress-strain curves, stress nephograms, strain nephograms, and other visual representations.

2. Displacement and deformation analysis: This technique focuses on analyzing and comparing the displacement and deformation of the structure to evaluate its stiffness and deformation capacity. It includes generating displacement-time curves, deformation nephograms, displacement nephograms, and other visual representations.

3. Fatigue analysis: This technique involves analyzing and comparing the structural behavior under repeated stress to assess its life and safety. It includes generating stress-cycle number curves, predicting life expectancy, and assessing damage.

4. Modal analysis: This technique is used to analyze and compare the vibration modes of the structure to evaluate its vibration characteristics and stability. It includes calculating natural frequencies, modal shapes, and vibration modes.

5. Multiphysics analysis: This technique involves analyzing and comparing the mechanical behavior of the structure in conjunction with other physical fields such as thermal, electricity, and fluids. It assesses the overall performance and reliability of the structure through coupling analysis and multiphysics optimization.

These analysis and validation techniques can be used individually or in combination to achieve more accurate and comprehensive results. For example, stress and strain analysis can be combined with fatigue analysis and displacement deformation analysis to assess the stability and safety of the structure. Modal analysis can be combined with stress and strain analysis to assess the vibration characteristics and stability of the structure.

7.5 Co-simulations of Thermal and Structural Properties

7.5.1 Requirements and Methods of Structural Co-simulations

Structural simulation plays a vital role in mechanical design and engineering analysis, enabling engineers to assess the performance and reliability of mechanical and structural components. However, in real-world applications, these components often interact with other systems, such as electrical, hydraulic, and thermodynamic systems. To accurately simulate the behavior of such integrated systems, structural co-simulations are required. This section will outline the requirements and methods for structural co-simulations and explore their practical applications in engineering.

Structural co-simulations involve integrating structural simulation with simulations in other physical domains to accurately model the behavior of mechanical and structural components in complex environments. This approach requires considering the interactions between mechanical components and systems in fields like electrical, hydraulic, and thermodynamics. The key requirements for structural co-simulations include:

1. Coupling mechanical and electrical systems: accounting for interactions between motors and gear systems, sensors and controllers, and other components.
2. Coupling mechanical and hydraulic systems: capturing interactions between hydraulic cylinders and valve systems, ensuring a comprehensive simulation.
3. Coupling mechanical and thermodynamic systems: modeling interactions between engines and cooling systems, considering thermal effects.
4. Coupling across different physical domains: capturing interactions among mechanical, electrical, hydraulic, and thermodynamic systems.

The primary objective of structural co-simulations is to simulate the interactions between mechanical and structural components across various physical domains, enabling more accurate and comprehensive simulation results.

To facilitate structural co-simulations, several commonly used methods are employed:

1. Finite Element Analysis (FEA) and Multiphysics Simulations:

Finite element analysis is a widely utilized method for mechanical and structural simulations, enabling the evaluation of component performance and reliability. In structural co-simulations, FEA is combined with simulation methods from other physical domains, such as electrical, hydraulic, and thermodynamic systems. This integration, known as multiphysics simulation, enables the simulation of interactions between mechanical and structural components across different physical fields. To achieve this, mechanical and structural components are decomposed into finite element models and combined with models from other physical domains. Numerical solvers are then employed to solve for the interactions between all physical domains, resulting in accurate behavior predictions for the mechanical and structural components.

2. Multibody Dynamics Simulations (MBD):

Multibody dynamics simulation is a method used to simulate the motion and interaction of mechanical and structural components. Mechanical and structural components are represented as rigid bodies, and their interactions are modeled using forces and torques. This approach is suitable for simulating the interactions between mechanical and structural components in diverse physical fields. It involves decomposing the mechanical and structural components into rigid bodies and using various force and torque models to describe their interactions. A numerical solver is used to solve for the interactions among all rigid bodies.

3. Discrete Element Simulations (DEM):
Discrete element simulation is employed to simulate the motion and interaction of discrete media, such as particles, dust, and grains. The discrete media is represented by a discrete set of particles, with their interactions modeled using forces and torques. This method allows for simulating the interaction between mechanical and structural components and media like dust or particles. In DEM, mechanical and structural components are decomposed into discrete particles, and various force and torque models are used to describe their interactions with the media.
4. Coupled Simulations:
Coupled simulation involves the integration of multiple simulation methods to simulate the behavior of mechanical and structural components. This method typically requires specialized simulation software and tools, such as ANSYS, ABAQUS, and SIMULIA.
5. Experimental Simulations:
Experimental simulation utilizes experimental data to verify simulation results. In the context of structural co-simulations, experimental simulation can be employed to validate the interaction between mechanical and structural components across different physical domains. For instance, experimental data can be used to verify the interaction between mechanical and hydraulic systems, ensuring the accuracy and reliability of simulation results. By comparing and validating experimental data with simulation results, accurate behavior predictions for mechanical and structural components can be obtained.

7.5.2 Framework and Process of Structural Co-simulations

Structural co-simulations is a comprehensive approach that integrates multidisciplinary knowledge to simulate and analyze complex systems. It encompasses various fields such as structural dynamics, material mechanics, fluid mechanics, and control systems. By modeling and simulating the coupling relationships between different domain models, the overall system behavior can be analyzed through simulation. The framework of structural co-simulations typically consists of the following main components: multi-domain coupling modeling, simulation calculations, data processing, and result analysis. Among these, multi-domain coupling modeling is the central component, where the coupling relationships between models from different domains, including structural dynamics, material mechanics, fluid mechanics, and control systems, are established and integrated.

Multi-domain coupling modeling is a crucial step in structural co-simulations as it involves establishing the coupling relationships between models from different domains. During the modeling process, it is essential to consider the interactions and influences among various fields and their impacts on the entire system.
1. Structural dynamics model
The structural dynamics model is fundamental in structural co-simulations. It encompasses the characteristics of mass, stiffness, damping, as well as the dynamic response

and deformation behavior of the structure under external loads. When developing the dynamic model, factors such as geometry, material properties, support conditions, and load conditions of the structure need to be taken into account. Common structural dynamic models include the finite element model, finite-difference model, and spring-mass model.

2. Material mechanics model

The material mechanics model describes the stress-strain relationship of materials under various loads. In structural co-simulations, the material mechanics model is primarily employed to characterize the elastic and plastic deformation behavior, as well as fatigue life and fracture behavior of materials. Common material mechanics models include linear elastic models, nonlinear elastic models, plastic models, and damage models.

3. Fluid mechanics model

The fluid mechanics model is employed to describe the motion and deformation behavior of fluids under forces, encompassing parameters such as velocity, pressure, density, and temperature of the flow field. In structural co-simulations, the fluid mechanics model is often coupled with the structural dynamics model to consider the interaction and influence between the fluid and the structure. Common hydrodynamic models include Reynolds-Averaged Navier-Stokes (RANS), Large Eddy Simulations (LES), and Direct Numerical Simulations (DNS).

4. Control system model

The control system model describes the operational mechanism, control algorithms, and the influence of the control system on the system performance. In structural co-simulations, the control system model is typically coupled with the structural dynamics model to consider the impact of the control system on the structural response. Common control system models include PID controllers, fuzzy controllers, neural network controllers, and others.

After establishing the multi-domain coupling model, simulation calculations are performed to simulate the dynamic response and deformation behavior of the system under different loads and working conditions. To ensure accurate and stable simulation results, it is important to select the appropriate numerical calculation method and solver.

Numerical calculation methods are commonly used in structural co-simulations and involve techniques such as the finite element method, finite-difference method, and boundary element method. These methods are utilized to establish models for structural dynamics, material mechanics, fluid mechanics, and other aspects, enabling simulation calculations.

A solver is a computational program used to solve mathematical equations, particularly in the fields of structural dynamics, material mechanics, fluid mechanics, and other models. Common solvers include direct solutions, iterative solutions, and multi-grid methods. These solvers play a crucial role in obtaining solutions for the equations involved in the simulation process.

7.5.3 Data Transfer and Interface in Structural Co-simulations

Structural co-simulations involve the integration of multiple simulation tools to analyze complex systems. To ensure accurate and reliable simulations, data transfer and interface settings are crucial between these tools. This section outlines the basic principles and methods of data transfer and interface in structural co-simulations.

In structural co-simulations, various types of data, including geometric information, physical parameters, and boundary conditions, need to be transferred between different simulation tools. Here are some common methods for data transfer:
1. File transfer
File transfer is a fundamental method for data exchange. SimulationsâŁ™ tools can export or import data using file formats such as IGES, STEP, STL, and DXF. File transfer is simple to use but may require manual import and export, resulting in slower data transfer.
2. Interface transmission
Interface transmission involves connecting different simulation tools directly to enable data transfer. Middleware or drivers are often used to implement this method. For example, the Abaqus-MATLAB interface allows data transfer between Abaqus and MATLAB. Interface transmission offers fast data transfer speeds and automated data exchange.
3. Database transmission
Database transfer involves storing data in a database, which can be accessed by different simulation tools. This method typically utilizes a database management system (DBMS). For instance, a database transfer can facilitate data exchange between Simulink and LabVIEW. Database transmission provides fast data transfer and enables data sharing among multiple simulation tools.

In structural co-simulations, interface settings are necessary to enable data transfer and sharing between different simulation tools. Here are some common methods for interface setup:
1. Interface script
Interface script involves writing script code to facilitate data transfer and interface settings between different simulation tools. Interface scripts can be implemented using various programming languages such as Python or MATLAB. The advantage of interface script is its flexibility, allowing customization based on specific requirements.
2. Interface plug-in
Interface plug-in involves installing plug-ins to achieve data transfer and interface settings between different simulation tools. These plug-ins are typically provided by manufacturers or third parties and enable automated data exchange and interface configurations. Interface plug-ins are user-friendly and expedite data sharing between different simulation tools.

Interface standard involves using a unified interface specification to facilitate data transfer and interface settings between different simulation tools. These standards are developed by organizations like standardization bodies or vendor consortia, including Modelica and FMI. The advantage of interface standard is its universality and extensibility, enabling data exchange and sharing among multiple simulation tools.

In structural co-simulations, different simulation tools often employ different data formats. To achieve data transmission and sharing, data format conversion is required.

Data conversion tools are software tools that facilitate conversion between different data formats. These tools typically offer plug-ins supporting various data formats such as IGES, STEP, STL, and DXF. The advantage of data conversion tools is their ability to automatically convert between different data formats. Interface plug-ins can also be used for data format conversion, as they often support conversion between the data formats required by different simulation tools. Interface plug-ins are easy to use and enable swift data format conversion. Additionally, interface standards are commonly employed for data format conversion. For example, Modelica's data format is supported by multiple simulation tools, allowing data sharing and format conversion between them.

7.6 Co-simulations of Structural Simulation Software with ADS

7.6.1 Structural Co-simulations of SIMULIA Abaqus with ADS

SIMULIA Abaqus is a kind of advanced finite element analysis (FEA) software developed by Dassault Systèmes, widely used in various engineering fields such as automotive, aerospace, civil engineering, and biomedicine. It offers powerful capabilities for linear and nonlinear static and dynamic structural analysis, enabling the solution of complex engineering problems. Abaqus also supports data interaction with other CAE software, facilitating multiphysics coupling analysis. To ensure compatibility, Abaqus and Advanced Design System (ADS) should be installed on the same computer.

ADS is a kind of electronic design automation (EDA) software developed by Keysight Technologies for high-frequency and high-speed electronic system design. To achieve co-simulations between Abaqus and ADS, an interface between the two types of software needs to be established.

The main functions of structural co-simulation include:

1. Multiphysics field coupling analysis: Enables two-way coupling analysis between the electromagnetic field and structural field, considering the influence of the electromagnetic field on structural response and vice versa.

2. Optimized design: Through co-simulations, equipment performance can be optimized during the design stage, reducing the number of design iterations and improving design efficiency.

3. Higher simulation accuracy: Co-simulations allow for the consideration of interaction between various physical fields, thereby enhancing the accuracy of simulation results.

4. Support for a wide range of material models: Abaqus offers a comprehensive library of material models supporting various common engineering materials.

The operation process is as follows:

1. Build the model: Construct the structural model in Abaqus CAE, defining geometry, material properties, boundary conditions, etc.

2. Build an electromagnetic model in ADS: Create an electromagnetic model in ADS, including devices and electromagnetic components.

3. Simulations settings: Configure simulation parameters in Abaqus, such as solver type, analysis steps, and output options.

4. Configure the co-simulations interface: Ensure the correct configuration of the interface between Abaqus and ADS, establishing correlation between the two models.

5. Run co-simulations: Execute the co-simulations, where Abaqus and ADS automatically exchange data and collaborate in the background. Abaqus focuses on structural analysis, while ADS focuses on electromagnetic field analysis.

6. Result post-processing: Use post-processing tools in Abaqus CAE and ADS to analyze and visualize simulation results, including stress, strain, displacement fields, and electromagnetic field distributions.

7. Optimization design: Based on the simulation results, optimize the design by modifying geometric dimensions, material parameters, or boundary conditions, and perform co-simulations again.

8. Verification and evaluation: Verify and evaluate the optimized design to ensure compliance with design requirements and performance indicators.

To enable co-simulations between Abaqus and ADS, the following interfaces should be established:

1. Data exchange interface: Abaqus and ADS need to exchange various data, such as structural deformation, stress distribution, and electromagnetic field distribution. This data can be exchanged through files or memory. Common data exchange formats include CSV, TXT, and HDF5.

2. Control interface: During co-simulations, Abaqus and ADS must coordinate to control the simulation process. This can be achieved using scripts or plug-ins. For example, Python scripts can automate specific tasks in Abaqus, while VBScript scripts in ADS can control the simulation process.

3. Event-triggered interface: In co-simulations, it may be necessary to trigger specific actions based on certain events, such as reaching a threshold or meeting a condition. An event-triggered interface facilitates this functionality. For instance, a stress threshold set in Abaqus can trigger an action in ADS when the stress exceeds the threshold.

7.6.2 Structural Co-simulations of MSC Software with ADS

MSC Software is a globally recognized provider of high-quality computer-aided engineering software and solutions. Their comprehensive product portfolio encompasses various domains, including structural mechanics, fluid mechanics, acoustics, electromagnetism, and multiphysics fields.

The MSC CoSim engine offers a direct co-simulations interface that allows for the coupling of different solvers/disciplines within a multiphysics framework. It enables engineers to build co-simulations models between Adams, Marc, and scFLOW. In addition to the CoSim engine, MSC Software supports other co-simulations approaches, such as the Functional Model Interface (FMI) and the Adams Marc co-simulations Interface (ACSI).

The simulation workflow using MSC Software's federated architecture consists of the following steps:

1. Structural model development

Utilize software tools like MSC Patran or MSC Nastran to create the structural model. Define the geometry, material properties, boundary conditions, etc.

2. Load and constraint definition

In MSC Patran or MSC Nastran, specify the loads, constraints, and the type of analysis to be solved for the structure.

3. Import structure model into ADS

Import the structure model from MSC Patran or MSC Nastran into ADS (Advanced Design System) and define the corresponding circuit model within ADS.

4. Simulations parameter setup

Configure simulation parameters in ADS, such as simulation time, step size, and the interface between MSC Software and ADS.

5. Perform co-simulations

Conduct co-simulation in ADS, where the structure and circuit are Co-simulated.

6. Analyze simulation results

Evaluate the simulation results in MSC Patran or MSC Nastran, including analyzing stress, deformation, vibration of the structure, as well as voltage, current, and power characteristics of the circuit.

7. Structural design optimization

Based on the simulation results, perform structural optimization design techniques like topology optimization, shape optimization, parameter optimization, etc., to achieve the optimal structural design.

7.7 Co-simulations Structural Simulation Software with HFSS

7.7.1 Structural Co-simulations of ANSYS Mechanical with HFSS

ANSYS Mechanical is a kind of professional finite element analysis software widely used for solving structural mechanics problems in various engineering applications. It offers advanced modeling capabilities, automated solution processes, and comprehensive post-processing functions, enabling efficient analysis of complex engineering problems. ANSYS also facilitates multi-physical domain system-level collaborative simulation analysis through the integration of ANSYS Maxwell, Twin Builder, ANSYS Workbench (ANSYS Mechanical and ANSYS CFD) software. Co-simulations analysis can be achieved by directly mapping ANSYS Workbench.

The operation process for multi-physical field coupling combined structural simulation using ANSYS Mechanical and HFSS (High-Frequency Structure Simulator) is as follows:

1. Create a new project and add HFSS and ANSYS Mechanical modules to the project space.
2. Set the simulation parameters in HFSS, including the geometric model, material properties, and boundary conditions.
3. Perform the electromagnetic field simulation in HFSS and generate the simulation results.
4. In ANSYS Workbench, connect the Solution port of HFSS to the HFSS Solution port of ANSYS Mechanical.
5. Open ANSYS Mechanical, import the structural model, and define the material properties, boundary conditions, and loads.
6. Configure the parameters coupled with HFSS in ANSYS Mechanical. This includes selecting the electromagnetic field data to be coupled, setting the coupling mode (such as electromagnetic-thermal coupling and electromagnetic-structural coupling), and specifying the coupling coefficient.
7. Run the multi-physical field coupling structural co-simulation in ANSYS Mechanical. The software will automatically import the simulation results from HFSS and perform coupling analysis with the structural model.
8. After the simulation is completed, utilize the post-processing capabilities of ANSYS Mechanical to analyze the simulation results. This includes visualizing the distribution of stress, strain, displacement, and other parameters, as well as evaluating the influence of the electromagnetic field on the structure. Based on the analysis results, optimize the design and iterate through the simulation process until the design requirements are met.

7.7.2 *Structural Co-simulations of COMSOL Multiphysics with HFSS*

COMSOL Multiphysics is a kind of highly powerful finite element analysis software specifically designed for simulating complex physical phenomena. It offers a wide range of functionalities across various fields such as electromagnetics, structural mechanics, thermal conduction, fluid mechanics, acoustics, and more. The key strength of COMSOL Multiphysics lies in its capability to perform multiphysics field coupling analysis, allowing engineers and researchers to simulate multiple interconnected physical processes within a unified environment. This provides a comprehensive simulation solution for obtaining accurate design results. By combining the simulation capabilities of COMSOL Multiphysics and HFSS (High-Frequency Structure Simulator), users can achieve coupling analysis of multi-physical fields, such as electromagnetics and structural mechanics, enabling more accurate design outcomes. The installation and configuration process is as follows:

1. Configure COMSOL and HFSS co-simulations: After installation, import the HFSS model into the COMSOL Multiphysics software and set the relevant coupling parameters. The specific steps are as follows:

2. Open COMSOL Multiphysics, navigate to the File menu, click on Import, and select the HFSS model file to import.

3. Add the appropriate physical fields and boundary conditions for the imported model in COMSOL Multiphysics software.

4. In the physical field settings, select the coupling parameters related to HFSS and define the coupling relationship between the electromagnetic field and the structural field.

The operational process is as follows:

1. Preparation: Ensure that the COMSOL Multiphysics and HFSS co-simulations environment is installed and configured.

2. Import the HFSS model: Launch COMSOL Multiphysics and import the HFSS model file.

3. Set physical fields and boundary conditions: Add the necessary physical fields (e.g., electromagnetic field, structural field) and specify the appropriate boundary conditions for the imported model.

4. Configure coupling parameters: In the physical field settings, select the coupling parameters relevant to HFSS and define the coupling relationship between the electromagnetic field and the structural field.

5. Meshing: Generate a mesh for finite element calculations.

6. Run the simulation: Set the solver parameters and initiate the simulation, then wait for the calculation to complete.

7. Result analysis: Utilize the post-processing functionalities of COMSOL Multiphysics to visualize the simulation results and export the data.

8. Optimization design: Based on the simulation results, adjust and optimize the model parameters as necessary.

By employing the combined structural simulation of COMSOL Multiphysics and HFSS, users can conveniently conduct multi-physical field coupling analyses, thereby enhancing the accuracy and efficiency of the design process.

7.7.3 Structural Co-simulations of Ansys Multiphysics with HFSS

Ansys Multiphysics is an advanced engineering simulation software that integrates multiphysics field analysis. It finds applications in various engineering fields, such as structural analysis, fluid dynamics, and electromagnetic field analysis. Developed by ANSYS, this software offers engineers a comprehensive, efficient, and reliable solution for multiphysics simulations. Ansys Multiphysics enables engineers to conduct in-depth analysis of complex engineering problems, optimize designs, and verify product performance. It provides a range of solvers and modules that allow for multiphysics simulations on a unified platform. These modules cater to different physical fields.

The Structural Analysis module allows for the examination of static and dynamic responses of structures. Static analysis encompasses linear statics, nonlinear statics, steady-state thermal conduction, and more. Dynamic analysis includes linear dynamics, nonlinear dynamics, transient thermal conduction, and others. The module also supports advanced functions such as fatigue analysis, optimization, and failure analysis.

The Fluid Dynamics module enables the analysis of fluid flow characteristics under various working conditions. It supports different fluid models, including laminar flow, turbulent flow, and multiphase flow. Additionally, advanced features such as thermal transfer, turbulence models, and chemical reactions are also supported. The Electromagnetic Field Analysis module is utilized for analyzing static electric fields, static magnetic fields, time-varying electromagnetic fields, and more. This module facilitates solutions for electromagnetic fields, propagation, radiation of electromagnetic waves, and offers multiphysics field coupling functions like thermal and structural coupling.

Ansys Multiphysics also encompasses other modules such as acoustic analysis, optical analysis, and manufacturing process simulation, among others. These modules can be used independently or in conjunction with other modules to achieve multiphysics simulations.

To conduct structural co-simulations using Ansys Multiphysics and HFSS, the following configuration is required: Launch Ansys Workbench and add a System Coupling component in the Project Schematic window. Then, drag and drop a Structural component and an HFSS component onto the engineering canvas, connecting them to the System Coupling component. This allows Ansys Multiphysics and HFSS to share data and Co-simulate. Each component requires separate settings while ensuring simulation parameter consistency. Additionally, it is essential to select the appropriate coupling mode based on the specific engineering problems, such as force coupling or thermal coupling.

The key feature of combined structural simulation using Ansys Multiphysics and HFSS is the ability to couple multiphysics fields. Through co-simulations, the interaction between the electromagnetic field and the structural field can be considered, resulting in more accurate simulation results. Furthermore, multiphysics field coupling aids in discovering phenomena that may be challenging to observe in a single physical field analysis, ultimately enhancing product performance and safety.

Ansys Multiphysics and HFSS employ efficient solvers and algorithms specifically optimized for different physical fields and engineering problems. These solvers and algorithms deliver reliable simulation results in a short amount of time. Additionally, both software packages support parallel computing, further improving simulation efficiency.

7.8 Static Simulations of Structures by Combining Hypermesh and Ansys

In the modern field of engineering, Finite Element Analysis (FEA) has become an indispensable part. This method can solve complex engineering problems and provide detailed performance evaluations, enabling design optimization. Hypermesh and Ansys are two widely used software applications for finite element analysis, with the former primarily used for pre-processing and the latter for solving and post-processing. This chapter will provide a detailed introduction on how to use Hypermesh in conjunction with Ansys for static structural simulations. Firstly, it is important to understand the basic process of finite element analysis, which consists of three main stages: pre-processing, solving, and post-processing. In the pre-processing stage, we mainly focus on constructing the geometric model, defining materials, and building the finite element system model. Constructing the geometric model is the first step in pre-processing. We can use third-party modeling software such as Creo or SolidWorks for modeling. Once the modeling is completed, the geometric file needs to be saved in a neutral format and then imported into Hypermesh, as shown in Fig. 7.1.

Next, we need to define the material properties of the model. In Hypermesh, we can create corresponding material options and input the material properties, which

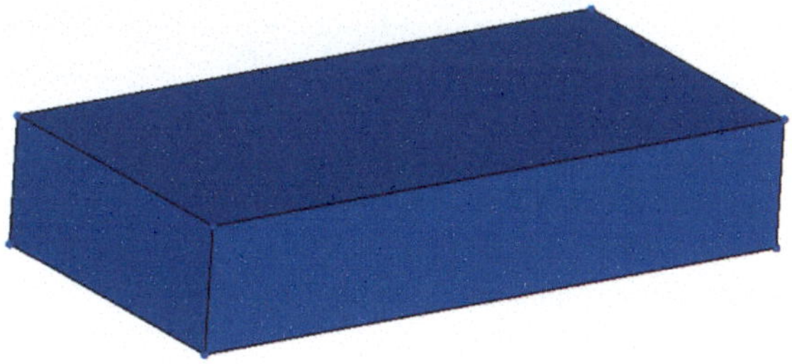

Fig. 7.1 Geometric model construction

Fig. 7.2 Elements of geometric model construction

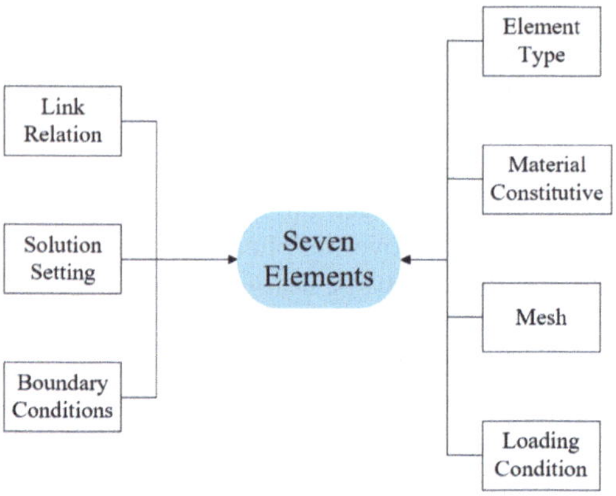

mainly include density (7850 kg/m^3), elastic modulus (210e9 Pa), and Poisson's ratio (0.3). The construction of the finite element system model is the final step in pre-processing, which mainly consists of seven elements: connection relationships, solution settings, boundary conditions, element types, material constitutive behavior, mesh, and loading conditions, as shown in Fig. 7.2. For link relationships and solving configurations, we configure them in Hypermesh. We select the Solid186 element type, with default keywords and material properties. In the boundary condition settings, we need to specify where the model is subjected to external forces or constraints. Meshing is a critical step in constructing the finite element model, and for regular geometries, we can perform meshing all at once, as shown in Fig. 7.3. During the solving phase, we need to set loading conditions and boundary conditions, and then proceed with the solving setup. Setting loading conditions and boundary conditions refers to determining where and how forces or constraints are applied within the model. In this example, we apply fixed constraints at the left two ends of the cuboid and apply a force of 1N in the negative Z-axis direction at each node on the right end of the cuboid, as shown in Fig. 7.4. In the solving setup, we need to insert a solving control card and then export the file as a .cdb format, saving it in a fully English path. Finally, we enter the post-processing phase. We open the Ansys Launcher and set

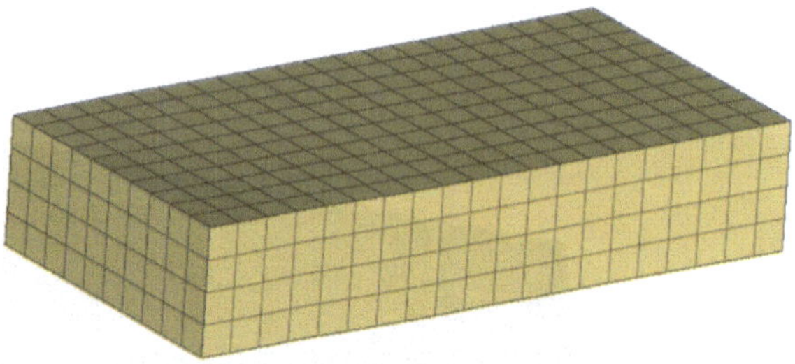

Fig. 7.3 Mesh partitioning of geometric model

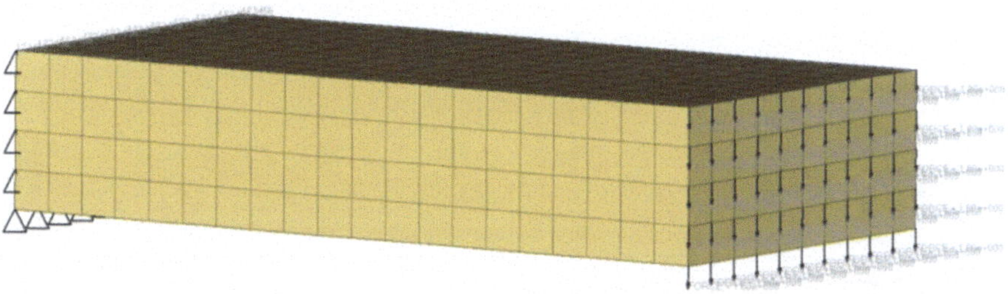

Fig. 7.4 Application of force to a cuboid geometric model

the working directory to the previously saved English path. Then, we import the .cdb format file exported from Hypermesh for direct solving and visualization of deformation and stress results, as shown in Figs. 7.5 and 7.6. In general, the process of performing static analysis using Hypermesh and Ansys is a complex procedure that involves several steps, including geometric model construction, material definition, finite element model creation, solution through applied loads and boundary conditions, and finally, result visualization and evaluation in post-processing. Throughout this process, it is essential to have a deep understanding and mastery of each step in order to obtain accurate and effective simulation results.

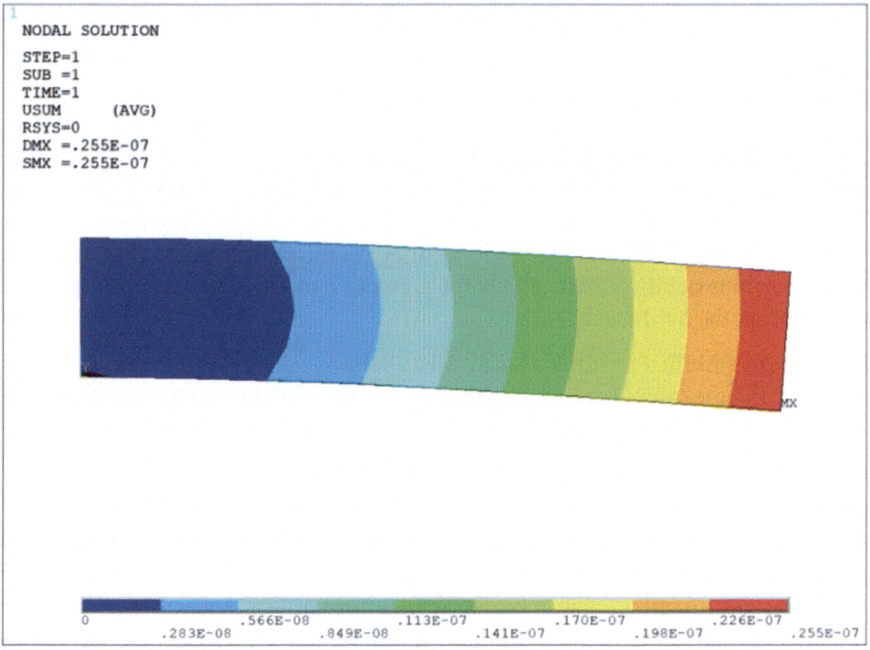

Fig. 7.5 Simulation results of deformation

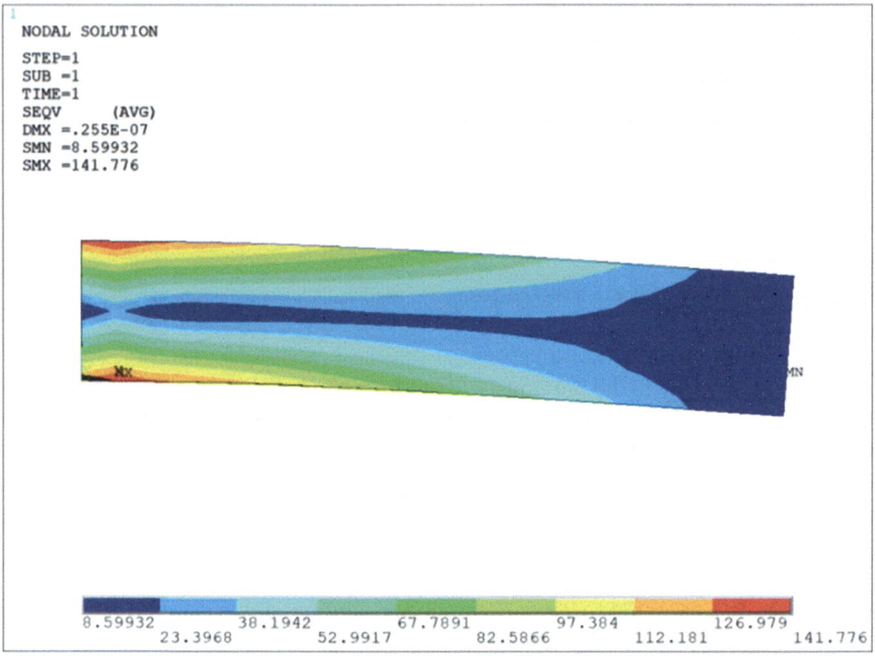

Fig. 7.6 Simulation results of stress

Problems

P7.1 ADS Multi-Physics Simulation Technology
Explore the enhanced multiphysics simulation capabilities of ADS to assess the stability of power amplifiers under various operating conditions. Utilize the RFPro electromagnetic visualization feature to identify the physical locations and frequencies that cause instability. This allows for corrective measures to be taken prior to manufacturing the hardware.

P7.2 Mapping Multi-Physics Fields to Ansys HFSS
In this exercise, the example project of a waveguide filter in EDT (Electronic Desktop) is used.

- Step 1: Solve the coupled Icepak-HFSS model in EDT.
- Step 2: Connect the multiphysics model in Workbench.
- Step 3: Ensure pressure feedback and/or thermal feedback are enabled based on project objectives.
  For stress/deformation, enable feedback only for entities that will appear in the mechanical simulation. For thermal feedback, enable feedback only for bodies with specified material properties (including thermal modifiers). Therefore, make sure to define the materials correctly in EDT first.

- Step 4: Match geometry and coordinate systems.
- Step 5: Create a workflow in Workbench.
- Step 6: Map Loads.
- Step 7: Update the thermal model.
  Right-click > Update Settings for Steady-State Thermal. If HFSS has not been solved yet, this will solve the initial non-deformed isothermal situation and enable future connections to the thermal/mechanical solver.
- Step 8: Repeat Step 7, but for Static Structural.
  For imported loads, there will be one from steady-state thermal and one from HFSS. The steady-state thermal connection should be fine by default.
- Step 9: Enable HFSS Update.
- Step 10: Update HFSS.
- Step 11: Repeat until your Multiphysics Simulation converges.
  The Workbench environment is necessary to access the relevant options and settings for bi-directional coupling. In HFSS, the fields requiring bi-directional coupling, such as thermal and structural, need to be set. Data transfer is required between HFSS and thermal/structural, regardless of whether there is thermal/force transfer. The data transfer is bi-directional.

References

1. Guo Z (2005) The deformation and processing of structural materials. Elsevier
2. Natsuki T, Endo M (2004) Stress simulation of carbon nanotubes in tension and compression. Carbon 42(11):2147–2151
3. Shin HV et al (2016) Reconciling elastic and equilibrium methods for static analysis. ACM Trans Graph (TOG) 35(2):1–16
4. Shimada S et al (1994) Structure of micromachined surface simulated by molecular dynamics analysis. CIRP Ann 43(1):51–54
5. Brown SB, Song H (1992) Implications of three-dimensional numerical simulations of welding of large structures. Weld J 71(2):55–62
6. Nakayama H, Washino K (2001) Simulations based optimization using. In: Knowledge-based intelligent information engineering systems & allied technologies: KES'2001, vol 5, p 466
7. Sun H, Burton HV, Huang H (2021) Machine learning applications for building structural design and performance assessment: state-of-the-art review. J Build Eng 33:101816
8. Zhao L et al (2010) Lightweight design of mechanical structures based on structural bionic methodology. J Bionic Eng 7(4):S224–S231
9. Messner MC (2020) Convolutional neural network surrogate models for the mechanical properties of periodic structures. J Mech Des 142(2):024503
10. Yan W et al (2018) Modeling process-structure-property relationships for additive manufacturing. Front Mech Eng 13:482–492
11. Tuegel EJ et al (2011) Reengineering aircraft structural life prediction using a digital twin. Int J Aerosp Eng 2011:1–12
12. Tuegel EJ et al (2011) Reengineering aircraft structural life prediction using a digital twin. Int J Aerosp Eng
13. Guang M, Xubin Z, Jun M (2016) Mechanical problems in momentous projects of aerospace engineering. Adv Mech 46(1):201606

14. Noor AK et al (2000) Structures technology for future aerospace systems. Comput Struct 74(5):507–519
15. Zhu P, Zhang Y, Chen GL (2009) Metamodel-based lightweight design of an automotive front-body structure using robust optimization. Proc Inst Mech Eng Part D J Automob Eng 223(9):1133–1147
16. Müller U et al (2020) Crash simulation of wood and composite wood for future automotive engineering. Wood Mater Sci Eng 15(5):312–324
17. Kim KJ, Won ST (2008) Effect of structural variables on automotive body bumper impact beam. Int J Automot Technol 9:713–717
18. Gheorghe V et al (2021) New design of composite structures used in automotive engineering. Symmetry 13(3):383
19. Gao F et al (2020) Structural deformation monitoring and numerical simulation of a supertall building during construction stage. Eng Struct 209:110033
20. Beer M, Liebscher M (2008) Designing robust structures-a nonlinear simulation based approach. Comput Struct 86(10):1102–1122
21. Khodabandelu A et al (2020) Developing a simulation model for lifting a modular house. In: Construction research congress 2020: computer applications. American Society of Civil Engineers, Reston, VA
22. Gao F et al (2020) Structural deformation monitoring and numerical simulation of a supertall building during construction stage. Eng Struct 209:110033
23. Schmid M, Tomek P, Hanus P (2022) Multi-physical contact simulation in vehicle applications. Prod Eng Arch 28(4):369–374
24. Sarvar F, Poole NJ, Witting PA (1990) PCB glass-fibre laminates: thermal conductivity measurements and their effect on simulation. J Electron Mater 19(12):1345–1350
25. Keyes DE et al (2013) Multiphysics simulations: challenges and opportunities. Int J High Perform Comput Appl 27(1):4–83
26. Bazant ZP (2010) Can multiscale-multiphysics methods predict softening damage and structural failure? Int J Multiscale Comput Eng 8(1)
27. Stone HS (1990) High-performance computer architecture. Addison-Wesley Longman Publishing Co., Inc
28. Klenke S, Baca T (1996) Structural dynamics test simulation and optimization for aerospace components. Test Eval Int Aerosp Forum
29. Henning F et al (2019) Fast processing and continuous simulation of automotive structural composite components. Compos Sci Technol 171:261–279
30. Xiang JH et al (2013) Construction of macromolecular structural model of anthracite from Chengzhuang coal mine and its molecular simulation. J Fuel Chem Technol 41(4):391–400
31. Mayer N et al (2016) Structural analysis of composite components considering manufacturing effect. Compos Struct 140:776–782
32. Reddy JN (2019) Introduction to the finite element method. McGraw-Hill Education
33. Cheng AH, Cheng DT (2005) Heritage and early history of the boundary element method. Eng Anal Boundary Elem 29(3):268–302
34. Moczo P, Kristek J, Halada L (2004) The finite-difference method for seismologists. In: An introduction, p 161
35. Cebeci T et al (2005) Computational fluid dynamics for engineers: from panel to Navier-Stokes methods with computer programs. Springer, Long Beach, California
36. Constantin P, Foias C (2020) Navier-stokes equations. University of Chicago Press
37. Bertalmio M, Bertozzi AL, Sapiro G (2001) Navier-stokes, fluid dynamics, and image and video inpainting. In: Proceedings of the 2001 IEEE computer society conference on computer vision and pattern recognition. CVPR 2001, vol 1. IEEE
38. Knoedel P, Gkatzogiannis S, Ummenhofer T (2017) Practical aspects of welding residual stress simulation. J Constr Steel Res 132:83–96
39. Yaghi A et al (2006) Residual stress simulation in thin and thick-walled stainless steel pipe welds including pipe diameter effects. Int J Press Vessels Pip 83(11–12):864–874

40. Salmanpour AH, Mojsilovic N, Schwartz J (2013) Deformation capacity of unreinforced masonry walls subjected to in-plane loading: a state-of-the-art review. Int J Adv Struct Eng 5:1–12
41. Marrett R, Peacock DCP (1999) Strain and stress. J Struct Geol 21(8–9):1057–1063
42. Umemoto M et al (2000) Tensile stress-strain analysis of single-structure steels. Metall and Mater Trans A 31:1785–1794
43. Greek S, Ericson F (1998) Young's modulus, yield strength and fracture strength of microelements determined by tensile testing. MRS Online Proc Libr (OPL) 518:51
44. Demczyk BG et al (2002) Direct mechanical measurement of the tensile strength and elastic modulus of multiwalled carbon nanotubes. Mater Sci Eng A 334(1–2):173–178
45. Higson GR (1964) Recent advances in strain gauges. J Sci Instrum 41(7):405
46. Haecker CJ et al (2005) Modeling the linear elastic properties of Portland cement paste. Cem Concr Res 35(10):1948–1960
47. Sathikh S, Moorthy MBK, Krishnan M (1996) A symmetric linear elastic model for helical wire strands under axisymmetric loads. J Strain Anal Eng Des 31(5):389–399
48. Huang F (1988) On the mathematical model for linear elastic systems with analytic damping. SIAM J Control Optim 26(3):714–724
49. Preston DL, Tonks DL, Wallace DC (2003) Model of plastic deformation for extreme loading conditions. J Appl Phys 93(1):211–220
50. Prasatya P, McKenna GB, Simon SL (2001) A viscoelastic model for predicting isotropic residual stresses in thermosetting materials: effects of processing parameters. J Compos Mater 35(10):826–848
51. Harmathy TZ (1967) A comprehensive creep model, pp 496–502
52. Hsieh SS, Ting EC, Chen WF (1982) A plastic-fracture model for concrete. Int J Solids Struct 18(3):181–197

Chapter 8
Automation Interfaces and Script Programming Methods for Co-simulations

8.1 Introduction

In the fields of electronic and electromagnetic engineering, advanced simulation tools such as Automated Design System (ADS) and High-Frequency Structure Simulator (HFSS) have become indispensable for optimally analyzing and designing products through simulations. With the development of computer technologies and numerical methods, the software tools have been significantly improved and become very powerful, providing researchers or engineers with much convenience in use. To further enhance the efficiency of use, reduce human errors, and explore the functions of software tools, learning how to use the programming interfaces and scripts for automatic designs has become a necessary skill for professionals in the fields. To this end, the basic functions and features of ADS and HFSS softwares will be reviewed. The ADS is a kind of mainstream software for the simulations of radio frequency (RF) and microwave circuits and has been extensively applied for the designs of RF and microwave circuits, analyses of signal integrity, analyses of power integrity, etc. The HFSS, on the other hand, is a sort of simulation software for electromagnetic fields based on the well-known numerical method, i.e., Finite Element Method (FEM), and has been widely used for the analyses and designs of antennas, RF/microwave devices, electronic packaging, and military targets. In practical applications, the ADS and HFSS are often used in combination with other kinds of simulation software to meet different requirements.

Based on the knowledge on the ADS and HFSS, we will first provide a detailed introduction to the automation interfaces [1–3] and script programming methods [4] in the ADS. The ADS software offers rich programming interfaces and script supports, such as Advanced Extension Language (AEL) interface, Python Data Link (PDL), and ADS-MATLAB interface, among others. By utilizing these interfaces and scripts, users can achieve a personalized customization of the software and data interaction with other kinds of software such as Python and MATLAB. The detailed usage methods of these interfaces and scripts will be narrated in incorporation with

M. S. Tong and X. Y. Li, *Co-simulations of Microwave Circuits and High-Frequency Electromagnetic Fields*, https://doi.org/10.1007/978-981-99-8307-0_8

principles and examples. In a similar way, we will then give an introduction to the automation interfaces and script programming methods in the HFSS. The HFSS software supports various scripting languages like VBScript and IronPython. Through these scripting languages, users can achieve an automation of design processes in the HFSS, including the geometry modeling, simulation parameter setting, simulation execution and optimization, and result extraction. The detailed usage guidance of HFSS-based automation interfaces and script programming methods will be described along with application examples. Finally, we will offer some useful tips and considerations for using the automation interfaces and script programming methods based on the ADS and HFSS so that some common pitfalls and issues can be avoided to ensure a better user experience. With a deep understanding of the automation interfaces and script programming methods in the two sorts of software, users can fully make use of their capabilities, improve the analysis and design efficiencies, and reduce the human errors.

This chapter aims to provide a comprehensive guide for the above automation interfaces and script programming methods in the ADS and HFSS and help readers efficiently make use of these two sorts of software for their optimal analyses and designs. However, before diving into the contents of this chapter, it is recommended that readers first acquire some basic operational experiences with the ADS and HFSS as well as a certain level of programming skills with the Python and MATLAB. In addition, the examples and cases used in this chapter may involve certain specialized knowledge, and we suggest that readers be able to incorporate their own professional backgrounds when reading.

8.2 Advanced Extension Language (AEL) Interfaces in ADS

The Advanced Design System (ADS) is a kind of leading Electronic Design Automation (EDA) software widely used in the simulations for RF circuits [5, 6], microwave circuits [7, 8], signal integrity [9], and power integrity [10], etc. The ADS provides powerful functions to help users tackle complex analysis and design challenges. However, users may need to customize the functions for specific tasks and this is where the Advanced Extension Language (AEL) interface comes into play. The AEL interface offers a powerful scripting language that allows users to customize the ADS functions. In this section, we will provide a detailed introduction to the basic concepts of the AEL interface, including how to write, load, and run the AEL scripts, as well as how to use the AEL interfaces to implement customized ADS functions. We will also illustrate how to leverage the AEL interfaces to solve real-world problems through examples.

8.2.1 Basic Concept of AEL Interfaces

AEL is a script language based on the C language, with a syntax structure similar to C [11, 12]. AEL allows users to create custom functions, procedures, and data types to implement custom functionality in ADS. The following are the basic components of the AEL interface:

1. AEL Code Structure: Similar to the C language, AEL code consists of functions and procedures, with statements separated by semicolons (;). AEL code is typically saved in files with the extension ".ael".
2. Data Types: AEL supports various data types, including integers, real numbers, strings, lists, and structures.
3. Variables and Constants: AEL supports local variables and global variables. Local variables are only visible within the function or procedure where they are defined, while global variables are visible throughout the AEL code. Constants are immutable values that can be used in the code.
4. Operators and Expressions: AEL supports various operators, including arithmetic operators, relational operators, and logical operators. These operators can be used to create expressions for calculating values and controlling program flow.
5. Control Structures: AEL supports various control structures, including conditional statements (if-else), loop statements (for, while), and branching statements (switch-case).
6. Functions and Procedures: AEL supports user-defined functions and procedures. Functions are used to compute and return a value, while procedures are used to perform operations without returning any value.
7. Error Handling: AEL provides error-handling capabilities to capture and handle runtime errors, ensuring stable program execution.

8.2.2 Writing, Loading, and Running AEL Scripts

The steps to write an AEL script begin by opening a new file with a text editor such as Notepad + + and Sublime Text. Save it as a file with the extension ".ael". Write AEL code to define required functions, procedures, variables, etc. Here is an example of a simple AEL script:

```
// Calculate the sum
of two numbers
function int
add(int a, int b)
{
    return a + b;
}
//Main procedure
```

```
procedure main() {
    int num1 = 5;
    int num2 = 10;
    int sum = add(num1, num2);
    printf("The sum of \%d
    and \%d is \%d.\n",
    num1, num2, sum);
}
```

To save an AEL script file and load/run an AEL script, these steps should be followed:

1. Open ADS software.
2. From the menu bar, select File > Customize PCell. Change the type from "Use this layout" to "AEL Macro" and enter the name of the function (not the filename). This will open the AEL debugger window.
3. In the AEL debugger window, select "File" > "Load AEL file" and choose the previously written AEL script file.
4. After loading the AEL file, select "Run" > "Run Procedure." In the pop-up window, enter the name of the procedure to run (e.g., "main") and click the "OK" button. This will execute the AEL script and display the output results in the AEL debugger window.

Using AEL interface to implement custom ADS functionality:

Through the AEL interface, various custom functionalities can be implemented to meet specific design task requirements. Here are a few examples of custom functionalities implemented using the AEL interface:

1. Parameter sweeping: Designers can write AEL scripts to sweep one or more parameters within a certain range and analyze the effects of parameter variations on design performance. For example, an AEL script can be created to sweep the bias voltage of an RF amplifier and analyze its impact on gain, stability, and other metrics.
2. Data processing and analysis: AEL can be used to process and analyze simulation results to extract useful information. For instance, an AEL script can be written to compute various performance metrics such as insertion loss and group delay, and generate reports.
3. Automated design workflows: AEL scripts can enable automated design workflows, thereby enhancing design efficiency. For example, an AEL script can be written to automate various tasks such as device modeling, parameter optimization, layout generation, and so on.

8.2.3 A Simple Example

During simulation, it is easy to encounter a phase jump of 360 °, which brings inconvenience to adjustments [5]. To address this issue, the AEL function in ADS software can be utilized to eliminate the periodic phase jumps. The specific method is as follows:

First, create an AEL file in the \expressions\ael directory of the ADS installation directory, for example, ABC.AEL. Then, save the following code in ABC.AEL:

```
defun phaseshift_cor (v)
{
    return v−(ceil(v/360−1))∗360;
}
```

Add the following line in the user_defined_fun.ael file: load ("ABC.ael"), then restart the software to use the custom function. Then, write the following expressions in the simulation results (Fig. 8.1):

$$phaseshifter = phaseshift\_cor(phase(S(2, 1)) - phase(S(4, 3))) \quad (8.1)$$

$$phaseshift = phaseshift\_cor(phaseshifte) \quad (8.2)$$

This section details the AEL interface in ADS. The AEL interface is a powerful scripting language that allows the user to customize the functionality of the ADS software. We will discuss the basic concepts of the AEL interface, including how to

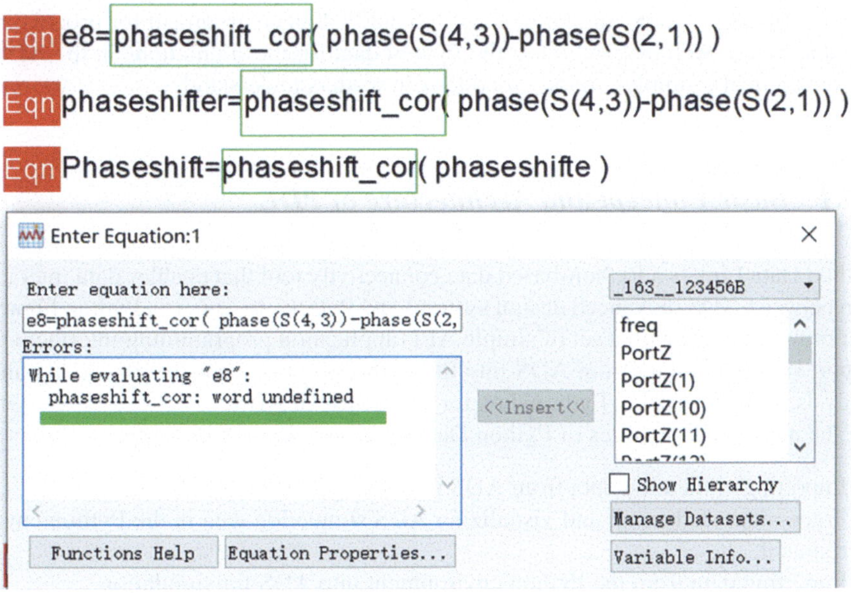

Fig. 8.1 Debugging in phase shift equation input

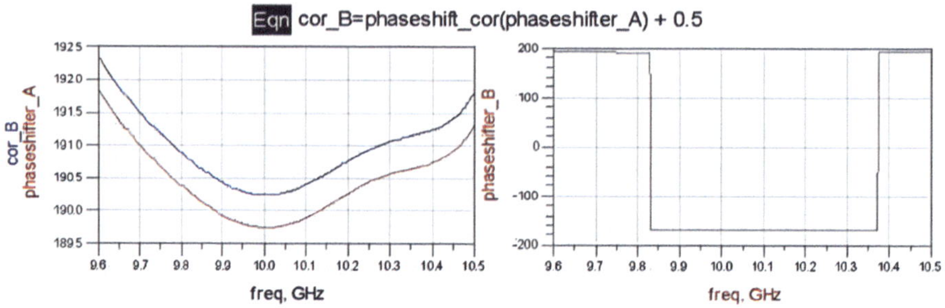

Fig. 8.2 Comparison of simulation results

write, load, and run AEL scripts, and how to use the AEL interface to implement custom ADS functionality. Another great resource is the ADS online help, chapter Design Tools > AEL. Here you will find instructions and command reference documentation. The Search tab is useful if you are looking for more details about the features you see in the log files (Fig. 8.2).

8.3 Python Data Link (PDL) for ADS

In this section, we will provide a detailed explanation on how to achieve a data interaction between the ADS and Python using the PDL [13, 14]. We will first explore the fundamental concepts and architecture of the PDL, and then demonstrate how to import the ADS simulation data into the Python environment for processing and analysis as well as how to use the Python data for the simulations in the ADS. Some practical examples are provided to help readers understand.

8.3.1 Basic Concept and Architecture of PDL

Python Data Link is a Python-based data connectivity tool that enables data interaction between ADS (advanced design system) and Python. Its core is a Python library that provides users with a set of simple API (application programming interface) to import simulation data from ADS into the Python environment for processing and analysis, as well as using Python data for simulation in ADS.

The main functionalities of Python Data Link include the following:

1. Importing simulation data from ADS into the Python environment.
2. Processing, analyzing, and visualizing ADS simulation data in the Python environment.
3. Importing data from the Python environment into ADS for simulation.

The architecture of Python Data Link consists of the following components:

1. Data import and export: PDL (Python data link) provides a series of interfaces for transferring data between ADS and Python.
2. Data processing and analysis: PDL supports users to use various data processing and analysis tools (such as NumPy, SciPy, and matplotlib) in the Python environment to process and analyze ADS simulation data.
3. Data visualization: PDL supports users to visualize ADS simulation data using various visualization tools (such as matplotlib and seaborn) in the Python environment.
4. ADS simulation: PDL supports importing data from the Python environment into ADS for simulation purposes.

8.3.2 Layout Simulations by Python-Calling ADS

In the official documentation of ADS, the article "Running a Momentum Simulation from Command Line" provides a comprehensive explanation of how to use the ADS interface, specifically for conducting layout simulations through the command line. Before invoking ADS, it is necessary to modify the system environment variables. The configuration of environment variables on Windows is as follows:

```
set HPEESOF_DIR=<path_to_ADS_installation_directory>
# for example: set HPEESOF_DIR
=C:\Progam Files\Keysight\ADS
set ADS_LICENSE_FILE
=<path_to_ADS_license_file_or_server>
# for example: set ADS_LICENSE_FILE
=C:\Program Files\Keysight\EEsof_License_Tools
\licenses\license.lic
set PATH=%HPEESOF_DIR%\bin;%PATH%
```

Please note that you should not directly use these commands in the cmd console as it will only create temporary variables. If you need to configure environment variables, you should go to the environment variable settings yourself. After configuring, you will need to restart your computer for the settings to take effect. You can test if the configuration was successful by running the command "adsMomWrapper—version."

Next, we need to generate the simulation initialization file. Open the simulation interface in the layout file where simulation is needed, and then generate the simulation input file. This file will be saved in the simulation folder of your workspace, for example: "D:/<your ads workspace>/simulation///layout."

Before starting the simulation, you need to navigate to the specified target folder in the cmd and then invoke the corresponding command. Here are the relevant code instructions:

```
os import math import numpy as np import pandas as pd

# "liner" corresponds to ".cti"
and "adapt" corresponds to ".afs".
Extracting S−parameters and reading S−parameters.
def read_cti(path: str):
print(path)
read_c = open(path + "/proj.cti")
data = read_c.read()
read_c.close()
data = data.split("BEGIN\n")
s_str = data[2].split("END")
s11_str = s_str[0].replace("\t", "")
s11_str = s11_str.split("\n")
s_str = data[4].split("END")
s21_str = s_str[0].replace("\t", "")
s21_str = s21_str.split("\n")
s11_r = []
s11_i = []
s21_r = []
s21_i = []
s21_phase = []
for i in range(len(s11_str) − 1):
s11_ri = s11_str[i].split(",")
s21_ri = s21_str[i].split(",")
s11_r.append(float(s11_ri[0]))
s11_i.append(float(s11_ri[1]))
s21_r.append(float(s21_ri[0]))
s21_i.append(float(s21_ri[1]))
s21_phase.append(math.atan2(s21_i[i],
s21_r[i]) * 180 / math.pi)
return s21_phase

# Modify the configuration file before simulation,
that is, modify the device design parameters.

def generate_s21(proj_path: str, w1):
i_hang = 0:
with open(proj_path + r"\proj_a") as read_f,
open(proj_path + r"\.proj_a", 'w') as write_f:
for line in read_f:
if i_hang == 2:
print (line)
line = "ADD P1 :W0.000000 −0.9500,−0.2750:
%.4f,−0.2750 %.4f,0.4350 −0.9500,0.4350;\r" % (w1, w1)
```

```
write_f.write(line)
else:
write_f.write(line)
i_hang = i_hang + 1:

os.remove(proj_path + r"\proj_a")
os.rename(proj_path + r"\.proj_a",
proj_path + r"\proj_a")

i_hang = 0:
w1_p = w1 * math.pow(10, -3):
with open(proj_path + r"\proj.pin") as read_f_1,
open(proj_path + r"\.proj.swap.pin", 'w') as write_f_1:
for line in read_f_1:
if i_hang == 25:
print(line)
line = "<x>%.5f</x>\r" % w1_p
write_f_1.write(line)
else:
write_f_1.write(line)
i_hang = i_hang + 1

os.remove(proj_path + r"\proj.pin")
os.rename(proj_path + r"\.proj.swap.pin",
proj_path + r"\proj.pin")

# os.system("cd %s && adsMomWrapper -O
-3D proj proj" % proj_path)
os.chdir(proj_path)
print(os.getcwd())
# Simulations Command
os.system("adsMomWrapper -T proj proj")
os.system("adsMomWrapper -DB proj proj")
os.system("adsMomWrapper -M proj proj")
os.system("adsMomWrapper -3D proj proj")

def Delete_S21(proj_path: str, fileList):
for parent, dirnames, filenames in os.walk(proj_path):
goodfilelist = []
fullfilelist = []
for x in fileList:
goodfile = proj_path + x
goodfilelist.append(goodfile)

for filename in filenames:
```

```
fullpath = parent + '/' + filename
fullfilelist.append(fullpath)

for xlist in fullfilelist:
if xlist not in goodfilelist:
os.remove(xlist)
pass

if __name__ == '__main__':
# proj_path = "D:/project_all/ADS/xxx
/layout/emSetup_MoM"
proj_path = "D:/em"
w1 = 4.0500
s21_csv = pd.DataFrame()
S21_Phase = np.zeros([1, 402])
w1_list = np.arange(0.65, 139.05, 0.01)
# Files that need to be kept for simulation,
files not listed here will be deleted.
fileList = [r'/emStateFile.xml',
r'/momentum.cfg', r'/proj.cfg', r'/proj.lcf',
r'/proj.ltd', r'/proj.opt',
r'/proj.params', r'/proj.pin',
r'/proj.plan', r'/proj.prt', r'/proj.sti', r'/proj_a']

for i in range(len(w1_list)):
generate_s21(proj_path, w1_list[i])
S21_Phase_hang = read_cti(proj_path)
S21_Freq = np.array(w1_list[i])
S21_Freq = np.reshape(np.append(S21_Freq,
np.reshape(np.array(S21_Phase_hang),
[1, 401])), [1, 402])
S21_Phase = np.append(S21_Phase, S21_Freq, axis=0)
Delete_S21(proj_path, fileList)

S21_data = pd.DataFrame(S21_Phase)
S21_data.to_csv("D:/python/phase/xxx.csv")
print("save complete")
```

The file of the S-parameters after simulation is saved in the .cti file. Note: S-parameter files generated by different types of simulation are different.

8.4 ADS-MATLAB Interfaces

The ADS-MATLAB interface is an important interface for facilitating data interaction between ADS (advanced design system) and MATLAB [15–17]. Through the ADS-MATLAB interface, users can easily access the simulation results from ADS in MATLAB and perform data analysis and processing using MATLAB. This article will provide a detailed introduction on how to utilize the ADS-MATLAB interface to exchange data between ADS and MATLAB, as well as how to use MATLAB to process ADS simulation results. Additionally, it will provide some practical examples to demonstrate how to achieve automated design using the ADS-MATLAB interface.

8.4.1 Data Transfer Based on ADS-MATLAB Interface

The ADS-MATLAB interface provides several methods for data transfer, including MAT-file transfer, COM transfer, TCP/IP transfer, etc. Among them, MAT-file transfer is the simplest method. In ADS, users can use the built-in MATLAB functions in ADS to save simulation results to a MAT-file. For example, the following ADS script saves the S-parameter matrix to a MAT-file:

```
save('S_params.mat', 'S_params');
```

In MATLAB, users can use the built-in load function to read data from a MAT-file. For example, the following MATLAB code reads the S-parameter matrix from a MAT-file:

```
load('S_params.mat');
```

With this method, the user can transfer the ADS simulation results to MATLAB for data analysis and processing.

8.4.2 Processing of ADS Simulation Results with MATLAB

In MATLAB, users can use various data analysis and processing tools, such as drawing, curve fitting, data statistics, and so on. Here are some common MATLAB tools and how to use them to process ADS simulation results.

MATLAB provides a powerful drawing tool, which can be used to draw a variety of graphics, such as linear chart, scatter plot, histogram, and so on. When dealing with ADS simulation results, plotting is a very useful tool to help users better understand the simulation results. For example, the following MATLAB code plots the frequency response of the S-parameter:

```
plot(freq, 20*log10(abs(S_params)));
xlabel('Frequency (GHz)');
ylabel('S-parameters (dB)');
```

When dealing with ADS simulation results, it is sometimes necessary to curve fit the data in order to better understand the data. MATLAB provides various curve fitting tools, such as polyfit, fit, and cftool. For example, the following MATLAB code will fit a quadratic polynomial to the frequency response curve of the S-parameter:

```
p = polyfit(freq, 20*log10(abs(S_params)), 2);
fit_curve = polyval(p, freq);
plot(freq, 20*log10(abs(S_params)), freq, fit_curve);
xlabel('Frequency (GHz)');
ylabel('S-parameters (dB)');
```

When dealing with ADS simulation results, data statistics, such as mean value and variance, are sometimes needed. MATLAB provides various statistical tools, such as mean, STD, and var. For example, the following MATLAB code calculates the mean and standard deviation of the S-parameters:

```
avg = mean(abs(S_params));
std_dev = std(abs(S_params));
```

8.4.3 Examples of ADS-MATLAB Interfaces

Below, two practical cases will be presented to demonstrate how to achieve automated design using the ADS-MATLAB interface.

Impedance matching is an important issue in radio design. Through impedance matching, the maximum energy can be transmitted when the signal is transmitted in the circuit [6]. In ADS, the Impedance Match Toolbox can be used for impedance matching design. The following describes how to implement a matching design using ADS-MATLAB interface.

1. Create ADS project

Create a new ADS project, then create a schematic CELL 1, and add a microstrip line to make a simple matching circuit, as shown in Fig. 8.3. Then we need to run an ADS simulation once to generate the netlist.log file, otherwise the following program will throw an error.

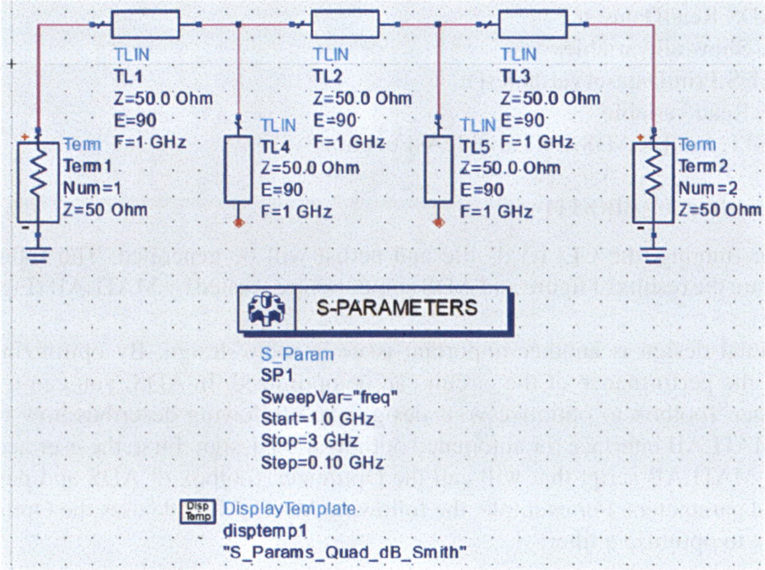

Fig. 8.3 Matching circuit of microstrip

2. MATLAB program, copy the following program to the MATLAB page:

```
%%I.Clear the screen
clc
clear
close all
%%II.Build a co−simulations environment
% Create TADS Interface object
ADS=TADSInterface();
% Set ADS installation path
ADSInstallationDir
= 'C:\Program Files\K Keysight\ADS';
ADS.SetADSPaths(ADSInstallationDir);
%Convert S parameter units to dB
to_dB = @(s) 20*log10(abs(s));
% Set Dataset File Path and Netlist File Path
ADS.NetlistFile = 'C:\Users\Administrator
\MyWorkspace2_wrk\netlist.log';
ADS.NetlistFile = 'C:\Users\Administrator
\MyWorkspace2_wrk\CELL1.ds';
% Run the simulation
ADS.RunSimulation ();
% ReadDataset();
```

```
ADS.ReadDataset();
% Show all variables
ADS.PrintDatasetVariables();
% Read variable
[S11, freq] = ADS.GetVariableAsFunction
('S[1,1]', 'freq');
plot(freq, to_dB(S11));
```

After running, the CELL1.ds file and netlist will be generated. The following figures are the results of figure and ADS simulation generated by MATLAB (Figs. 8.4 and 8.5).

Optimal design is another important issue in radio design. By optimizing the design, the performance of the circuit can be optimized. In ADS, you can use the Optimizer Toolbox to optimize your design. The following describes how to use ADS-MATLAB interface for automated optimization design. First, the user needs to write a MATLAB script that will call the Optimizer Toolbox of ADS and pass the required parameters. For example, the following MATLAB code uses the Optimizer Toolbox to optimize a filter:

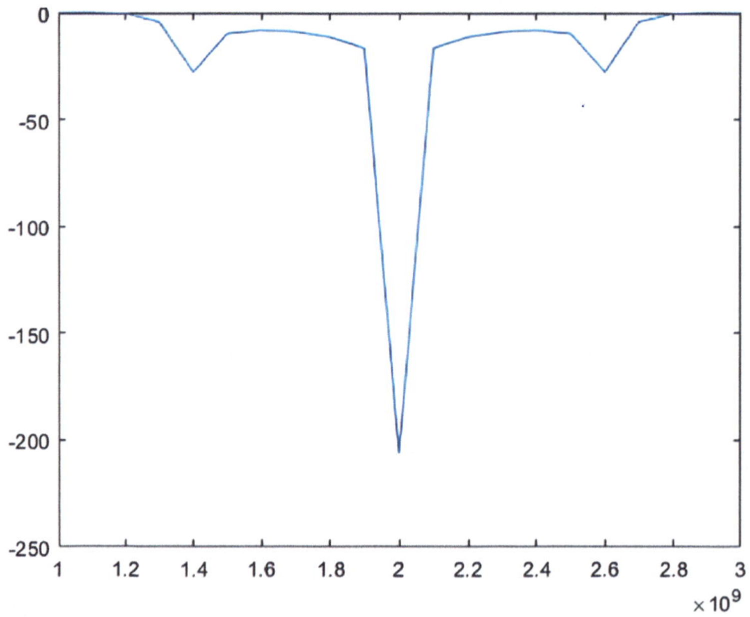

Fig. 8.4 MATLAB simulation results of S parameters, where the abscissa is Freq, the unit is Hz, and the ordinate is dB(S(1,1))

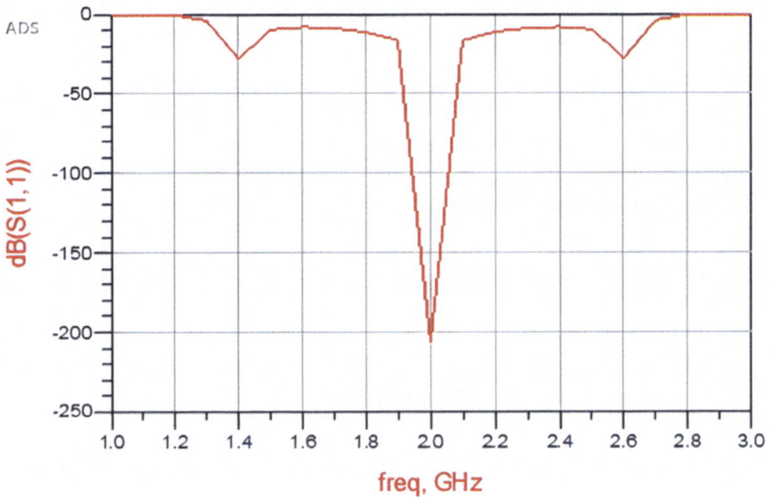

Fig. 8.5 ADS simulation results of S parameters

```
% connect to ADS
adsh = actxserver('AgilentADS.Simulator');
adsh.Init();

% create a new design
adsh.NewProject();
adsh.SetActiveDesign('MyDesign');

% create a new schematic
adsh.NewSchematic('MySchematic');

% add a filter component
filter = adsh.AddComponent('my_filter', 'FILTER');
filter.SetParameter('Freq', '1GHz');
filter.SetParameter('R', '50ohm');

% add an optimizer
opt = adsh.AddComponent('my_optimizer', 'OPTIMIZER');
opt.SetParameter('Objective', 'abs(S21)');
opt.SetParameter('Algorithm', 'quasi-newton');

% set the optimization variables opt.SetParameter
('Variables', {'C1', 'C2', 'L1', 'L2'});
```

```
opt.SetParameter
('Lower', [1e−12, 1e−12, 1e−9, 1e−9]);
opt.SetParameter
('Upper', [100e−12, 100e−12, 10e−9, 10e−9]);
% connect the components adsh.ConnectComponents
(filter, opt, 1);
% run the optimizer opt.Run();
% get the optimized values
C1 = opt.GetVariable('C1');
C2 = opt.GetVariable('C2');
L1 = opt.GetVariable('L1');
L2 = opt.GetVariable('L2');
% save the optimized values to a MAT−file
save('optimized_values.mat', 'C1', 'C2', 'L1', 'L2');
```

In the MATLAB script, we first use the "actxserver" function to connect to ADS. Then, we create a new ADS design and create a new schematic within that design. Next, we add a filter and an optimizer, and set the objective function and algorithm for the optimizer, as well as the optimization variables and their upper and lower bounds. Finally, we call the "Run" function of the optimizer to perform the optimization and use the "GetVariable" function to retrieve the optimization results. The results are then saved to a MAT-file. Next, we can write an automated MATLAB script that invokes the optimization design script and passes different parameters based on various design requirements, enabling the automation of the optimization process. For example, the following MATLAB code will automate multiple optimization designs to find the best filter:

```
% set the frequency range
freq = linspace(1e9, 2e9, 101);

% initialize the best filter
best_C1 = 0;
best_C2 = 0;
best_L1 = 0;
best_L2 = 0;
best_loss = Inf;

% loop over all possible values of C1, C2, L1, and L2
for C1 = linspace(1e−12, 100e−12, 11)
for C2 = linspace(1e−12, 100e−12, 11)
for L1 = linspace(1e−9, 10e−9, 11)
for L2 = linspace(1e−9, 10e−9, 11)

% run the optimization design
run_optimization_design(C1, C2, L1, L2, freq);

```

```
% load the results from the MAT−file
load('optimized_values.mat');

% calculate the loss
loss = 20∗log10(abs(filter.GetParameter('S21')));

% update the best filter
if loss < best_loss
best_C1 = C1;
best_C2 = C2;
best_L1 = L1;
best_L2 = L2;
best_loss = loss;
end

end
end
end
end

% print the best filter
fprintf('Best filter: C1=%g F, C2=%g F,
L1=%g H, L2=%g H, loss=%g dB\n',
best_C1, best_C2, best_L1,
best_L2, best_loss);
```

In the MATLAB script, we first set the frequency range and initialize the optimal filter. Then, we iterate through all possible values of C1, C2, L1, and L2, and call the "run _ optimization _ design" function to perform the optimization design. This function will invoke the optimization design script and pass the required parameters. Next, we use the "load" function to read the optimization results from the MAT-file and calculate the loss. Finally, we update the optimal filter and print the results. This approach allows for a fast and efficient execution of the optimization design and enables the discovery of the best filter.

8.5 VBScript for HFSS

VBScript or VBS (Visual Basic Script) is a language used in HFSS to write automation scripts [18, 19]. This section will introduce how to use VBScript to achieve automated design in HFSS, including creating geometric models, setting simulation parameters, running simulations, and obtaining simulation results. HFSS provides script functionality based on VBScript, allowing users to write scripts for automating the design process and improving efficiency.

VBScript is a lightweight scripting language, which is a simplified version of Visual Basic (VB) and primarily used for scripting in the Windows operating system. VBScript is widely used in many applications, and HFSS provides support for VBScript so that users can utilize scripts for automation design. This section will provide detailed information on the basic syntax, data types, control structures, functions, and objects of VBScript, serving as foundational knowledge for subsequent HFSS automation design.

We can click the mouse on the tool menu, select "record script" command, choose a saving path and name, and click "save." When executing the VBS, select the "run script" command. We can use VBScript to connect multiple single-point calculations using the "for...next" loop statement, forming a sweep operation. This ensures the convergence of each frequency point within the frequency band when performing broadband simulations, avoiding the deficiencies in HFSS's own frequency sweeping accuracy.

```
Dim oAnsoftApp
Dim oDesktop
Dim oProject
Dim oDesign
Dim oEditor
Dim oModule
Set oAnsoftApp = CreateObject
("AnsoftHfss.HfssScriptInterface")
Set oDesktop = oAnsoftApp.GetAppDesktop()
oDesktop.RestoreWindow
Set oProject = oDesktop.SetActiveProject("Project2")
Set oDesign = oProject.SetActiveDesign("HFSSDesign1")
Set oModule = oDesign.GetModule("AnalysisSetup")
oModule.InsertSetup "HfssDriven",
Array("NAME:Setup1", "Frequency:=",
"10MHz", "MaxDeltaE:=", _
0.02, "MaximumPasses:=", 6,
"MinimumPasses:=", 1, "MinimumConvergedPasses:=", _
1, "PercentRefinement:=", 20,
"ReducedSolutionBasis:=", false, "DoLambdaRefine:=", _
true, "DoMaterialLambda:=", true,
"Target:=", 0.3333, "UseConvOutputVariable:=", _
false, "IsEnabled:=", true,
"ExternalMesh:=", false, "UseMaxTetIncrease:=", _
false, "MaxTetIncrease:=", 100000)
Dim frequency,sfrequency
frequency=Array(30,40,60,100,150,200,300)
' Define frequency range/MHz
For i = 0 To 7 sfrequency = CStr(Frequency(i)) & "MHz"
oModule.EditSetup "Setup1",
```

```
Array("NAME:Setup1", "Frequency:=", sfrequency,
"MaxDeltaE:=", _ 0.02, "MaximumPasses:=", 6,
"MinimumPasses:=", 1, "MinimumConvergedPasses:=",
_ 1, "PercentRefinement:=", 20,
"ReducedSolutionBasis:=", false, "DoLambdaRefine:=",
_ true, "DoMaterialLambda:=", true,
"Target:=", 0.3333, "UseConvOutputVariable:=",
_ false, "IsEnabled:=", true,
"ExternalMesh:=", false, "UseMaxTetIncrease:=",
_ false, "MaxTetIncrease:=", 100000)
Next
```

This is only the frequency control part and doesn't involve anything else. The highlighted part is what we need to rewrite, while the rest is automatically generated by the system. The drawback of using VBS for such frequency scanning simulation is that the post-processing can be more complex. Since the results of individual point calculations are overwritten, only the result of the last frequency point is retained. Therefore, post-processing calculations need to be performed after completing iteration for each frequency point.

8.5.1 Modification of Model Parameters with VBS

Using VBS, we can modify model parameters. More precisely, it can perform all the operations that we can manually do. For complex models or models that require frequent modifications, using VBS for control is more convenient than manual control, especially in design and optimization. I will now explain how to use VBS to assign electromagnetic parameters to a dispersive medium using the following example. In this case, we will use the frequency sweep control program mentioned above, primarily focusing on controlling material parameters. Create a new material, Material1, and set variables ε_r and tan, representing the real part of the permittivity and the loss tangent, respectively. Using VBS, assign values to these variables at each frequency point. This enables the assignment of different electromagnetic parameters for a dispersive material at different frequency points. The relevant program is as follows:

```
Dim oAnsoftApp
Dim oDesktop
Dim oProject
Dim oDesign
Dim oEditor
Dim oModule
Set oAnsoftApp = CreateObject
("AnsoftHfss.HfssScriptInterface")
Set oDesktop = oAnsoftApp.GetAppDesktop()
```

```
oDesktop.RestoreWindow
Set oProject = oDesktop.SetActiveProject("Project2")
oProject.ChangeProperty Array
("NAME:AllTabs", Array("NAME:ProjectVariableTab",
Array("NAME:PropServers", _
"ProjectVariables"), Array("NAME:NewProps", Array
("NAME:$epr", "PropType:=", _
"VariableProp", "UserDef:=", true, "Value:=", "10"))))
oProject.ChangeProperty Array("NAME:AllTabs", Array
("NAME:ProjectVariableTab",
Array("NAME:PropServers", _
"ProjectVariables"), Array("NAME:NewProps",
Array("NAME:$etan", "PropType:=", _
"VariableProp", "UserDef:=", true, "Value:=", "1"))))
Set oDefinitionManager = oProject.GetDefinitionManager
()
oDefinitionManager.AddMaterial Array
("NAME:Material1", "CoordinateSystemType:=", _
"Cartesian", Array("NAME:AttachedData"),
"permittivity:=", "$epr", "permeability:=", _
"1", "dielectric_loss_tangent:=", "$etan")
Dim epr,epi,etan,i
epr=Array(37.9600,32.3400,28.2100,
25.0900,22.6100,20.6000,18.9300)
epi=Array(58.1900,49.1600,43.1800,
38.8300,35.4200,32.7200,30.4600)
For i=0 To 7
...
etan=epi(i)/epr(i)
oProject.ChangeProperty Array("NAME:AllTabs",
Array("NAME:ProjectVariableTab",
Array("NAME:PropServers", _
"ProjectVariables"),
Array("NAME:ChangedProps",
Array("NAME:$epr", "Value:=",
CStr(epr(i))))))
oProject.ChangeProperty Array("NAME:AllTabs",
Array("NAME:ProjectVariableTab",
Array("NAME:PropServers", _
"ProjectVariables"),
Array("NAME:ChangedProps",
Array("NAME:$etan", "Value:=", CStr(etan)))))
...
Next
```

This is just a partial program for controlling material parameters, and the highlighted parts are what we need to modify. By employing some programming techniques flexibly, we can use the same method to control any place where variables can be set using VBS. For example, when introducing broadband scanning to a half-wave dipole, we can define the length of the dipole as a variable L, and then modify L = 150/frequency(i) at each frequency point.

8.5.2 Post-processing Operations with VBS

The advantage of VBS is its repetitiveness. When the simulation is repeated, the following results will overwrite the previous results. Therefore, it is necessary to export the results of each simulation. At this time, relatively more VB statement calling techniques will be used, such as text creation and reading and writing operations, which can be found in the VBS examples provided by HFSS. The use of the field calculator is also essential, we can get most of the results through the field calculator, and then put these results through the program control output, and finally in the summary. The following paragraph is an example of using the field calculator to find the power of the scattered field:

```
dim sfilesfile="E:xx.txt"
Set oModule = oDesign.GetModule("Solutions")
oModule.EditSources "ScatteredFields",
Array("NAME:SourceNames", "IncWave1"),
Array("NAME:Modes", _ 1),Array("NAME:Magnitudes",1),
Array("NAME:Phases",0), Array("NAME:Terminated"),
Array("NAME:Impedances")
Set oModule = oDesign.GetModule("FieldsReporter")
oModule.EnterQty "Poynting"
oModule.CalcOp "Real"
oModule.EnterSurf "Plane1"
oModule.CalcOp "NormalComponent"
oModule.CalcOp "Integrate"
oModule.ClcEval "Setup1 : LastAdaptive",
Array("Freq:=", sfrequency, "Phase:=", _ "0deg")
oModule.CalcOp "Abs"oModule.CalcWrite sfile,
"Setup1: LastAdaptive", Array("Freq:=",
_ sfrequency, "Phase:=", "0deg")
```

8.6 IronPython Scripts for HFSS

In addition to VBScript, HFSS also supports automation design using IronPython scripts [20, 21]. This section will provide a detailed introduction on how to write HFSS automation scripts using IronPython and leverage the powerful capabilities of the Python language to achieve more advanced automation design tasks. We will explore the application of IronPython scripts in HFSS and provide practical examples to help readers better understand and master the usage of IronPython scripts. In the field of scientific research and engineering design, electromagnetic simulation technology has become an indispensable tool. Among them, HFSS (high-frequency structure simulator) is one of the widely used three-dimensional electromagnetic field simulation software with a high market share, extensively applied in the design and analysis of various RF, microwave, antenna, and other high-frequency structures. However, for complex design tasks, using the graphical user interface for operations may result in inefficiency and potential errors. Therefore, combining HFSS with IronPython can provide a highly automated design approach, enabling engineers and researchers to efficiently accomplish simulation tasks. In this chapter, we will briefly introduce the backgrounds of IronPython and HFSS and analyze the advantages of their integration.

8.6.1 Introduction to IronPython

IronPython is a Python implementation based on the .NET Framework, which allows users to write .NET applications using the Python language [22, 23]. This enables IronPython to interact with many existing .NET libraries while also taking advantage of Python's conciseness, readability, and powerful features. One of the main advantages of IronPython is its seamless integration with other .NET programming languages such as C# and VB.NET, enabling cross-language code interoperability. This allows developers to leverage the strengths of various languages and choose the most suitable tools for their projects.

8.6.2 Generation of Automation Scripts for HFSS

In this section, we describe how to write HFSS automation scripts using IronPython. This includes importing relevant libraries, creating HFSS projects, creating geometric models, setting boundary conditions, adding material properties, setting simulation parameters, and saving and exporting files.

8.6.2.1 Import Related Libraries

First, you need to import the libraries used to interact with HFSS. These libraries include the ones that come with IronPython as well as the API libraries for HFSS. The following is a typical import example:

```
import os
import sys
import clr
sys.path.append(r'C:\Program Files\AnsysEM\xxx\Win64')
sys.path.append(r'C:\Program Files\AnsysEM\xxx\Win64
\common\IronPython\DLLs')
sys.path.append(r'C:\Program Files\AnsysEM\xxx\Win64
\common\IronPython\Lib')
clr.AddReference('Ansys.Ansoft.Edb')
clr.AddReference('Ansys.Ansoft.Hfss')
import Ansys.Ansoft.Edb as Edb
import Ansys.Ansoft.Hfss as Hfss
```

Note that you need to change the path to the actual installation path of HFSS on your computer.

8.6.2.2 Creation of HFSS Projects

Create a new HFSS project and set the project name:

```
hfss_app = Hfss.HfssApplicationClass()
hfss_proj = hfss_app.CreateNewProject()
hfss_proj.SetName("HFSS_Project")
```

8.6.2.3 Creation of Geometric Models

Create a new design and set the design name, type, and version:

```
design = hfss_proj.CreateDesign("HFSS_Design",
"HFSS", "Version2022")
```

Next, create the geometric model. In this example, we will create a simple rectangular microstrip antenna:

```
modeler = design.GetModeler()
editor = modeler.GetActiveEditor()
patch = editor.CreateRectangle(
["Name:RectangleParameters", "IsCovered:True",
"XStart:0mm", "YStart:0mm", "ZStart:0mm",
"Width:20mm", "Height:10mm", "WhichAxis:Z"]
)
```

8.6.2.4 Setup of Boundary Conditions

Set the boundary condition of the rectangular antenna, for example, add a radiation boundary condition:

```
boundary = design.GetModule("BoundarySetup")
boundary.AssignRadiation(
["Name:Radiation", "Faces:=", [patch.Id]]
)
```

8.6.2.5 Addition of Material Properties

Add material properties for the rectangular antenna, for example, set the material of the antenna to copper:

```
editor.AssignMaterial(
["Name:=", "copper", "Component:=",
"patch", "MaterialValue:=", "copper"]
)
```

8.6.2.6 Setup of Simulation Parameters

Setup simulation parameters, such as frequency range and mesh type:

```
analysis = design.GetModule("AnalysisSetup")
analysis.InsertFrequencySweep(
"Setup1",
["Name:=", "Sweep", "StartValue:=", "1GHz",
"StopValue:=", "10GHz", "StepSize:=", "0.1GHz",
"Type:=", "Discrete", "SaveFields:=", False]
)
```

Finally, save the project to the specified directory and export it as an HFSS script file:

```
hfss_proj.SaveAs(r"C:\Users\YourUsername
\Documents\HFSS_Project.hfss")
hfss_proj.ExportScript(r"C:\Users\YourUsername
\Documents\HFSS_Project.vbs")
```

At this point, we have completed the writing of a basic HFSS automation script. In the following sections, we will describe how to leverage the power of the Python language for more advanced automated design tasks.

8.6.3 Advanced Automation Designs

In this section, we'll look at how to leverage the power of the Python language for more advanced automated design tasks. This includes parametric design, optimization algorithm and simulation, result processing and visualization.

8.6.3.1 Parametric Designs

Parametric design allows engineers and researchers to perform design optimization and sensitivity analysis by modifying parameters to adjust geometry and simulation settings. The following is a simple parametric design example where we modify the width and height of a rectangular microstrip antenna:

```
Width _ list = [10, 15, 20, 25, 30]
# width parameter list
Height _ list = [5, 7.5, 10, 12.5, 15]
# height parameter list
for width in width_list:
for height in height_list:
# Modify the width and height
editor.Modify(
["NAME:patch",
"Width:=", "{}mm".format(width),
"Height:=", "{}mm".format(height)]
)
# Simulate
analysis.Analyze("Setup1")
# Obtain simulation results
report = design.GetModule("ReportSetup")
s11_data = report.GetSolutionDataPerVariation
("Sweep", ["Freq:=", "All", "X Component:=", "S11"])
# Process and analyze the results
...
```

8.6.3.2 Optimization of Algorithms and Simulations

Combined with Python's powerful scientific computing libraries, such as NumPy and SciPy, users can introduce various optimization algorithms, such as genetic algorithm and particle swarm algorithm, into the HFSS simulation process. The following is a simple example of genetic algorithm optimization:

```
from scipy.optimize import differential_evolution
# Define the optimization objective function
def objective_function(x):
width, height = x
# Modify the width and height
editor.Modify(["NAME:patch","Width:=",
"{}mm".format(width),
"Height:=", "{}mm".format(height)])
# Simulate
analysis.Analyze("Setup1")
# Obtain simulation results,
such as obtaining S parameters
report = design.GetModule("ReportSetup")
s11_data = report.GetSolutionDataPerVariation
("Sweep", ["Freq:=", "All", "X Component:=", "S11"])
# Compute an optimization objective,
such as minimizing S11
return -max(s11_data)
# Set the parameter range
bounds = [(10, 30), (5, 15)]
# Use genetic algorithms for optimization
result = differential_evolution
(objective_function, bounds)
best_width, best_height = result.x
# Design using optimization results
...
```

8.6.3.3 Result Processing and Visualization

Using Python's rich data processing and visualization libraries, such as pandas and matplotlib, users can conduct in-depth analysis and visualization of HFSS simulation results.

8.6.4 Design of Logarithmic Periodic Antennas

In this section, we will give the log-periodic antenna design for the co-simulation of PYTHON and HFSS. The Python code is programmed by the calculation formula to realize the automation of the complex modeling process. Some API interface calls are required. Some sample codes are as follows:

```
oDesktop = oAnsoftApp.GetAppDesktop()
oProject.InsertDesign("HFSS",
"HFSSDesign1", "DrivenModal", "")
oDesign = oProject.SetActiveDesign("HFSSDesign1")
oEditor = oDesign.SetActiveEditor("3D Modeler")
oModule = oDesign.GetModule("BoundarySetup")
oModule = oDesign.GetModule("AnalysisSetup")
oModule.InsertSetup("HfssDriven",
[
"NAME:Setup1",
"AdaptMultipleFreqs:=" , False,
"Frequency:=" , "500MHz",
"MaxDeltaS:=" , 0.02,
"PortsOnly:=" , False,
"UseMatrixConv:=" , False,
"MaximumPasses:=" , 6,
"MinimumPasses:=" , 1,
"MinimumConvergedPasses:=", 1,
"PercentRefinement:=" , 30,
"IsEnabled:=" , True,
"BasisOrder:=" , 1,
"DoLambdaRefine:=" , True,
"DoMaterialLambda:=" , True,
"SetLambdaTarget:=" , False,
"Target:=" , 0.3333,
"UseMaxTetIncrease:=" , False,
"PortAccuracy:=" , 2,
"UseABCOnPort:=" , False,
"SetPortMinMaxTri:=" , False,
"UseDomains:=" , False,
"UseIterativeSolver:=" , False,
"SaveRadFieldsOnly:=" , False,
"SaveAnyFields:=" , True,
"IESolverType:=" , "Auto",
"LambdaTargetForIESolver:=", 0.15,
"UseDefaultLambdaTgtForIESolver:=", True,
"RayDensityPerWavelength:=", 4,
"MaxNumberOfBounces:=" , 5,
```

```
"InfiniteSphereSetup:=" , -1,
"SkipSBRSolveDuringAdaptivePasses:=", True

])
oEditor.Delete(
[
"NAME:Selections",
"Selections:=" , "spacing"
])
oModule = oDesign.GetModule("BoundarySetup")
oModule.AssignPerfectE(
[
"NAME:PerfE1",
"Objects:=" , ["element1"],
"InfGroundPlane:=" , False
])
oDesign.SetDesignSettings(
[
"NAME:Design Settings Data",
"Use Advanced DC Extrapolation:=", False,
"Use Power S:=" , False,
"Export After Simulations:=", False,
"Allow Material Override:=", True,
"Calculate Lossy Dielectrics:=", False,
"Perform Minimal validation:=", False,
"EnabledObjects:=" , [],
"Port Validation Settings:=", "Standard"
],
[
"NAME:Model Validation Settings",
"EntityCheckLevel:=" , "Strict",
"IgnoreUnclassifiedObjects:=", False,
"SkipIntersectionChecks:=", False
])
oModule = oDesign.GetModule("RadField")
oModule.InsertFarFieldSphereSetup(
[
"NAME:Infinite Sphere1",
"UseCustomRadiationSurface:=", False,
"ThetaStart:=" , "0deg",
"ThetaStop:=" , "180deg",
"ThetaStep:=" , "2deg",
"PhiStart:=" , "-180deg",
"PhiStop:=" , "180deg",
"PhiStep:=" , "2deg",
```

```
"UseLocalCS:=", False
])
oModule = oDesign.GetModule("ReportSetup")
oModule.CreateReport("S Parameter Plot 1",
"Modal Solution Data",
"Rectangular Plot", "Setup1 : Sweep",
[
"Domain:=", "Sweep"
],
[
"Freq:=", ["All"],
"hh:=", ["Nominal"]
],
[
"X Component:=", "Freq",
"Y Component:=", ["dB(S(1,1))"]
], [])
oModule.CreateReport("Gain Plot 1",
"Far Fields", "3D Polar Plot",
"Setup1 : LastAdaptive",
[
"Context:=", "Infinite Sphere1"
],
[
"Phi:=", ["All"],
"Theta:=", ["All"],
"Freq:=", ["0.5GHz"],
"hh:=", ["Nominal"]
],
[
"Phi Component:=", "Phi",
"Theta Component:=", "Theta",
"Mag Component:=", ["dB(GainTotal)"]
], [])
Save()
AnalyzeAll()
```

The simulation result is shown in the figure below, which is the current direction on the surface of the assembly line, and the two flow in opposite directions, that is, the transmission line mode. According to the distribution of electric field, it can be seen that the oscillator with corresponding frequency participates in resonance at different frequencies. The distribution of the electric field at low and high frequencies is shown dynamically, from which it can be seen that the low-frequency band is mainly longer oscillators, while the high-frequency band is concentrated on short oscillators. However, the content described in this section is merely an introductory

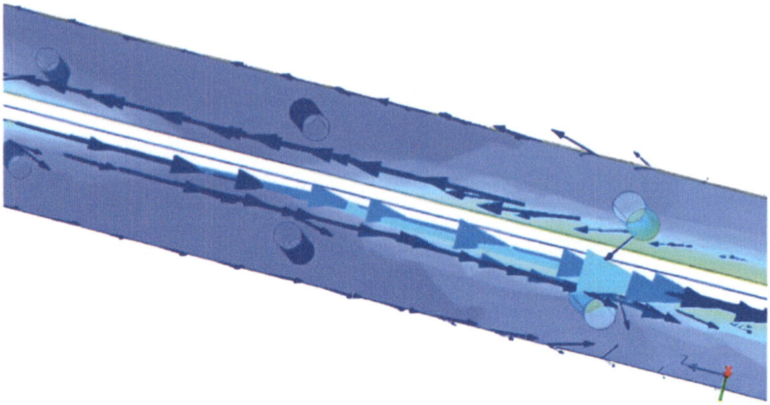

Fig. 8.6 Current flow direction of the assembly line

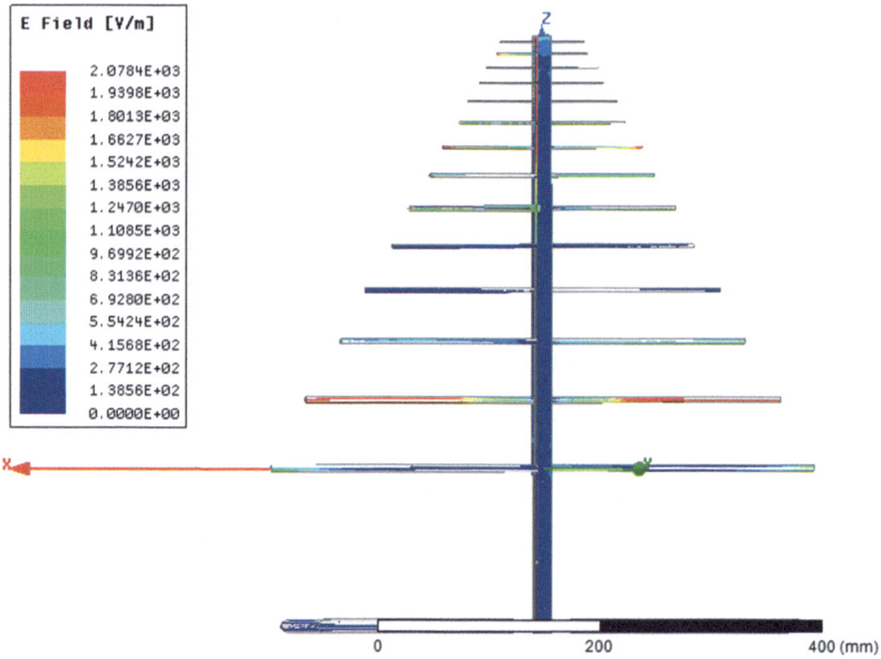

Fig. 8.7 Electric field distribution at the frequency of 3 GHz

tutorial on the application of IronPython in HFSS. In actual engineering, users may encounter more complex issues and challenges. Therefore, it is recommended that readers continue to learn and explore in practical applications, fully leveraging the advantages of Python and HFSS, to bring greater convenience and benefits to their engineering projects (Figs. 8.6, 8.7, 8.8, 8.9, and 8.10).

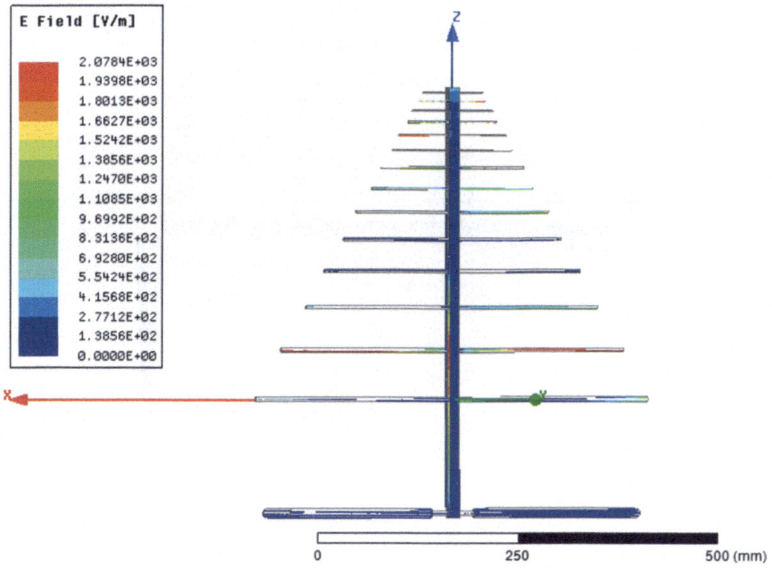

Fig. 8.8 Electric field distribution at the frequency of 500 MHz

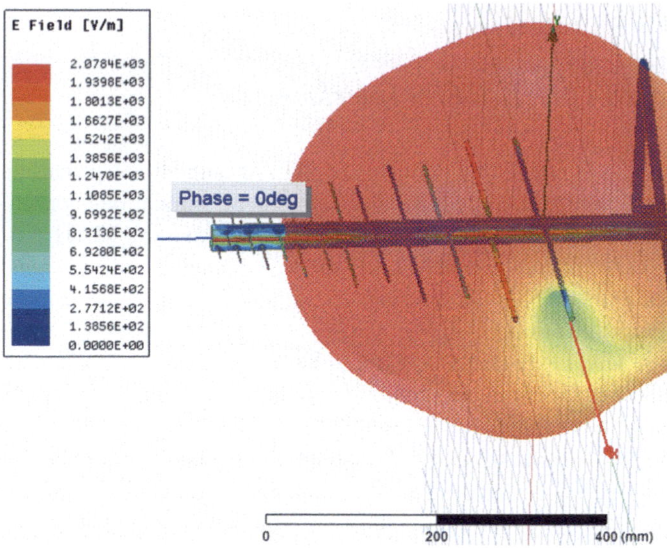

Fig. 8.9 Electric field distribution at the frequency of 300 MHz

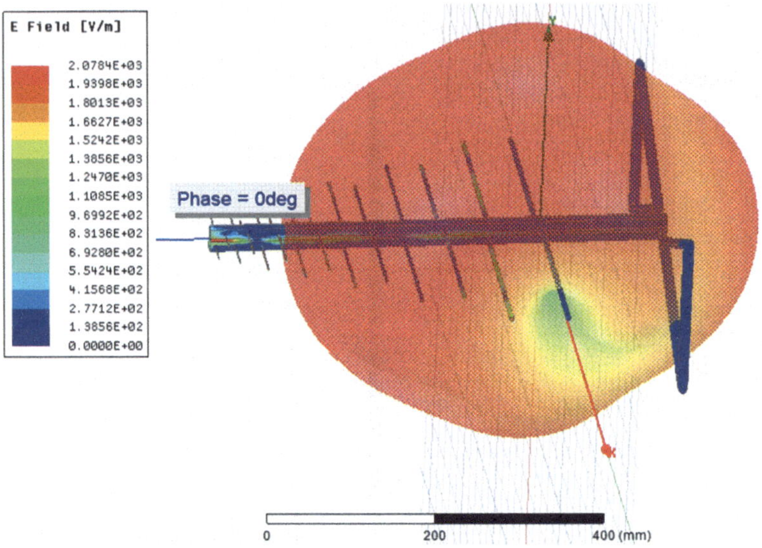

Fig. 8.10 Electric field distribution at the frequency of 1 GHz

8.7 MATLAB-HFSS Interfaces

8.7.1 Overview of MATLAB-HFSS Interfaces

MATLAB-HFSS API is an interface that connects MATLAB with HFSS, providing users with the ability to leverage the powerful electromagnetic simulation capabilities of HFSS within the MATLAB environment [24, 25]. By using the MATLAB-HFSS interfaces, designers can combine MATLAB's robust mathematical and signal processing capabilities with HFSS's electromagnetic simulations to achieve more advanced analysis and optimization.

The main features of MATLAB-HFSS API are as follows:

1. It provides MATLAB functions for creating, editing, and controlling HFSS projects.
2. It offers functions for defining materials, boundary conditions, excitations, and more within HFSS.
3. It supports HFSS data post-processing within the MATLAB environment.
4. It enables analysis of simulation results using MATLAB, such as parameter optimization and design sensitivity analysis.

8.7.2 Framework of MATLAB-HFSS Interfaces

The construction of the development framework mainly includes try..catch..end exception handling, generating vbs scripts without directly launching HFSS, and encapsulating framework functions. The main components of the construction are as follows:

```
clear;
clc;
try
\% Fill in the path
\% tmpPrjFile: pathname of the generated aedt
or HFSS project file
\% tmpScriptFile: pathname of the generated
VBS script file
tmpPrjFile = 'F:\vbsScript\Firstdemo.aedt';
tmpScriptFile = 'F:\vbsScript\xx.vbs';
\% hfssExePath: Path to HFSS software
hfssExePath = 'xx.exe';
\% create a readable and writable VBS script films
fid = fopen(tmpScriptFile, 'wt');
\% Create a new HFSS project
and insert a new design file..
hfssNewProject(fid);
Design_name='Draw_a box';
hfssInsertDesign(fid, Design_name);
\% Save the project file to the specified path
hfssSaveProject(fid, tmpPrjFile,1);
\% Close the HFSS Script File.
fclose(fid);
Disp ( 'VBS script generated);
catch
Disp (program exception);
fclose(fid);
end
```

8.7.3 MATLAB-Controlled File Preparation

Writing files in MATLAB can use fopen and fclose to control the VBS files we need to control, for example:

```
FID = fopen ( 'control. VBS', 'w +');\%
fclose all;
```

We can use fopen to open a script file and write the content we need into it. “fid” is the file handle, and you can check the usage of fopen in MATLAB by using “help fopen” or “doc fopen.” In order to control HFSS, we also need an initialization section. This initial section is placed at the beginning of each recorded script file.

Then, we write the initial section into our control VBScript (vbs) file. For example:

```
fid=fopen('control.vbs','w+');
fprintf(fid, 'Dim oAnsoftApp\n');
fprintf(fid, 'Dim oDesktop\n');
fprintf(fid, 'Dim oProject\n');
fprintf(fid, 'Dim oDesign\n');
fprintf(fid, 'Dim oEditor\n');
fprintf(fid, 'Dim oModule\n');
fprintf(fid, 'Set oAnsoftApp =
CreateObject("Ansoft.ElectronicsDesktop")\n');
fprintf(fid, 'Set oDesktop
= oAnsoftApp.GetAppDesktop()\n');
fprintf(fid, 'oDesktop.RestoreWindow\n');
fprintf(fid, 'Set oProject =
oDesktop.SetActiveProject("Project")\n');
\% Here Project is the name
of the project file in HFSS.
For different project files, you must modify this place
to find the relevant project in MATLAB.
Fprintf (FID, 'Set oDesign = oProject.
SetActive Design ( "HFSSdesigner") \ n');
\% HFSSd esigner is the design file name,
different design file names should also be modified
```

If you need other functions, you can use the fprintf function to write the contents of the previously recorded script files for different operations into our VBS script, and finally use dos to run our script files.

```
Dos' XX. VBS ';\%\% running script
```

In this way, running MATLAB can control HFSS to perform the desired operation and derive the corresponding results to complete the optimization task.

8.7.4 Two Examples

8.7.4.1 Calling HFSS to Model a Circular Microstrip Patch Antenna

```
clear;
clc;
try
\%\% Configuration Path
\% Fill in the path
\% tmpPrjFile: pathname of the generated aedt
```

```
or HFSS project file
\% tmpScriptFile: pathname of
the generated VBS script file
tmpPrjFile='D:\Installed
Software\Ansys_Electronic_2021\Ansys_Project_Directory
\Vbs_Practise_20220610\CircularPatchAntenna
\CircularPatchAntenna2. aedt';
tmpScriptFile='D:\Installed
Software\Ansys_Electronic_2021\Ansys_Project_Directory
\Vbs_Practise_20220610\CircularPatchAntenna
\CircularPatchAntenna2. vbs';

\% hfssExePath: Configure the path for
ANSYS Electronic software
hfssExePath='D:\Installed Software
\Ansys_Electronic_2021\Ansys_EM\AnsysEM21.2
\Win64\ansysedt.exe';\% Add the required m file path
addpath('D:\Installed Software\MATLAB
\toolbox\Matlab_HFSS_API\');
\% Create a read-write VBS script file
fid=fopen(tmpScriptFile,'wt');
\% Create a new Hfss project and
insert a new design file.
hfssNewProject(fid);
Design_name='CircularPatchAntenna2';
hfssInsertDesign(fid,Design_name);
\%\% parameter configuration, material selection,
parameter calculation, FR4 selected here,
can be changed to other materials,
modify the dielectric constant and board thickness
HSub_Cal=0.16;
er=4.4;
fr=2.45;
k=8.749/(fr*sqrt(er));
\% patch size;
r0=k/sqrt(1+2*HSub_Cal/(pi*er*k)*
(log10(pi*k/(2*HSub_Cal))+1.7726));
r1=r0*10;
r=vpa(r1,3);
\% ground dimension
GND_Lx=60;
GND_Wy=60;
\% Media Substrate Size
HSub=1.6;
WsubX=60;
```

```
WsubY=60;
\% Radiant Boundary Model
AirBox_X=120;
AirBox_Y=120;
AirBox_Z=70;
\%\% Start modeling, draw GND,
and set project boundary conditions;
hfssRectangle(fid,'GroundPlane','Z',
[-GND_Lx/2,-GND_Wy/2,0],GND_Lx,GND_Wy,'mm');
hfssSetColor(fid,'GroundPlane',[128,128,0]);
hfssSetTransparency(fid,{'GroundPlane'},0);
hfssAssignPE(fid,'PerfE_Ground',{'GroundPlane'});
\%\% draw the media substrate and
configure the material as Rogers 5880
hfssBox(fid,'Substrate',
[-WsubX/2,-WsubY/2,0],[WsubX,WsubY,HSub],'mm');
hfssAssignMaterial(fid,'Substrate','FR4_epoxy');
hfssSetColor(fid,'Substrate',[0,128,0]);
hfssSetTransparency(fid,{'Substrate'},0.6);
\%\% draw a circular radiation patch;
hfssCircle(fid,'Patch','Z',[0,0,HSub],16.8,'mm');
hfssAssignPE(fid,'PerfE_Patch',{'Patch'});
\%\% Draw coaxial line and set port
hfssCylinder(fid,'Feed','Z',[4.8,4.8,0],0.7,HSub,'mm');
hfssAssignMaterial(fid,'Feed','copper');
hfssCircle(fid,'Port1','Z',[4.8,4.8,0],1.4,'mm');
hfssSubtract(fid,{'GroundPlane'},{'Port1'});
\%\% draw the air box and
add the radiation boundary condition;
hfssBox(fid,'AirBox',[-AirBox_X/2,
-AirBox_Y/2,-AirBox_Z/2],
[AirBox_X,AirBox_Y,AirBox_Z],'mm');
hfssSetTransparency(fid,{'AirBox'},0.8);
hfssAssignRadiation(fid,'Radiation','AirBox');
\%\% Set Far-Field Spherical Coordinate System
hfssFarFieldSphere(fid, 'EHPlane',
-180, 180, 1, 0, 90, 90);
hfssFarFieldSphere(fid,'3D',0,180,1,0,360,1);
\%\% Draw coaxial feeder setting port
hfssCircle(fid,'Port','Z',[4.8,4.8,0],1.6,'mm');
hfssCircle(fid,'Port2','Z',[4.8,4.8,0],0.7,'mm');
hfssSubtract(fid,{'Port'},{'Port2'});
hfssAssignLumpedPort(fid,'1','Port',
[5.5,4.8,0],[6.4,4.8,0],'mm');
\%\% Set solution parameters and save
```

```
hfssInsertSolution(fid,'Setup2_45GHz',2.45,0.02,15);
hfssSolveSetup(fid,'Setup2_45GHz');
hfssInterpolatingSweep(fid,'Sweep2to3GHz',
'Setup2_45GHz',2,3,101);
\%\% Generate Report
hfssCreateReport(fid,
'E_H_Radiation_Plane','Far Fields',
'Rectangular Plot',
'Setup2_45GHz:LastAdaptive',
'dB(RealizedGainTotal)',fr);
\%\% Save File
hfssSaveProject(fid,tmpPrjFile,1);
\%\% Close the HFSS script films
fclose(fid);
Disp ( 'VBS script generated!' );
catch
Disp ( 'Program exception!' );
fclose(fid);
end
```

8.7.4.2 Design of PIFA Antenna Based on Genetic Algorithm

This section presents a method for designing a a planar inverted-F antenna (PIFA) by using a HFSS-MATLAB co-simulation [26, 27]. The antenna is intended for vehicular communication, specifically WLAN and 5G communication modes, with a working frequency range covering two bands: 2.4–2.484 and 4.8–5 GHz. The method consists of several modules, including the main function module, population initialization module, antenna modeling and fitness calculation module, selection operator, crossover operator, mutation operator module, and optimal solution selection module. The code for each module is shown below (Fig. 8.11).

The main function code is shown below:

```
function [pop,fitvalue,bestindividual,bestfit]
= self_main1()
clear;
clc;
\% population size
global popsize;
popsize = 50;
\% binary code length
chromlength = 542;
\% Crossover probability
pc = 0.6;
\% probability of mutation
```

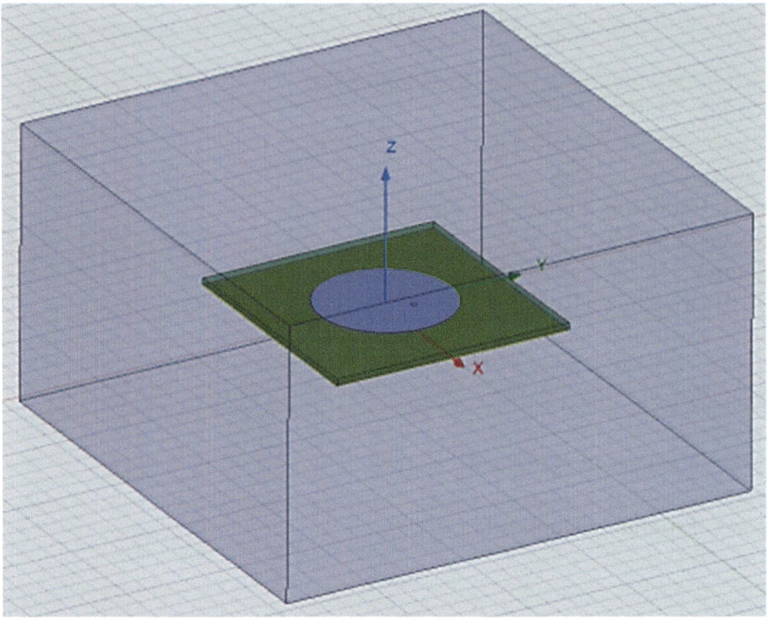

Fig. 8.11 Generated structure of microstrip antenna

```
pm = 0.005;
\% initial population
pop = initpop();

for i = 1:18
fprintf ( 'The generation is ---->> %1.1f\n',i );
\% Numeric value for calculating fitness
fitvalue = test_0330(pop);
\% Select an action
newpop = selection(pop,fitvalue);
\% Interleave operation
newpop = crossover(newpop,pc);
\% mutation operation
newpop = mutation(newpop,pm);
\% update population
pop = newpop;
\% Find the optimal solution
[bestindividual , bestfit] = best(pop,fitvalue);
\% x2 = binary2decimal(bestindividual);
\% x1 = binary2decimal(newpop);
\% y1 = cal_objvalue(newpop);
```

```
end

fprintf ( 'The best X is ---->> \%5.0f\n',
bestindividual );
fprintf ( 'The best Y is ---->> \%5.2f\n',bestfit);
end

The initialization population code is shown
in the following figure:
\% Initialize population
\% When the population is initialized,
a set of eight-digit decimal numbers
is randomly generated in the range 2-29,
arranged from smallest to largest.
\% Randomly generates two sets of
four-digit decimal arrays
arranged from small to large in the range 2-19
\% The number of populations is 50
and the number of iterations is 20.
function pop = initpop()

popsize =50;
chromlength = 542;
pop = ones(popsize,chromlength);
n=1;

while(n<50)
a = sort(round(rand(1,4)*15)+2);
b = sort(round(rand(1,8)*27)+2);
d = sort(round(rand(1,4)*15)+2);

wx = a(1,2);
while(wx<(a(1,3)+1))
lx = b(1,1);
while(lx<(b(1,4)+1))
c = 30*wx+lx;
pop(n,c) = 0;
lx = lx + 1;
end
wx = wx + 1;
end

wy = a(1,1);
while(wy<(a(1,4)+1))
```

```
ly = b(1,2);
while(ly<(b(1,3)+1))
c = 30*wy+ly;
pop(n,c) = 0;
ly=ly+1;
end
wy = wy+1;
end

wxx=d(1,2);
while(wxx < (d(1,3)+1))
lxx = b(1,5);
while(lxx<(b(1,8)+1))
c = 30*wxx+lxx;
pop(n,c) = 0;
lxx = lxx+1;
end
wxx = wxx+1;
end

wyy = d(1,1);
while(wyy<(d(1,4)+1))
lyy=b(1,6);
while(lyy<(b(1,7)+1))
c = 30*wyy+lyy;
pop(n,c) = 0;
lyy = lyy+1;
end
wyy = wyy+1;
end
n = n+1;
end
pop(n,541) = 0;
pop(n,542) = 0;

end

The code for modeling the antenna
and calculating the fitness is as follows:

\%model & simulation scripts
function [objvalue] = cal_objvalue(pop)
global popsize;
for n = 1 : popsize
chromlength = 542;
```

```
\% tmpPrjFile = 'F:\desktop\env\co_simulation_pifa
\HFSS\PIFAjoint.aedt';
\% tmpDataFile = 'C:\temp\tempdataa7.m';
tmpScriptFile = 'F:\Ansoft.ElectronicsDesktop.vb';

\%HFSS Executable
hfssExePath = '"F:\Ansoft.ElectronicsDesktop"';
fid = fopen(tmpScriptFile,'wt');
hfssNewProject(fid);
hfssInsertDesign(fid,'PIFAjoint');

\%Design 3Dmodel
hfssBOX(fid , 'BOX1' , [−9,0,0], [9,15,8] ,'mm');
hfssAssignMaterial(fid,'BOX1','air');

Axis = 'Z';
Width = −10;
Height = 15;
hfssRectangle(fid , 'GND' , Axis , [0,0,0],
Width, Height ,'mm');

Axis = 'Z';
Width = −10;
Height = 15;
hfssRectangle(fid , 'SUB' , Axis , [0,0,7],
Width, Height ,'mm');

a = 0;
b = 0;
m = 1;

\% This section should be holed on the sub board.
while(m<540)
if pop(n,m)==0
a = floor(m/30);
b = mod(m,30);
if b==0
sx = (0.5*a − 10.5);
sy = 14.5;
end
if b>0
sx = (0.5*a−10);
sy = (0.5*b);
end
```

```
nameslot = strcat('slot_',num2str(m));

Axis = 'Z';
Width = 0.5;
Height = 0.5;
hfssRectangle(fid , nameslot, Axis , [sx,sy,7],
Width, Height ,'mm');

hfssSubtract(fib,'SUB',nameflot);
end
m = m+1;
end

Axis = 'Y';
Width = −7;
Height = −2.5;
hfssRectangle(fid , 'SHORT' , Axis , [0,0,7],
Width, Height ,'mm');

Center = [−1.9 , 5 , 0];
Axis = 'Z';
Radius = 0.25;
Height = 7;
hfssCylinder(fid , 'FEED' , Axis ,Center ,
Radius ,Height , 'mm');

Center = [−1.9 , 5 , 0];
Axis = 'Z';
Radius = 0.5;
hfssCircle(fid , 'HOLE' , Axis , Center ,
Radius , 'mm');

hfssSubstract(fid , 'GND' , 'HOLE');
hfssAssignMaterial(fid,'FEED','copper');
hfssUnite(fit , 'SUB' , 'GND');
hfssUnite(fit , 'SUB' , 'SHORT');
hfssPE(fid,'SUB');

Center = [−1.9,5,0];
Axis = 'Z';
Radius = 0.5;
hfssCircle (fid , 'HOLE1' , Axis ,Center ,
Radius , 'mm');

\%Air box
```

```
Start_airbox = [−100 ; 100 ; 50];
Size_airbox = [200 , −200 , −100];
hfssBox(fid , 'AirBox' , Start_airbox ,
Size_airbox , 'mm');
\% Add radiation boundary condition
hfssAssignRadiaion(fid,'Radiation' , 'Airbox');

Setting of\% integral line
iLStart = [−1.9 , 5.12 , 0];
iLEnd = [−1.9 , 5.25 , 0];
Resistance = 50;
Reactance = 0;
hfssAssignLumpedPort(fid,'1','HOLE1',
iLStart,iLEnd,'mm',Resistance,Reactance);

%caculate solution
hfssInsertSolution(fid,'solve',2.5);
hfssSolveSetup(fid,'solve');
%sweeep
hfssInterpolatingSweep(fid,'sweep1','solve',0,1, ...
7,100);
fprintf(fid , 'oDesign.AnalyzeALL\n');


fprintf( fid , '\n');
fprintf( fid, 'Set oModule =
oDesign.GetModule("ReportSetup") \n');
fprintf( fid, 'oModule.CreateReport"S11",
"Modal Solution Data","Rectangular Plot",_\n');

fprintf( fid, ' "solve : Sweep1",
Array("Domain:=","Sweep1") ,
Array("Freq:=",Array("A11"), _\n');
fprintf( fid , 'Array("X Component:=",
"Freq","Y Component:=", Array("dB(S(1,1))")), _\n');

fclose(fid);

disp('Solving using HFSS..');
hfssExrcuteScript(hfssExePath,
tmpScriptFile,true,true);

tmpFid = fopen('test.csv','r');
tmpStr = fgetl(tmpFile);
tmpData = fscanf(tmpFid , '\%f\,%f\n' , [2 inf]);
```

```
fclose(tmpFid);

objvalue(n) = −[tmpData(2,45)+tmpData(2,46)+
tmpData(2,48)+tmpData(2,77)
+tmpData(2,78)+tmpData(2,80)];
d =−[tmpData(2,45)+tmpData(2,46)+tmpData(2,48)+
tmpData(2,77)+tmpData(2,78)+tmpData(2,80)];

fprintf('value of d : \%d\n',d);
datad1 = zeros(100,2);
tt = array2table(datad1,'VariableNames',
{'Freq_Ghz_','dB_S_1_1___'});
writetable(tt,'test.csv');

end
end
```

The selection operator code is as follows:

```
\% Input variable: pop binary population,
fitvalue: fitness value
\% Output variable: binary population
after newpop selection
function [newpop] = selection (pop,fitvalue)
\% Construct the wheel
[px,py] = size(pop);
totalfit = sum(fitvalue);
p_fitvalue = fitvalue/totalfit;
p_fitvalue = cumsum(p_fitvalue);
ms = sort(rand(px,1)); \% Generate a vector
of random numbers arranged from small to large
fitin = 1;
newin = 1;
while newin <= px
\%
newin
fitin
ms(newin)
p_fitvalue(fitin)

if(ms(newin)) < p_fitvalue(fitin)
newpop(newin,:) = pop(fitin,:);
newin = newin + 1;
else
fitin = fitin + 1;
```

```
end

end
end
```

The crossover operator code is as follows:

```
\% Crossover Operator Change
\% Input variables: pop: parent population
number of the binary; PC: probability of crossover
\% Output variable: newpop:
Population number after crossover
function [newpop] = crossover(pop,pc)
[px,py] = size(pop) ;
newpop = ones(size(pop));
for i = 1:2:px−1
if(rand < pc)
cpoint = round(rand∗py);
newpop(i,:) = [pop(i,1:cpoint) ,
pop(i+1,cpoint+1:py)];
newpop(i+1,:) = [pop(i+1,1:cpoint) ,
pop(i,cpoint+1:py)];
else
newpop(i,:) = pop(i,:);
newpop(i+1,:) = pop(i+1 , :);
end
end

end
```

The mutation operator code is as follows:

```
\% mutation operator
\% Input variables: pop: binary population,
pm: probability of mutation
\% Output variable: newpop: population after mutation
function [newpop] = mutation(pop,pm)
[px,py] = size(pop);
newpop = pop;
for i = 1:px
if(rand<pm)
mpoint = round(rand ∗ py);
if mpoint <= 0
mpoint =1;
end
newpop(i,:) = pop(i,:);
```

```
if newpop(i,mpoint) == 0
newpop(i,mpoint) = 1;
else
newpop(i,mpoint) = 0;
end
else
end
end
end
```

The code for finding the optimal solution is as follows:

```
\% Find the optimal fitness function
\% Input variables: pop: population,
fitvalue: population fitness
\% Output variable: bestindividual:
best individual, bestfit: best fit value
function [bestindividual,bestfit] = best(pop,fitvalue)
[px,py] = size(pop);
bestindividual = pop(1,:);
bestfit = fitvalue(1);
for i = 2:px
if fitvalue(i) > bestfit
bestindividual = pop(i,:);
bestfit = fitvalue(i);
end
end
end
```

The optimal output result is shown in Fig. 8.12, which is a binary two-dimensional array that reflects the structure of the antenna radiation patch. The antenna structure corresponding to the optimal array configuration output is shown in Fig. 8.13. It is a three-dimensional representation of the antenna, where we can see that the final antenna consists of a rectangular slot connected to a "T"-shaped slot. The simulation results of S-parameters before and after optimization are shown in Fig. 8.14 and Fig. 8.15, respectively. The graphs depict two frequency bands for the antenna, operating at 2.4 and 4.9 GHz, which fully meet our design requirements. The bandwidths for the two frequency bands are 0.05 and 0.4 GHz, respectively. The S_{11} values at both frequency points are below -10 dB, reaching -16.5 and -18 dB, respectively.

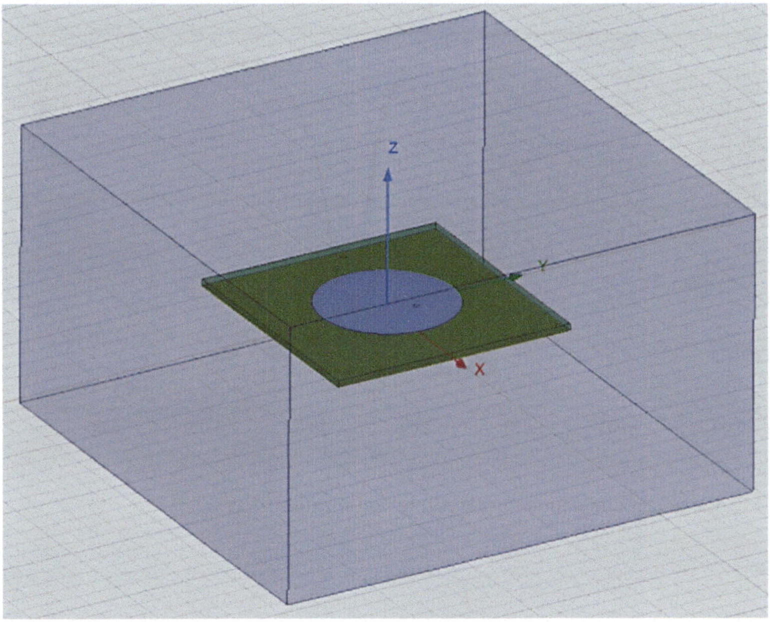

Fig. 8.12 Output optimal results

Fig. 8.13 3D diagram of optimized PIFA

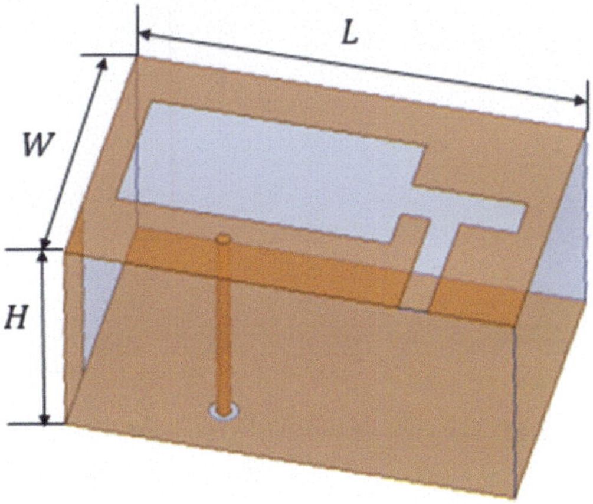

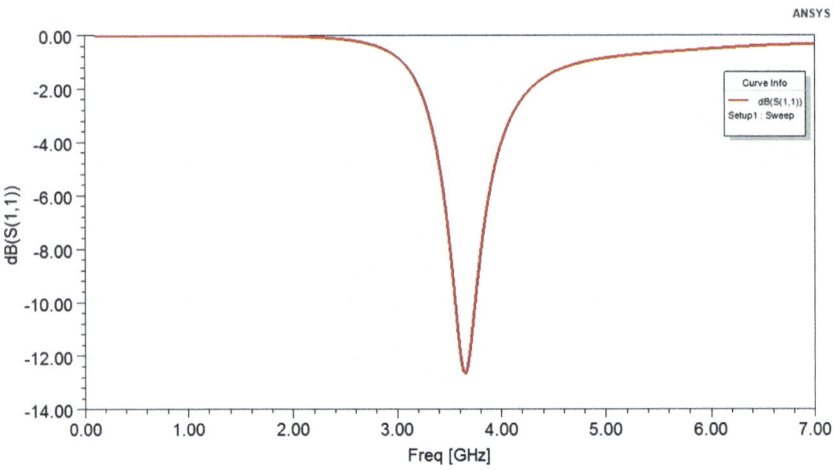

Fig. 8.14 S-parameter simulation results before optimization

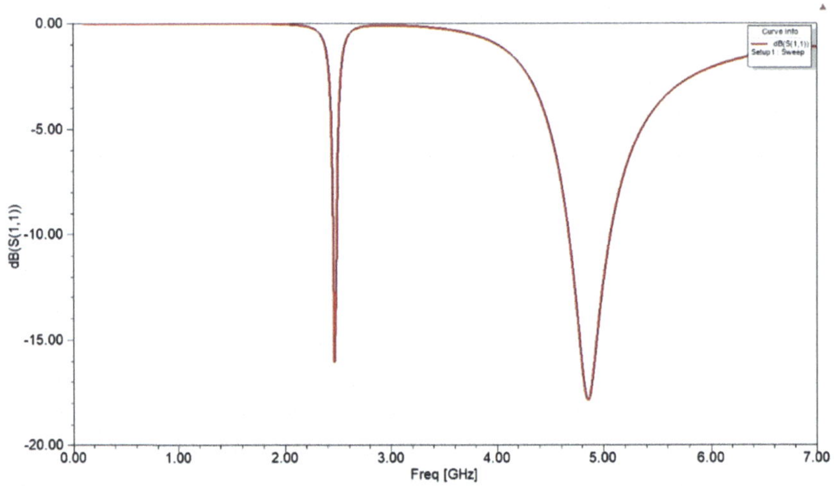

Fig. 8.15 S-parameter simulation results after optimization

Problems

P8.1 Advanced Data Processing in ADS
Practice using Datalink to import waveform files in the *.mat* format for RF simulation. Datalink is an advanced extension feature introduced in ADS 2019 and later versions. With Datalink, formulas can be used in the ADS data display window, and data can be processed using Python scripts. The .mat file is a data storage format file used by MathWorks' MATLAB software. By using the ADS Datalink feature, waveform

files stored in the *.mat* format can be converted to the ADS mdf format and used in the RF environment of ADS. The conversion function requires ADS version 2019 or later and the installation of the Datalink software module.

P8.2 Rapid Creation of Antenna Arrays in HFSS
Practice using MATLAB library functions to generate a vbs file and quickly create a 10×10 microstrip patch antenna array model in HFSS.

References

1. Kolberg D, Knobloch J, Zühlke D (2017) Towards a lean automation interface for workstations. Int J Prod Res 55(10):2845–2856
2. Ahn KU, Kim YJ, Park CS, Kim I, Lee K (2014) BIM interface for full versus semi-automated building energy simulation. Energy Build 68:671–678
3. Kim K, Kim Y, Shin Y, Choi K (1996) An integrated hardware-software cosimulation environment with automated interface generation. In: Proceedings seventh IEEE international workshop on rapid system prototyping. Shortening the path from specification to prototype, pp 66–71
4. Nembrini J, Samberger S, Labelle G (2014) Parametric scripting for early design performance simulation. Energy Build 68:786–798
5. Flühr H, Gangl S, Gruber M (2008) EDA toolbox for VHF data link system simulation. Aerosp Sci Technol 12(8):599–607
6. Deng H, Varanasi M, Swigger K, Garcia O, Ogan R, Kougianos E (2006) Design of sensor-embedded radio frequency identification (SE-RFID) systems. In: International conference on mechatronics and automation, pp 792–796
7. Guerrieri SD, Ramella C, Bonani F, Ghione G (2019) Efficient sensitivity and variability analysis of nonlinear microwave stages through concurrent TCAD and EM modeling. IEEE J Multiscale Multiphysics Comput Tech 4:356–363
8. You KY, Sim MS, Hamid F, Lee KY, Cheng EM, Esa F (2022) Simple MATLAB-based EDA for 5G phased-array antenna design. In: IEEE international RF and microwave conference (RFM), pp 1–4
9. Kashif R (2015) Lin F (2015) Signal integrity problems in electronic designing. In: Asia-Pacific microwave conference (APMC), vol 2, pp 1–3
10. Chen D, Hsiao J, Ling Y, Chen CT, Cai XK (2017) Unified and efficient power integrity impedance target flow for flexible PCB design. In: 2017 12th international microsystems, packaging, assembly and circuits technology conference (IMPACT), pp 292–294
11. Sood A, et al (1998) Modeling and mixed signal simulation of embedded passive components in high performance packages. In: Proceedings: 1998 international conference on multichip modules and high density packaging (Cat. No. 98EX154). IEEE, pp 1–4
12. Dan C, Kejun L, Longda Y, Xiaofei M, Fangfang Z (2015) Behavior model analysis of transceiver characteristic based on ADS automatic control. In: 2015 IEEE international conference on signal processing, communications and computing (ICSPCC). Ningbo, China, pp 1–4. https://doi.org/10.1109/ICSPCC.2015.7338898
13. Caine SH, Gordon EK (1975) PDL: a tool for software design. In: Proceedings of the 19–22 May 1975, national computer conference and exposition
14. Adrion WR, Branstad MA, Cherniavsky JC (1982) Validation, verification, and testing of computer software. ACM Comput Surv (CSUR) 14(2):159–192
15. Trinh J (2015) ADS och Matlab för optimering av pre-distortion av effektförstärkare

16. Otto S, et al (2009) Array and multiport antenna farfield simulation using EMPIRE, MATLAB and ADS. In: 2009 European microwave conference (EuMC). IEEE
17. Harding C, Griffiths A, Yu H (2007) An interface between MATLAB and OPNET to allow simulation of WNCS with MANETs. In: 2007 IEEE international conference on networking, sensing and control. IEEE
18. Kingsley-Hughes A, Kingsley-Hughes K, Read D (2004) VBScript programmer's reference. Wiley
19. Lomax P (1997) Learning VBScript. O'Reilly Media, Inc
20. Mueller JP (2010) Professional IronPython. Wiley
21. Foord M, Muirhead C (2009) IronPython in action. Manning Publications Co
22. Harris A (2009) Introduction to IronPython. Pro IronPython, pp 1–14
23. Harris A (2009) Pro IronPython. Apress
24. Tian W et al (2018) Application of genetic algorithm in M×N reconfigurable antenna array based on RF MEMS switches. Mod Phys Lett B 32(30):1850365
25. Hajri JBR, et al (2019) Fast and Automatic RF design based on MATLAB-HFSS control applied on magnetic absorber with metasurface. In: 2019 Photonics & Electromagnetics Research Symposium-Fall (PIERS-Fall). IEEE
26. Tian W et al (2018) Application of genetic algorithm in M×N reconfigurable antenna array based on RF MEMS switches. Mod Phys Lett B 32(30):1850365
27. Myers JC et al (2015) A multilayered metamaterial-inspired miniaturized dynamically tunable antenna. IEEE Trans Antennas Propag 63(4):1546–1553
28. Ghartemani MK, Khajehoddin SA, Jain PK, Bakhshai A (2011) Problems of startup and phase jumps in PLL systems. IEEE Trans Power Electron 27(4):1830–1838
29. Rathod VT (2019) A review of electric impedance matching techniques for piezoelectric sensors, actuators and transducers. Electronics 8(2):169
30. Ibnyaich S, Wakrim L, Hassani MMR (2021) Nonuniform semi-patches for designing an ultra wideband PIFA antenna by using genetic algorithm optimization. Wireless Pers Commun 117:957–969
31. Wakrim L, Yassini AE, Khabba A, Ibnyaich S, Hassani MMR (2021) Novel design of a triple band PIFA antenna by using a binary genetic algorithm. J Comput Electron 20:1373–1386
32. Pinho P, Pereira JR (2001) Design of a PIFA antenna using FDTD and genetic algorithms. In: IEEE antennas and propagation society international symposium, 2001 Digest: held in conjunction with: USNC/URSI National Radio Science Meeting (Cat. No. 01CH37229), vol 4, pp 700–703

Solutions

Problems of Chapter 1

P1.1

1. ADS start

   - Start ADS to enter the following interface.
   - Click "File->New->Workspace" to set the Workspace name (LNA_example) and storage path.
   - Click "Next."
   - In the New Workspace Wizard click Add User FavoriteLibrary/PDK and select the zipped component library in your ADS install directory ($HPEESOF\_DIR/oalibs/componentLib/$), select RF_Transistors_vendor_kit & S_Parameter_vendor_kit inyour install directory.
   - Select the directory where you want to unzip the library.
   - After the file is unzipped, enter the Design Kit name and click OK.
   - View the Vendor Libraries in the Component Library browser, and then hook the PDK name. You will see the window as shown below:

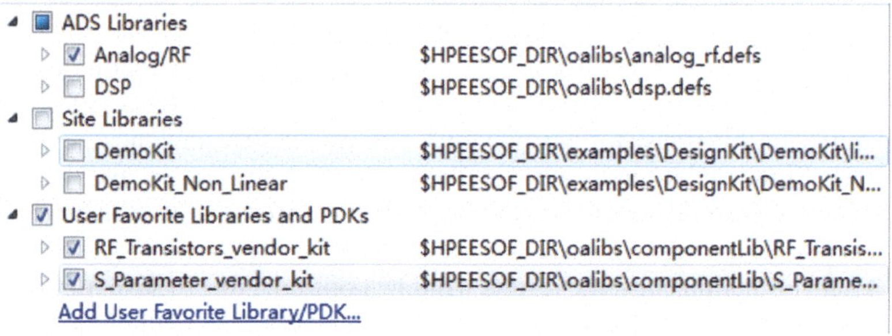

Fig. P1.1 Accessing vendor libraries and PDKs in the ADS component library browser

   - Click Finish to complete the workspace creation.

M. S. Tong and X. Y. Li, *Co-simulations of Microwave Circuits and High-Frequency Electromagnetic Fields*, https://doi.org/10.1007/978-981-99-8307-0

2. Transistor DC operating point scan

- In workspace, select File-New ... enter the following dialog box and name the new design as BJT_Curve. In "Schematic Design Templates," Select BJT_curve_tracer.
- In the new design, there will be components and icons pre-set by the system.

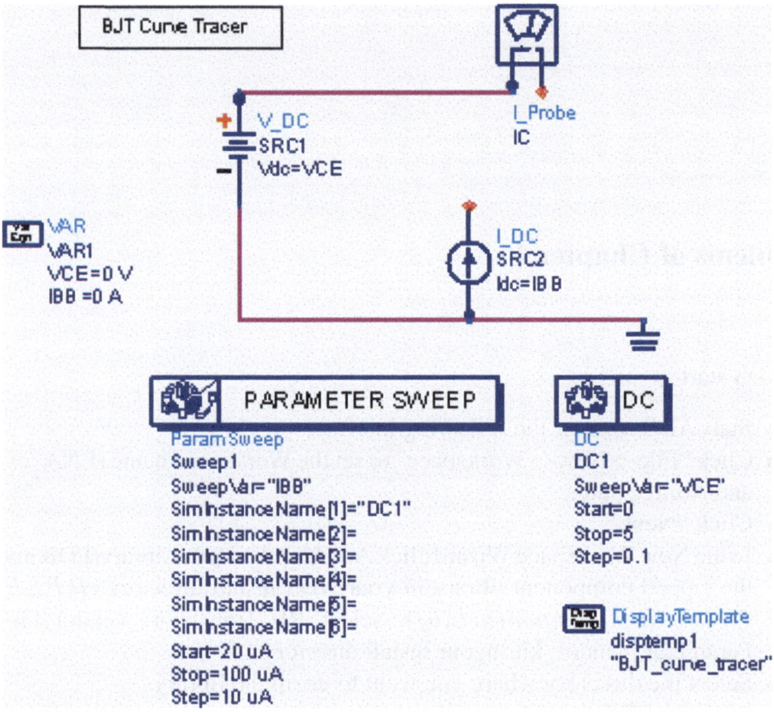

Fig. P1.2 Creating a BJT curve tracer design

- Click Component Library, open the library, search "41511."
- Choose the model "AT41511_19950125," switch to the schematic window, place the transistor, press Esc key to terminate the current operation.
- Access transistor, pay attention to confirm the line completely connected. The emitter of this transistor has two pins, you can choose any one.
- Press Simulate button to start simulation, a window pops up, and the window will display simulation or optimization process information. If an error occurs, it will give an error message.
- After simulation ends, the results window pops up. Note to save for the appropriate name. The "Marker" in the figure can be dragged with the mouse. As a result of the ADS design templates, the data here have been pre-set. In general, the data display needs a person to set it.
- In fact, the default scan parameters in the template and the requirement are usually not consistent, so you need to modify the schematic on the

schematic window. The method is relatively simple. Parameter sweep controller is very important, which will be frequently used in many cases.

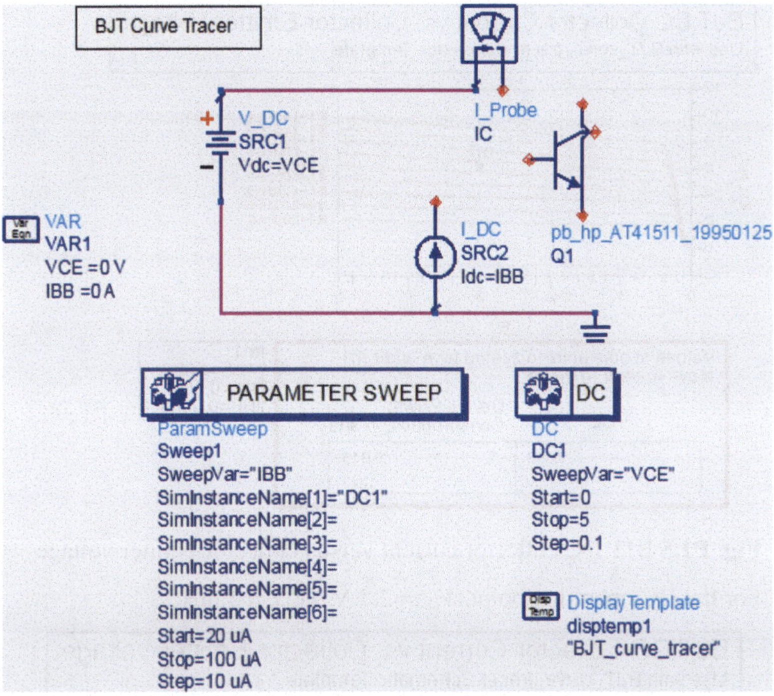

Fig. P1.3 Transistor DC operating point scan procedure using model AT41511

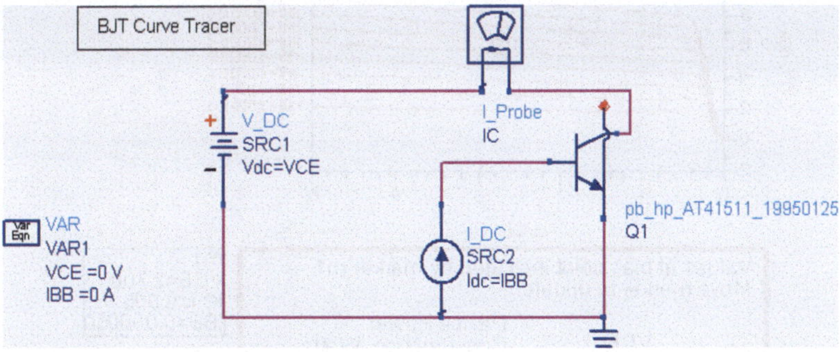

Fig. P1.4 Ensuring proper transistor connections for DC operating point scan

- To modify general parameters, one can either double-click on the target elements (components, controllers, etc.) or activate the parameter display in the design window. Advanced settings can only be modified by double-clicking on them.
- In this scenario, you can modify the scanning parameters and the simulation settings to select the optimal direct current (DC) operating point on the result

curve, enabling you to determine the corresponding DC bias voltage or current value accurately.

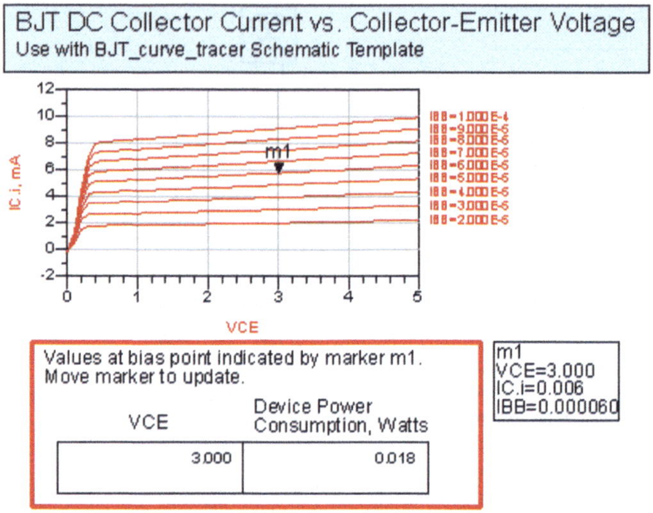

Fig. P1.5 BJT DC collector current versus collector-emitter voltage

- For the DC operating point, $V_{ce} = 2.7$ V, $IC.i = 5$ mA.

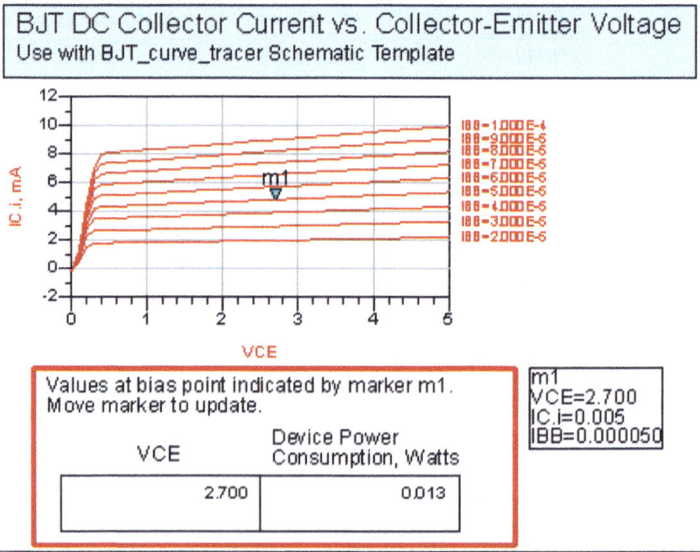

Fig. P1.6 BJT DC operating point analysis

3. S-parameter simulation

- Select File -> New -> Schematic enter the following dialog box, name cell asSP_of_spmod. Schematic Design Templates choose: S_Params. New schematic window is as follows:

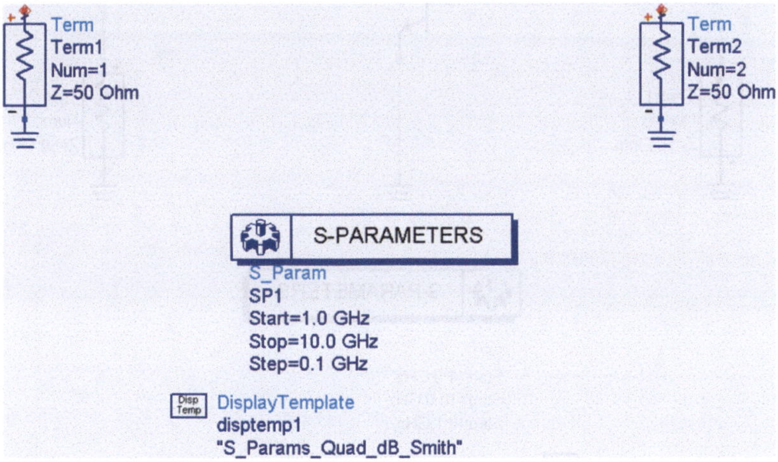

Fig. P1.7 Setting up S-parameter simulation

- Add the SP model (sp_hp_AT-41511_2_19950125) for transistors and connect the circuit. The term is commonly used in simulations to represent the characteristic impedance of the port.
- Since SP model itself corresponds to a certain operating point, S-parameter scan doesn't need DC bias.
- Observed parameters of the transistor sp model show that in this case, the working frequency range is 0.1 – 5.1 GHz.
- When simulating beyond this range, although the software can extrapolate the characteristics of the circuit using interpolation methods, the data is typically not reliable because the model becomes invalid.
- Modify parameters.
- Click the button to simulate, and the data output window pops up. Data output window as shown in the figure, which shows the output in different forms of S-parameters. Figure shows the input impedance matching of the transistor is not good.
- The tool bar used to control the display format is at the left side of the window.
- Click the appropriate button, you can output simulation results in accordance with the corresponding data format.
- Equation is used to customize the corresponding expressions, which can output simulation results after operations.
- Activate the graphical display. The parameters are listed in the left list, select the required parameters, such as: S (1,1). Add it to the right display list.

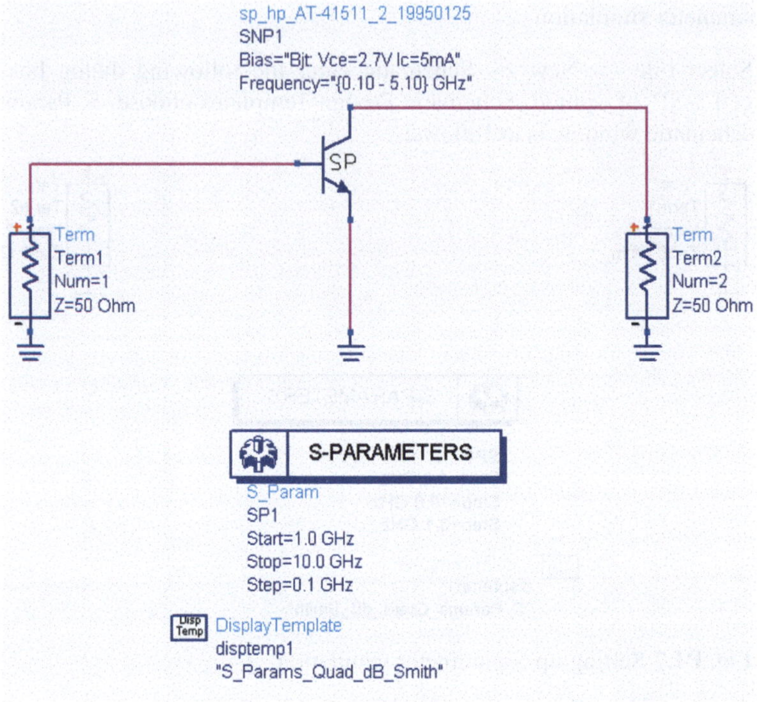

Fig. P1.8 Adding SP model and terminals for transistor S-Parameter simulation

- Then you will see the format of the data. For the S(1,1), select dB. S(1,1) is displayed as shown below:
- Activate the display mode of list of numbers. Add the parameters to the right of the display list. For the S (1,1) the default display is amplitude/angle format. Click $TraceOptions$, choose tab $TraceExpression$, the output format of the results can be edited, e.g., the bottom right is set to dB (S(1,1)), pay attention to the case of the letters.
- Select the list, the toolbar is activated, you can see all the data by turn page.
- Adding noise simulation: Back to Design window, double-click $S-PARAMETERS$. In the Noise tab, select $Calculatenoise$ and simulate.
- Select the appropriate format to display port 2 noise figure nf(2). nf(2) is noise figure (unit:dB) when port 2 is the output port.

4. S-parameter (SP) model design-impedance matching

- In many cases, before the simulation of the package model, through a pre-simulation of the sp model, the circuit approximate specifications can be obtained.
- Sp model design is often used as the primary stage of circuit design.

- 2 GHz input and output matching in SP model design.
- Create a new schematic window, named Spmod_LNA.
- Select S-parameter simulation controller in the left toolbar.

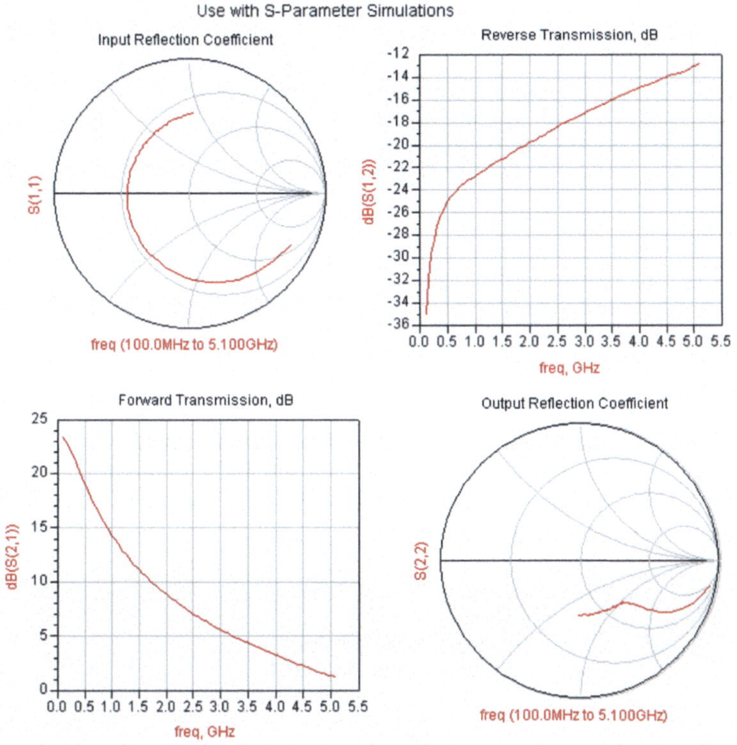

Fig. P1.9 S-Parameter simulation results

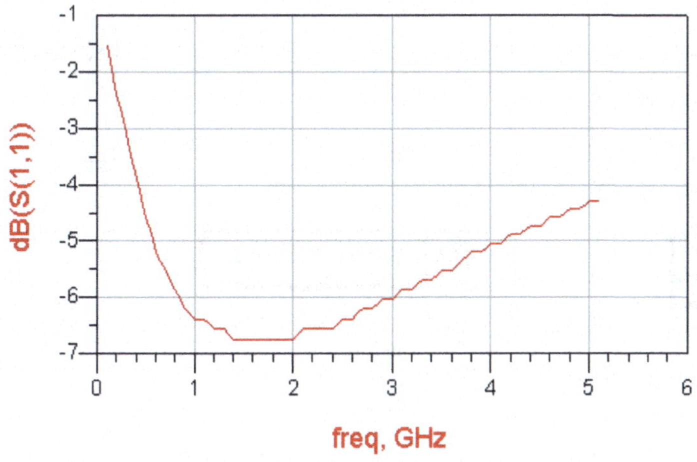

Fig. P1.10 Reflection coefficient frequency response graph

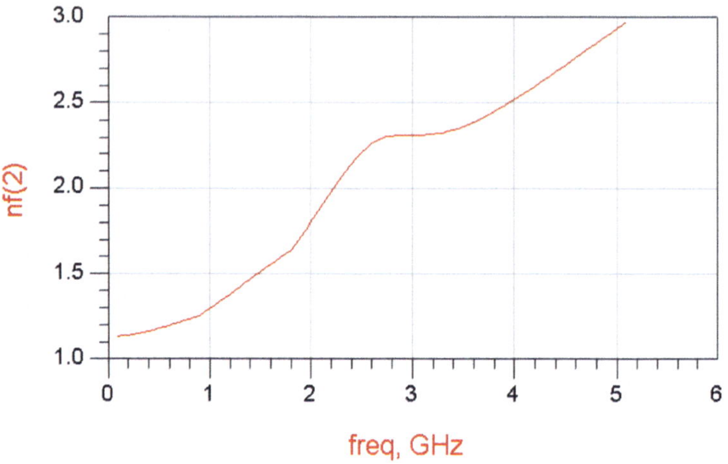

Fig. P1.11 Noise figure simulation results

- Options->Preferences->Units/Scale->Length: mil->mm.
- Select transistors *sp_hp_AT* − 41511_2_19950125 in the library, and place them in the schematic window. Modify the settings of the S-parameter simulation icon to the required value.

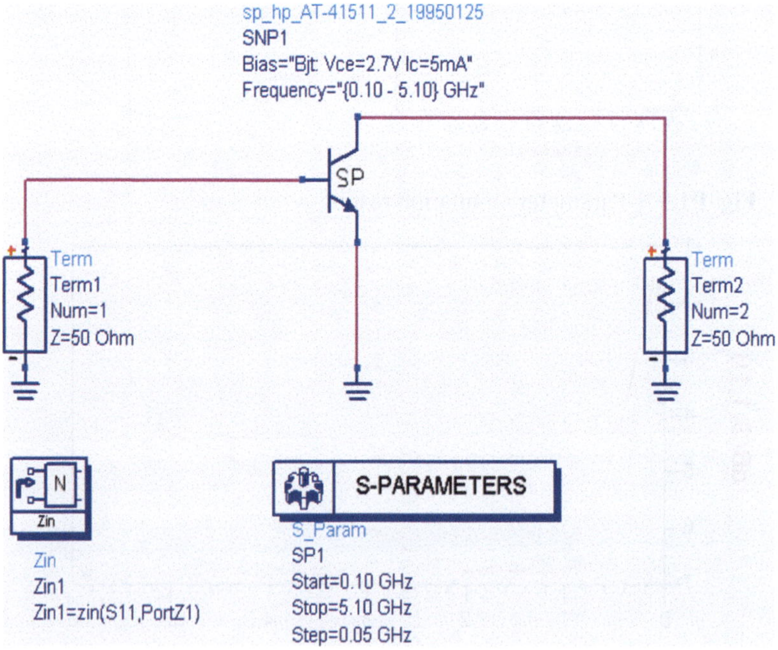

Fig. P1.12 Transistor S-Parameter test setup schematic

- Observation and simulation of the input impedance data in the output window.
- From the list available, we can see input impedance of 2 GHz point: 20.083/ 19.829.
- Translated into real/imaginary form: 18.89 + j * 6.81.

freq	Zin1
1.600GHz	18.498 / 1.167
1.650GHz	18.555 / 4.075
1.700GHz	18.674 / 6.954
1.750GHz	18.814 / 9.225
1.800GHz	18.994 / 11.455
1.850GHz	19.211 / 13.637
1.900GHz	19.465 / 15.765
1.950GHz	19.757 / 17.830
2.000GHz	20.083 / 19.829
2.050GHz	20.140 / 21.608
2.100GHz	20.224 / 23.383
2.150GHz	20.644 / 25.240
2.200GHz	21.095 / 27.017
2.250GHz	21.454 / 28.296
2.300GHz	21.830 / 29.528
2.350GHz	22.356 / 31.101
2.400GHz	22.911 / 32.593
2.450GHz	23.173 / 34.102
2.500GHz	23.461 / 35.583
2.550GHz	23.933 / 36.575
2.600GHz	24.419 / 37.523
2.650GHz	24.768 / 38.895
2.700GHz	25.140 / 40.234
2.750GHz	25.676 / 41.065
2.800GHz	26.227 / 41.856
2.850GHz	26.657 / 43.085
2.900GHz	27.111 / 44.279
2.950GHz	27.709 / 44.960
3.000GHz	28.320 / 45.606
3.050GHz	28.830 / 46.696
3.100GHz	29.362 / 47.751

Fig. P1.13 Simulation data of input impedance at 2 GHz frequency point

- Matching using microstrip line. The basic parameter setting of microstrip lines. $TLines - Microstrip$ toolbars are used to set up the microstrip line parameters. Click $HSUB$ to place microstrip substrate in the schematic. The meaning of the parameters: H: substrate thickness, Er: relative permittivity of substrate, Mur: permeability, Cond: metal conductivity, Hu: package height, T: thickness of metal layer, TanD: loss tangent, and Rough: surface roughness.
- In MSub, modify the parameters to the necessary values, as shown in the right figure.

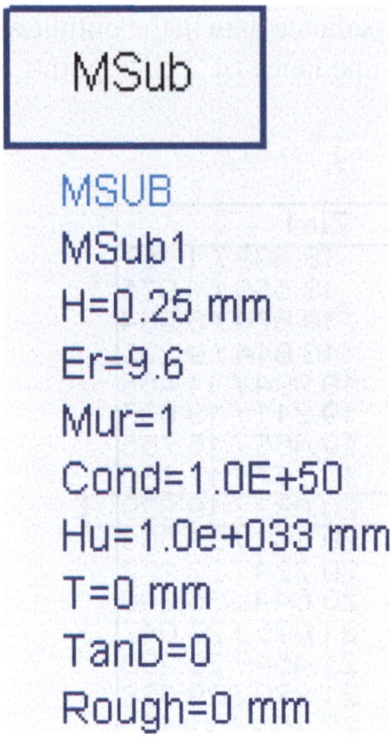

Fig. P1.14 Microstrip line parameters and substrate settings

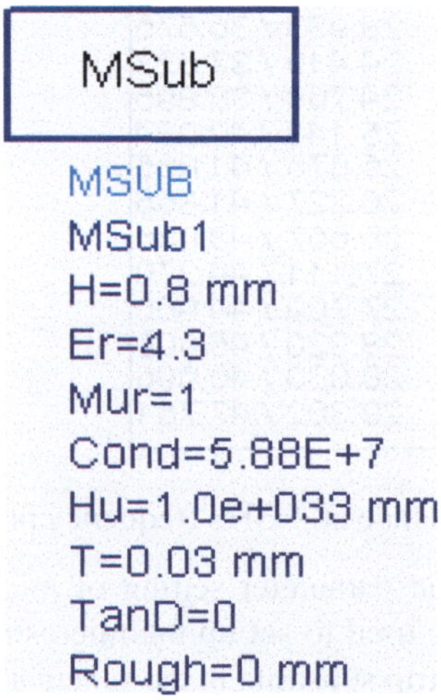

Fig. P1.15 Adjusting MSub parameters for circuit design

- Choose $PassiveCircuitDG - MicrostripCircuits$ toolbar.
- Use single branch line to match.
- Use synthesize tool in ADS to get the matching network.
- Double-click to edit parameter.
- The frequency is set to 2 GHz. Zin is set to the target value of 50 Ω. Zload is set as the simulated results of the input impedance of the transistor above.

- In the top row of the schematic window, choose *DesignGuide -> Passive Circuit -> PassiveCircuitControlWindow*, then pop-up window.
- Choose *Design*, after synthesis, we can get suitable matching network.
- Matching network is generated, click into the sub-circuits of matching network, as shown in the figure.
- T-shaped connector is introduced to take into account the resistance mutations. In practice, the T-shaped connector does not represent any actual length, but discontinuities of the circuit. About the specific meaning, see help documentation.

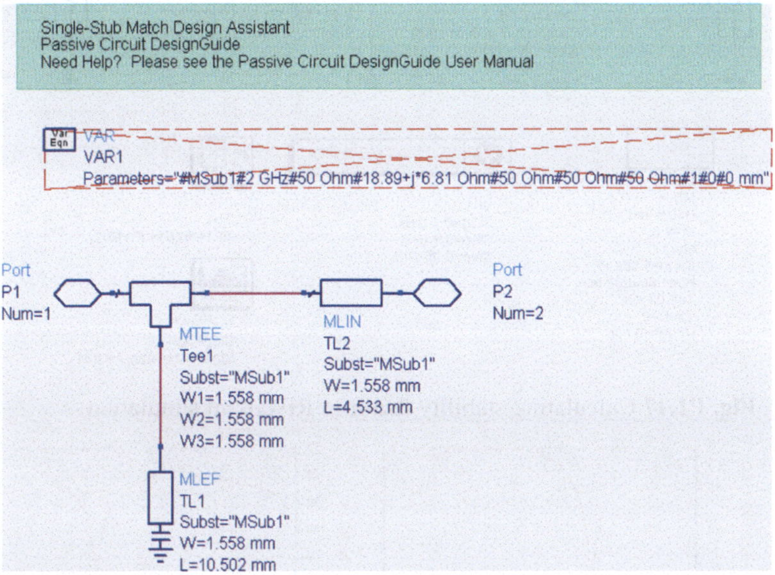

Fig. P1.16 Integration of T-Shaped connectors in RF matching network schematic

- Access matching network. Insert *Simulation – S_Param*, and use it to calculate stability factor.
- Simulated results (S-parameters).
- Simulated results Input impedance, Stability factor, Noise figure.
- From the simulation results above we can see, basically the circuit has reached a relatively good performance (2 GHz), such as good input match, high gain, and good stability factor and noise figure.
- On the other hand, the output matching is not good, the gain of the circuit may be further improved.

- It should be noted:
  In fact, the design of the output match can be the same as the one of input match. The software can synthesize it after calculating the output impedance. The following design approach is not the best way, only to introduce the use of optimization tools.

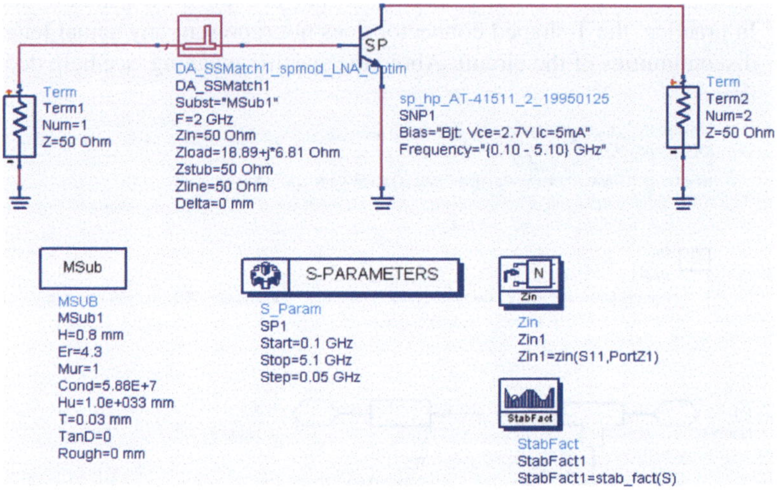

Fig. P1.17 Calculating stability factor in RF circuit simulation

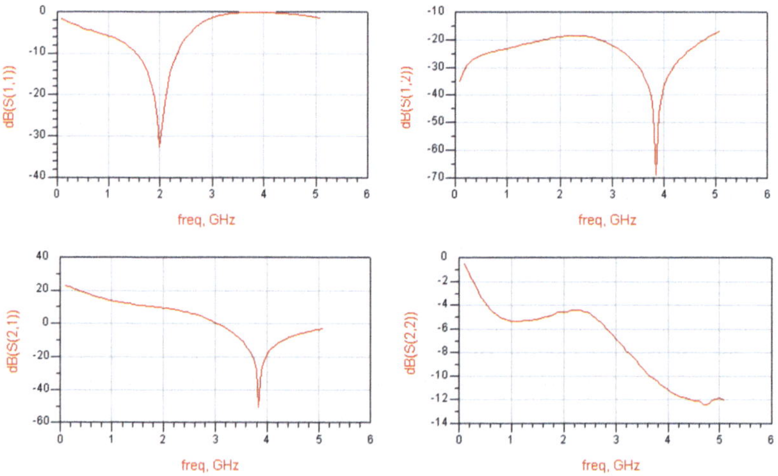

Fig. P1.18 S-Parameter frequency response graphs

- We also use a single branch line to match the output structure. Select *TLines – Microsrrip* and click microstrip tool and T-shaped connector tool. Then, connect the circuit as illustrated in the figure, ensuring components may be rotated as needed for correct orientation. The direction of components can be rotated.
- It needs to adjust the parameters of the microstrip and joint.
  From the above, we can see the width of the input matching network is 1.558 mm. (Of course, it is impossible to achieve this accuracy in the actual production of the circuit.) According to the settings when synthesis, this is actually the corresponding width of 50 Ωs characteristic impedance. Therefore, in the output matching circuit, all widths are set for this.
- Optimization tool bar *Optim/Stat/Yield*. Insert optimization icon and insert optimization goal icon.

freq	Zin1	StabFact1	nf(2)
1.600 GHz	28.785 / 14.057	1.003	1.946
1.650 GHz	30.917 / 14.047	1.013	1.942
1.700 GHz	33.267 / 13.512	1.025	1.934
1.750 GHz	35.481 / 12.410	1.034	1.922
1.800 GHz	37.847 / 10.819	1.044	1.906
1.850 GHz	40.342 / 8.702	1.051	1.915
1.900 GHz	42.919 / 6.021	1.059	1.919
1.950 GHz	45.505 / 2.742	1.074	1.919
2.000 GHz	47.990 / -1.150	1.090	1.915
2.050 GHz	50.212 / -4.715	1.089	1.907
2.100 GHz	52.226 / -8.858	1.088	1.897
2.150 GHz	53.677 / -14.457	1.088	1.884
2.200 GHz	54.357 / -20.423	1.089	1.873
2.250 GHz	53.983 / -26.056	1.097	1.863
2.300 GHz	52.889 / -31.667	1.106	1.860
2.350 GHz	51.197 / -37.585	1.109	1.867
2.400 GHz	48.890 / -43.146	1.113	1.888
2.450 GHz	46.616 / -47.925	1.106	1.919
2.500 GHz	44.038 / -52.367	1.100	1.960
2.550 GHz	41.010 / -56.275	1.109	2.010
2.600 GHz	38.024 / -59.752	1.118	2.072
2.650 GHz	35.318 / -63.010	1.115	2.147
2.700 GHz	32.676 / -65.919	1.112	2.239
2.750 GHz	30.022 / -68.324	1.112	2.351
2.800 GHz	27.536 / -70.448	1.113	2.483
2.850 GHz	25.280 / -72.493	1.110	2.638
2.900 GHz	23.145 / -74.311	1.109	2.825
2.950 GHz	21.087 / -75.776	1.111	3.046
3.000 GHz	19.160 / -77.083	1.113	3.303
3.050 GHz	17.376 / -78.382	1.105	3.607
3.100 GHz	15.685 / -79.544	1.097	3.962
3.150 GHz	14.089 / -80.418	1.098	4.377
3.200 GHz	12.569 / -81.213	1.099	4.858
3.250 GHz	11.107 / -82.075	1.092	5.415
3.300 GHz	9.706 / -82.845	1.085	6.061

Fig. P1.19 RF simulation data: input impedance, stability factor, and noise figure

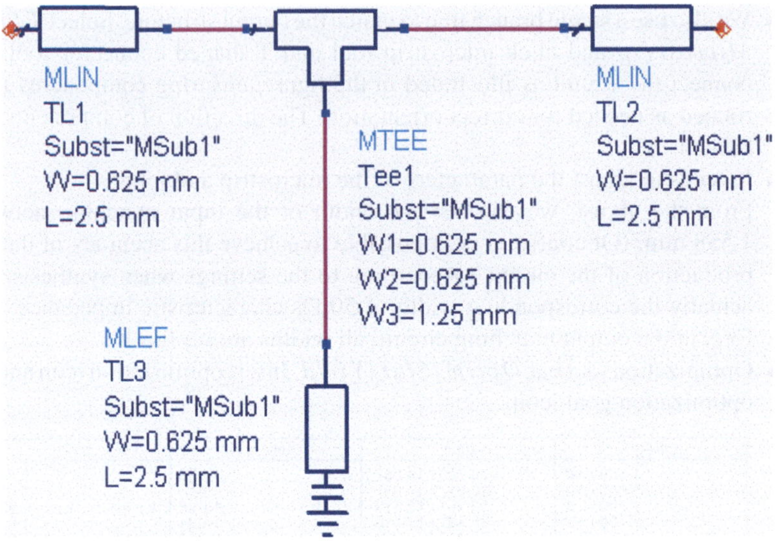

Fig. P1.20 Single-branch line and T-shaped connector in matching network

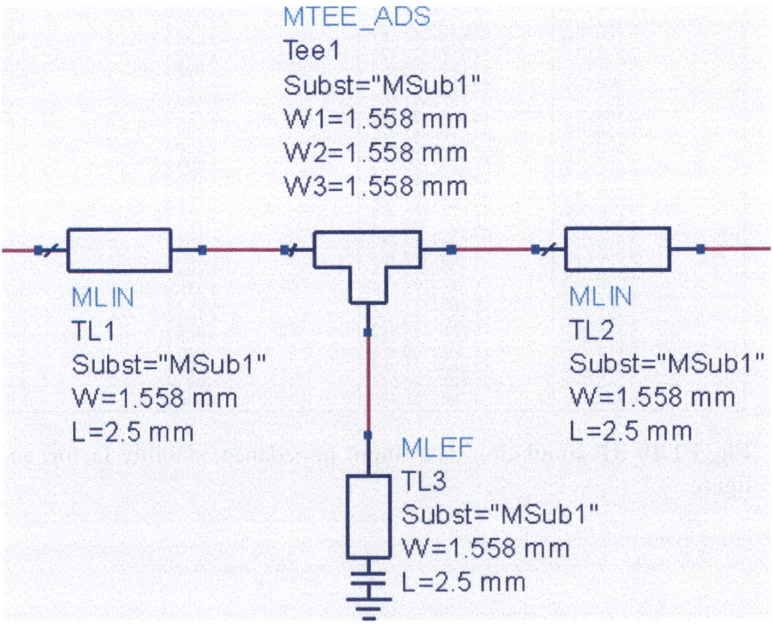

Fig. P1.21 Parameter adjustment for microstrip line

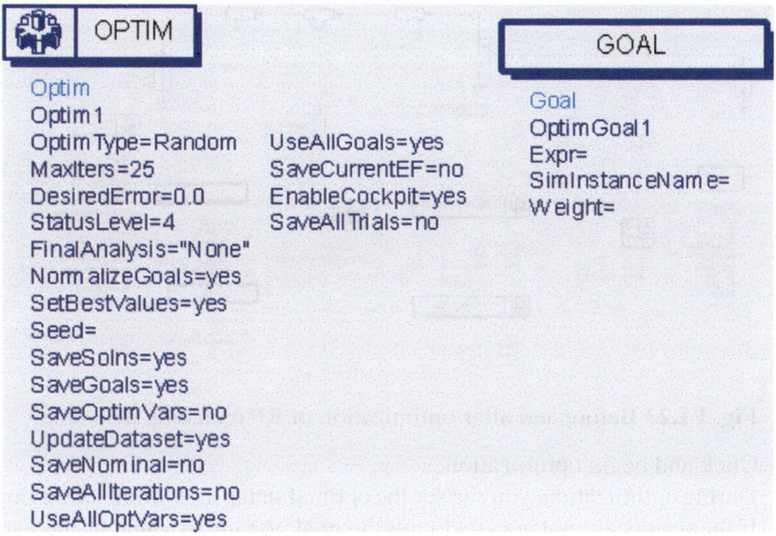

Fig. P1.22 Setting optimization goals

- According to optimization goal, add a new S-parameter controller, and set the frequency range around 2 GHz.
- Set the optimization goal. S(2,2) is small around 2 GHz. S(1,1) remain as small as possible around 2 GHz. As the current two goals are around 2 GHz, so the Siminstancename is set to "SP2."
- Active optimization of microstrip.
  Double-click the microstrip, which needs to optimize.
  Click *Tune/Opt/Stat/DOESetup....*
  Then activate the optimization, and set the optimal range (generally the smaller the better).
- Select the appropriate optimization method for optimizing.
  Methods usually include Random and Gradient.
  Random search method is usually used for large-scale gradient method for the local convergence.
- Schematic before optimization.

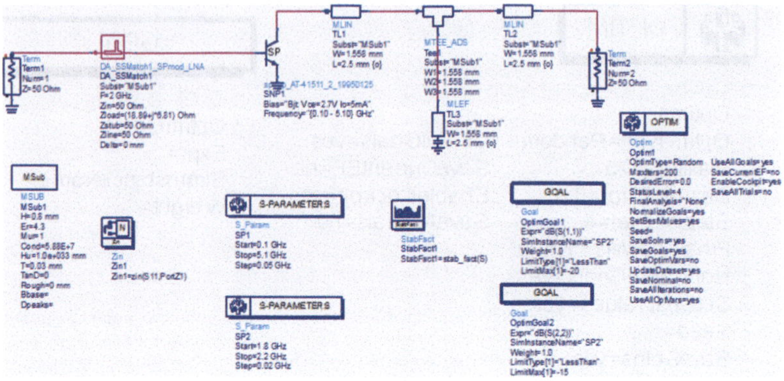

Fig. P1.23 Before and after optimization of RF matching network

- Click and begin optimization.
  During optimization, you can see the optimal status in "Optimization Cookpit." If the results are not satisfied with the goal, the initial value of the variables, optimal method, and number of iterations can be adjusted.
  After optimization, click *UpdateDesign* to update the optimized circuit parameters.
- Results show S(2,2) has been greatly improved, but at the same time S(1,1) worsened.
  Repeatedly adjust the optimization method and target weight, you can optimize the input matching network, and ultimately get a better results.

5. Achieve overall target

- Save as a new workspace LNA_example_overall_wrk.
- Adjust the input circuit's optimization method to align with that of the output circuits.

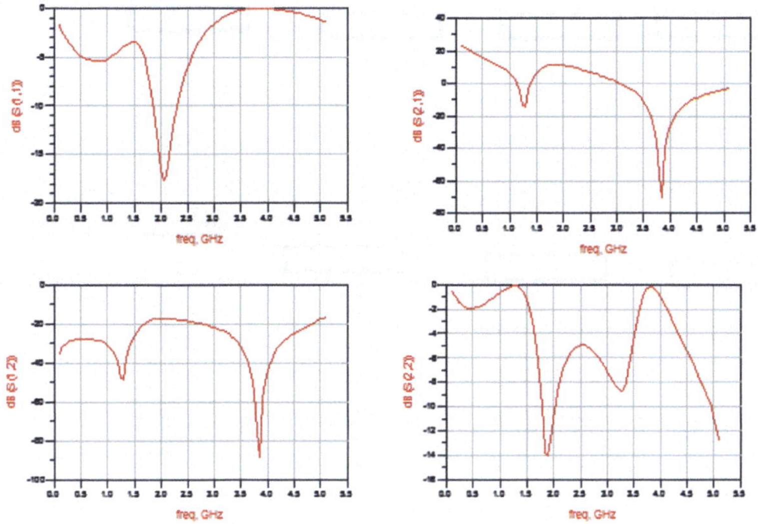

Fig. P1.24 Improved S-parameter results post RF network optimization

- Make the impedance matching in 1.8–2.2 GHz instead of only in 2 GHz.
- The line width is also changed to the variables to optimize.
- Optimization frequency changes to 1–3 GHz. The noise figure, amplifier gain, and stability factor are added to the optimization objectives, and through limited amplifier gain to meet gain flatness in band, and satisfy all requirements of the final target.
- The amplifier gain outside the working bandwidth is also added to the optimization objectives. Add the DC block capacitor in the input and output circuits.

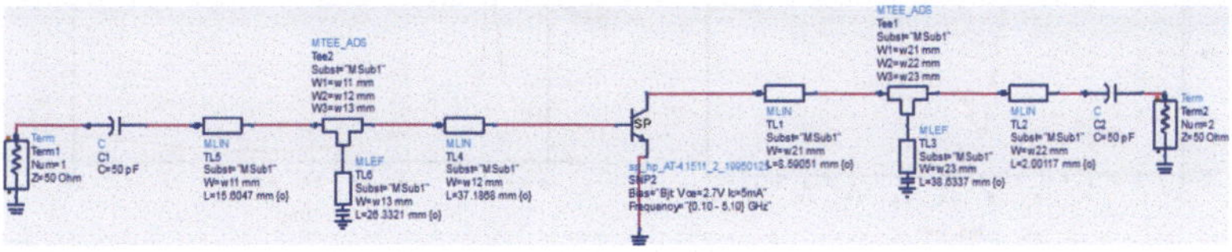

Fig. P1.25 The entire circuit schematic

- Simulated results.
- If the circuit stability factor becomes very small (less than 0.9), it is difficult to get optimization target. Or the values of S(1,1) of some of the whole frequency band are above the 0 dB, it needs to add negative feedback to improve the amplifier stability.
- Optimization on the part of the circuit goals may lead to the deterioration of some other goals, we can adjust the topology, add the match stages, and increase the number of optimization variables based on need.

Fig. P1.26 The whole goals in ADS

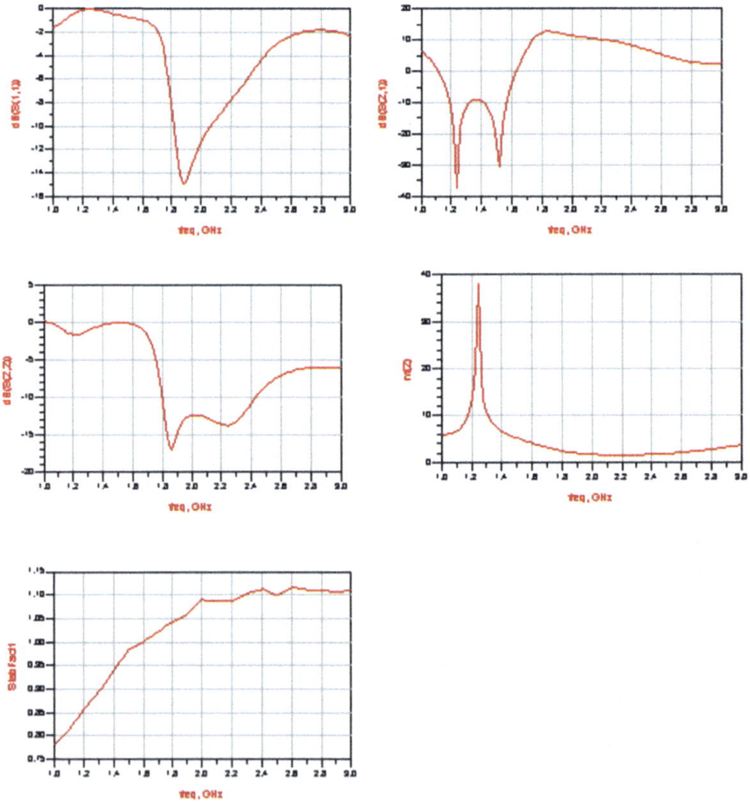

Fig. P1.27 Simulated results after an optimization process

P1.2

1. Starting Ansys HFSS.
2. Creating projects:
   On the File menu, click New. You specify the name of the project when you save it using the File>Save or File>Save As. Open a previously saved project using the File>Open command.

- Insert an HFSS design into a project. On the Project menu, click Insert HFSS Design.
- Selecting the solution type:
  On the HFSS menu, click Solution Type. Select Driven Modal in the solution types.
- Setting the model's units of measurement:
  To set the model's units of measurement:
  1. On the 3D Modeler menu, click Units. The Set Model Units dialog box appears.
  2. Select the new units for the model from the Select units pull-down list.
  3. Click OK to apply the new units to the model.
- Drawing a Model:
  To draw a WR-90 Rectangular waveguide,
  1. On the HFSS menu, click Draw. The Draw dialog box appears. Select Box. Dimensions of the box can be specified while drawing the box. At the lower end of the screen on the right is the Coordinate entry fields.
- Enter the Initial XYZ-coordinates and then enter the length in XY- and Z-directions in dX, dY, and dZ.
  For example, to draw the box with initial point to be origin and propagation along X-axis. Since it is a WR-90 rectangular waveguide the dimensions are a = 2.286 cm , b = 1.016 cm. Once you draw the box the properties window opens up, you can also specify the coordinates and size of the box here.
- Assigning materials:
  1. Right-click on the 3D Modeler Window to get the 3D Modeler menu.
  2. On the 3D Modeler menu, click Assign Material.
  3. The Select Definition window appears. By default, it lists all of the materials in Ansoft's global material library as well as the project's local material library.
  4. Select a material from the list. Select air or vacuum for the whole box as our rectangular waveguide is not filled with any dielectric.
  5. Click OK.
  6. The material you chose is assigned to the object.
- Assigning boundaries:
  Boundary conditions specify the field behavior at the edges of the problem region and object interfaces.
  1. Right-click on the 3D Modeler Window to select faces.
  2. Click on the faces to select the faces which are to be assigned to be a perfect conductor.
  3. On the HFSS menu, click Boundaries. Select Assign and choose Finite

conductivity. Assign Finite conductivity to 4 faces excluding the Port 1 and Port 2.

- Assigning excitations:
  Excitations in HFSS are used to specify the sources of electromagnetic fields and charges, currents, or voltages on objects or surfaces in the design. Assigning excitations is a two-step process.
- Assigning ports:
  1. Select the object face to which you want to assign the port.
  2. Click HFSS>Excitations>Assign>Wave Port.
  Wave port represents the surface through which a signal enters or exits the geometry. Hence two ports are required to be defined. HFSS assumes that each wave port you define is connected to a semi-infinitely long waveguide that has the same cross section and material properties as the port. HFSS generates a solution by exciting each wave port individually.
  3. The Wave Port wizard appears.
  4. Type the port's name in the Name text box or accept the default name, and then click Next.
  5. To specify more than one mode to analyze at the port, type a new value in the Number of Modes box, and then click Update. The mode spreadsheet is updated to include the total number of modes.
- Defining integration line:
  An integration line needs to be specified to define a port mode. Since we are analyzing the WR-90 waveguide for the first four modes we need to specify four integration lines.
  1. Select New Line from the mode's Integration Line list.
  2.The dialog box disappears while you draw the vector.
  3. Select the starting point of the vector by either clicking on the point on the screen or entering the coordinates in the X, Y, and Z fields at the bottom of the screen.
  4. Select the endpoint of the vector using the mouse or the keyboard. The endpoint defines the direction and length of the integration line.
  5. The Wave Port or Lumped Port dialog box reappears.
- Adaptive solution setup:
  1. On the HFSS menu, point to Analysis Setup, and then click Add Solution Setup.
  2. The Solution Setup dialog box appears. It is divided among the following tabs:
  General—Includes general solution settings;
  Advanced—Includes advanced settings for initial mesh generation and adaptive analysis;
  Ports—(if a port was defined) Includes mesh generation options for model ports;
  Defaults—Enables you to save the current settings as the defaults for future solution setups or revert the current settings to HFSS's standard settings.
  3. Click the General tab.

For Driven solution types, do the following:
a. Enter the Solution Frequency in the frequency units.
The minimum value for adaptive Mesh Frequency is two-third of the final frequency required. Although it is recommended to just adapt to the Final frequency.
Since we are analyzing the first 4 modes of the WR-90 waveguide, the cut-off frequency of the last mode is 16 GHz. Hence the final frequency is a value, which is higher than that. For example, 20 GHz.
b. Enter the Maximum Number of Passes = 10.
The Maximum Number of Passes value is the maximum number of mesh refinement cycles that you would like HFSS to perform. This value is a stopping criterion for the adaptive solution; if the maximum number of passes has been completed, the adaptive analysis stops. If the maximum number of passes has not been completed, the adaptive analysis will continue unless the convergence criteria are reached.
c. Enter the Delta s = 0.002.
The delta S is the change in the magnitude of the S-parameters between two consecutive passes.
4. Click Ok.

- Frequency Sweep setup:
  1. In the HFSS menu, select Analysis Setup and then select Add sweep.
  2. The Edit Sweep Dialog Box opens.
  3. Select Discrete and enter the start and stop frequency.
  Since we are analyzing the first 5 modes of the WR-90 waveguide, the cut-off frequency of the last mode is 16 GHz. Hence the stop frequency is a value, which is higher than that. For example, 20 GHz.
  4. Click OK.
- Running a simulation:
  To validate your model:
  1. Select HFSS menu > Validate Check.
  2. Click OK.
- To analyze:
  1. On the HFSS menu, click Analyze.
  While a simulation is running, you can monitor the solution's progress in the Progress window.
  You can also view the following solution data at any time during or after the solution:
  Convergence data—by clicking HFSS>Analysis Setup>Convergence.
  Matrices computed for the S-parameters, impedances, and propagation constants by clicking HFSS>Analysis Setup>Profile.
  Once the simulation is completed HFSS informs you in the message window.

3. Results:
HFSS > Results > Solution Data, the solution data window appears.
HFSS computes the following matrix data:

S-, Y-, and Z-Parameters.
VSWR.
Excitations: Gamma and Zo.

4. Plotting the results:
   HFSS> Results> Create Report.
   Or you can also go to the project tree and right-click on results and click create report. The Create Report window dialog box appears.
   (1) Select the report type you want to view from the pull-down list on the top of the dialog box.
   (2) Select the type of plot you want to create, from the display type pull-down list.
   (3) Click OK.
   The Traces dialog box appears.
   (4) In the Solution list, click the solution containing the data you want to plot.
   (5) In the domain list, click a domain. For modal and terminal S-parameter reports, the domain can be frequency or time. In this case we want frequency domain.
   (6) Click on Add Trace, click Done.

5. Analysis and results:
   (1) Analyze the propagation constant for the first five modes.
   (2) Determine the wavelength at different frequencies for the first five modes.
   (3) Determine the intrinsic impedance at different frequencies for the first five modes.
   (4) Analyze E and H field patterns for the first five modes.
   Plot of propagation constant versus frequency for TE10, TE20, TE01, TE11, and TM11 using HFSS.

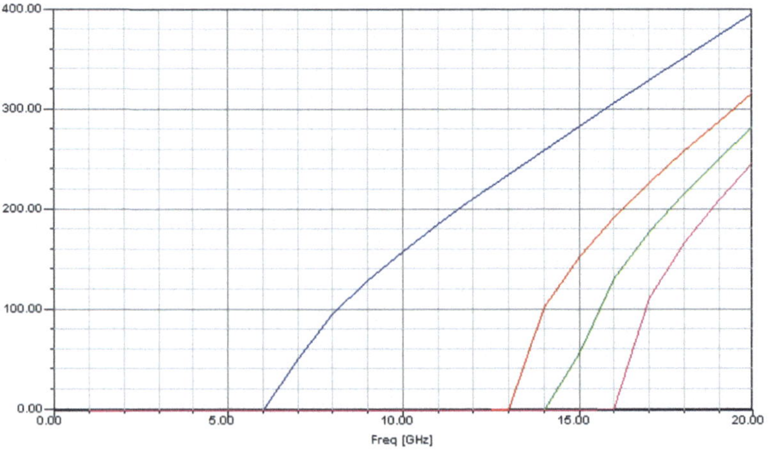

Fig. P1.28 Waveguide propagation constant analysis in HFSS

Note that a mode will only propagate if the propagation constant is real, which occurs when the operating frequency exceeds the mode's cut-off frequency. Since traveling waves are described by the function $(-j\beta z)$, the propagation constant 'B' must be real to result in a propagating wave with a phase shift described by the exponential term, which is inherently complex. As the traveling waves are functions of $\exp(-j\beta z)$, has to be real and make $\exp(-j\beta z)$ imaginary. Plot of propagation constant versus frequency for the first five modes using theoretical values similarly.

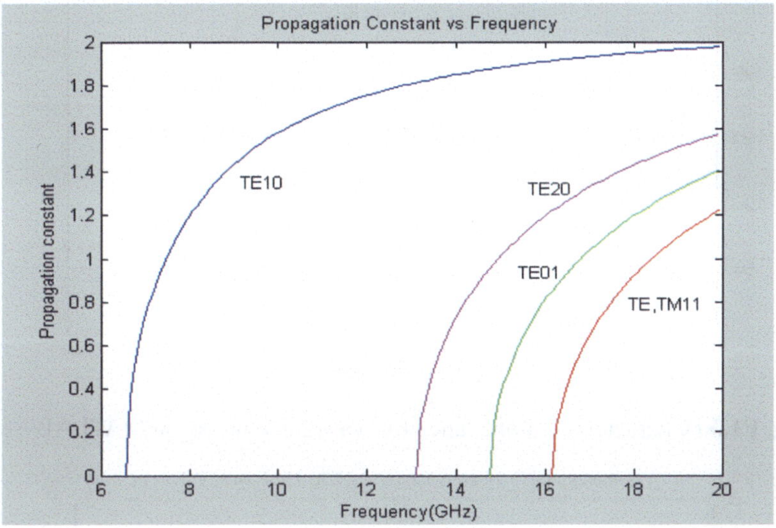

Fig. P1.29 Theoretical mode propagation characteristics in MATLAB

Similarly, you can draw plots for wavelength versus frequency and impedance versus frequency from the data given in the solution data box.

(1) In the project tree, right-click the solution setup of interest, and then click Matrix Data on the shortcut menu. The Solution Data dialog box appears. The Matrix Data tab is selected.

(2) In the Simulation pull-down list, click the solution setup and solved pass—adaptive, single frequency solution, or frequency sweep—for which you want to view matrices.

(3) Select the type of matrix you want to view: S, Y, and Z matrices or Zo (characteristic impedance).

The wavelength data is displayed when you check the gamma box. The available types depend on the solution type.

(4) Select the format—Magnitude/Phase, Real/Imaginary, dB/Phase, Magnitude, Phase, Real, Imaginary, or dB—in which to display the matrix information. The available formats depend on the matrix type being displayed.

Impedance versus frequency for the first five modes

Wavelength versus frequency for the first five modes

6. Field patterns for E and H fields:
   On the project tree click on Port Field Display>wave port 1>mode1.
   The TE10 mode will be displayed in the model.

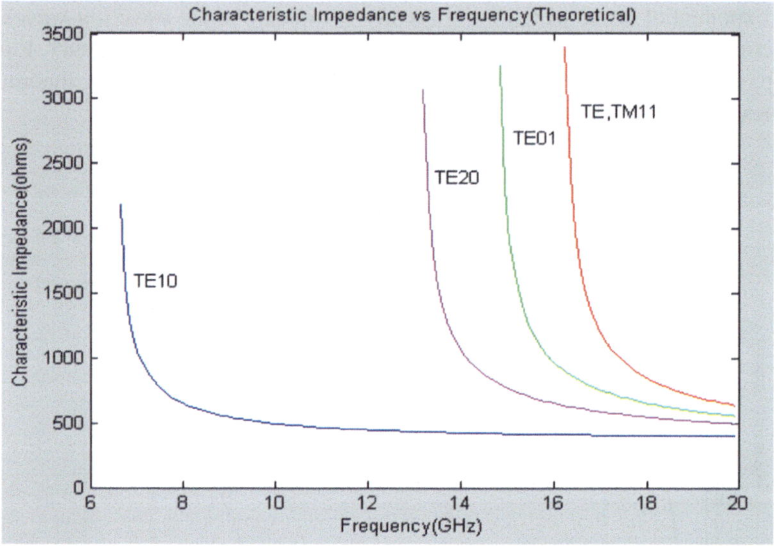

Fig. P1.30 Characteristic impedance for waveguide modes in MATLAB

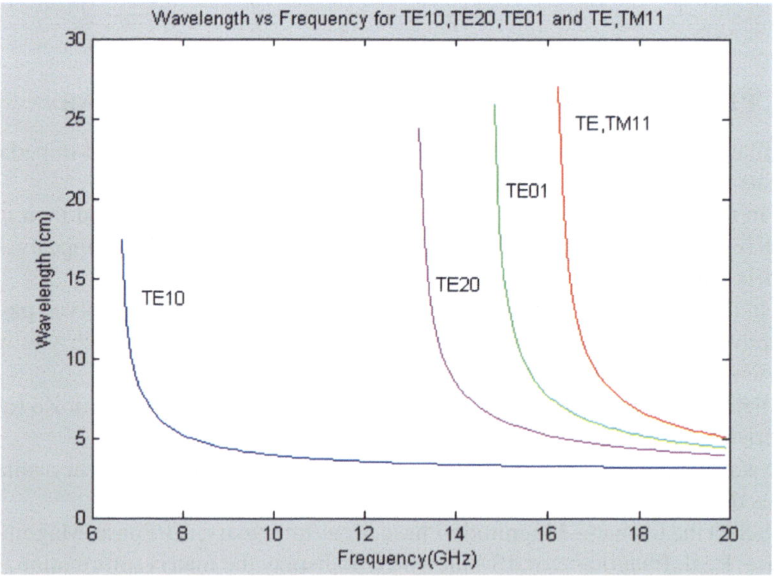

Fig. P1.31 Mode wavelength variation with frequency in MATLAB

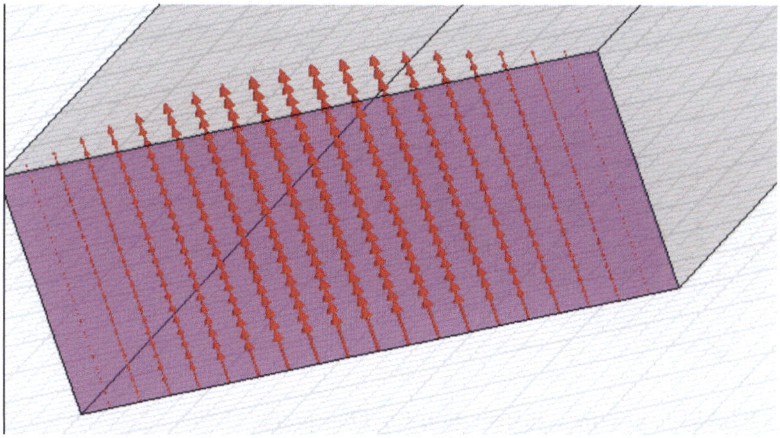

Fig. P1.32 Electric field distribution for TE_{10} mode

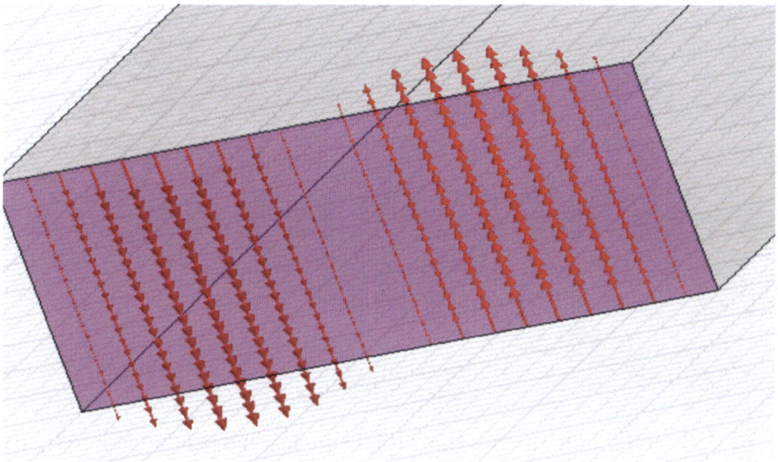

Fig. P1.33 Electric field distribution for TE_{20} mode

7. To correctly view the electric (E) and magnetic (H) field patterns in HFSS software, follow these steps:
(1) Select the surface on which you wish to visualize the field pattern.
(2) Navigate to the HFSS menu, and click on 'Fields > Plot Fields > Mag E' to view the magnitude of the electric field.
(3) For the magnetic field pattern, go to 'Fields > Plot Fields > Mag H'.
(4) When the 'Create Field Plot' dialog box appears, choose 'Mag E' under 'Quantity' and select either the specific model name under 'Volume' or opt to 'Select All Objects' if you want to see the fields for the entire model.
(5) The electric field pattern is displayed in volts per meter (V/m). The color scale illustrates the field intensity, which decreases from top to bottom.
(6) Please note, the magnetic field pattern should be indicated in ampere per meter (A/m), not V/m, as it represents the strength of the magnetic field, not the electric field.

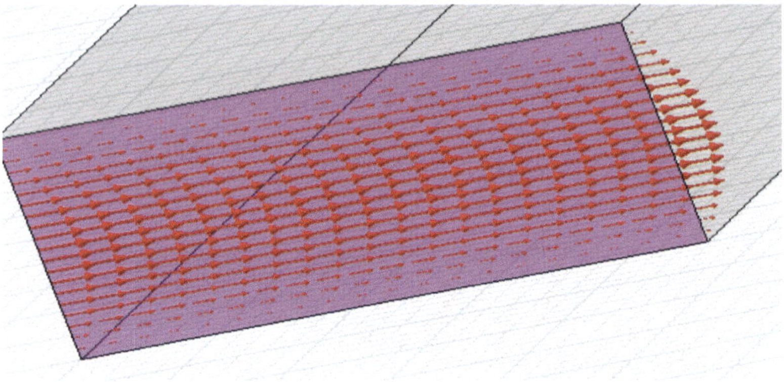

Fig. P1.34 Electric field distribution for TE_{01} mode

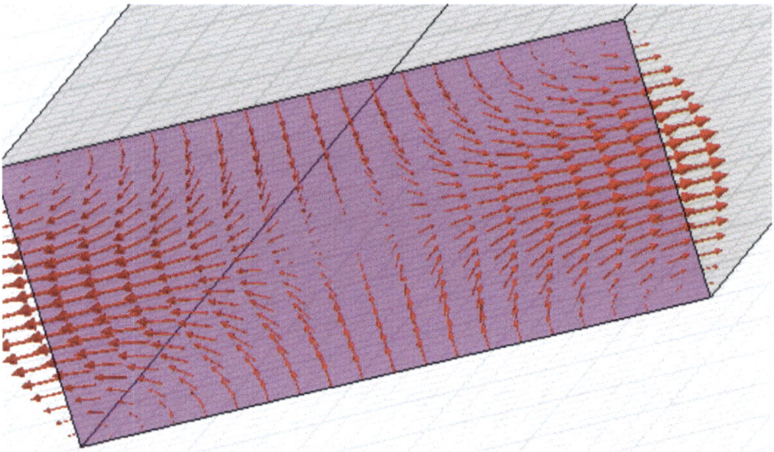

Fig. P1.35 Electric field distributions for TE_{11} (left-hand side) and TM_{11} (right-hand side) modes, respectively

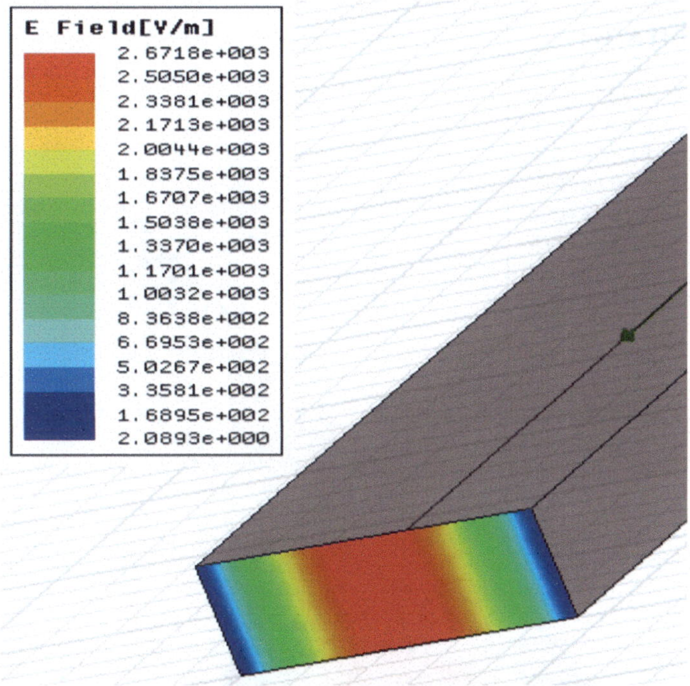

Fig. P1.36 E field in HFSS

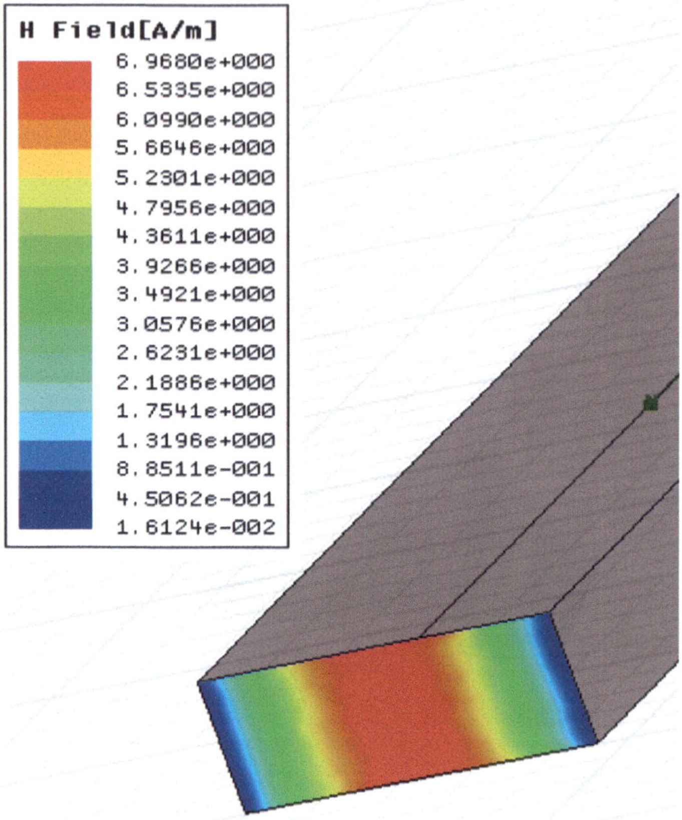

Fig. P1.37 H field in HFSS

Problems of Chapter 2

P2.1

- The core view of the power divider. The power divider can be decomposed into two separate two-port networks, the odd-mode network and the even-mode network, based on its symmetry, enabling individual analysis of each component.
  For the common port of a power divider, odd-mode excitation does not affect it, as long as we analyze the even-mode circuit. The return loss at Port 1 for the common mode and the even-mode circuit is equal. Consequently, the transmission line design of the power divider transforms into designing a $2Z_0$-to-Z_0 impedance converter.

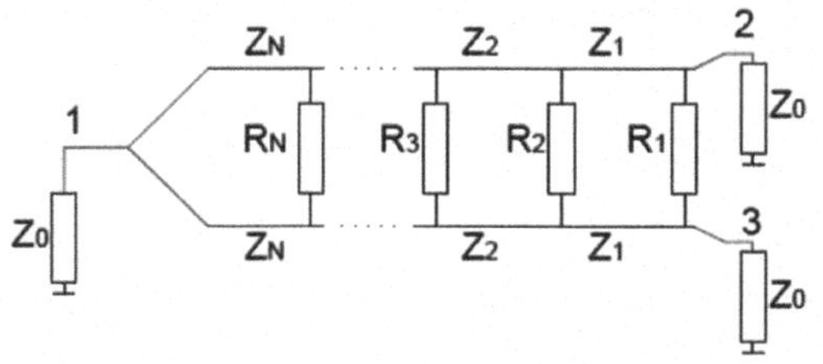

Fig. P2.1 N-section wideband power divider

The even-mode excitation circuit also functions as a power combiner. When Ports 2 and 3 have in-phase equal voltage inputs, the power can be transferred to Port 1 without any loss. Some individuals often confuse $S_{21} = -3$ dB, based on the reciprocity of passive circuits, with $S_{12} = -3$ dB, which makes it impossible to achieve power doubling. This situation occurs when there are only inputs at Ports 2 and 3, with no signal input at Port 3.
From the odd-mode circuit perspective, pure odd-mode excitation at the common port does not result in any power reaching it. The isolation resistor merely acts as an attenuator/load, fully absorbing the power fed in through the odd-mode excitation. Consequently, Ports 2 and 3 achieve low standing wave ratio and high isolation.

- Simulation and design of a power divider:
  Based on the theoretical analysis and conclusions, the design of a power divider essentially transforms into the design of an impedance transformer. There are various ways to implement impedance transformers, including stepped impedance transformers, transformers, and impedance transformation filters. Stepped impedance transformers are commonly encountered, but in reality, power dividers based on transformer and impedance transformation filter structures exhibit excellent performance in many specific scenarios.

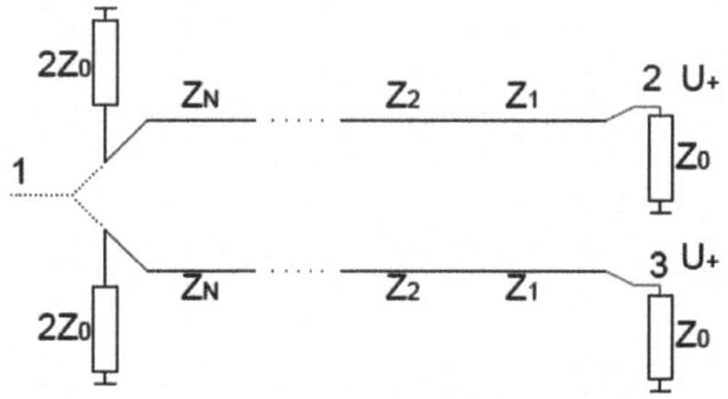

Fig. P2.2 Even-mode excitation equivalent circuit

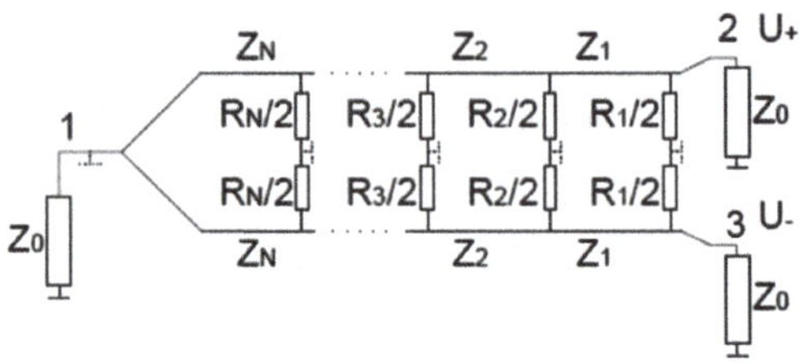

Fig. P2.3 Odd-mode excitation equivalent circuit

Below, we present a detailed design process for a power divider operating within the frequency range of 2 to 18 GHz. The chosen substrate material is Rogers 5880 with a thickness of 0.254 mm.

- Design procedure:

  Determination of the impedance transformation stages and the impedance values of each section.

  Confirmation of the isolation resistor value.

  Individual electromagnetic simulation of the impedance transformer.

  Electromagnetic simulation of the T-shaped section.

  Combined electromagnetic simulation of the entire power divider.
- Impedance transformer order and parameter acquisition:

  There are numerous software tools available for obtaining the order and parameters of an impedance transformer. In this study, we utilize modeling and simulation in ADS (advanced design system) to acquire the corresponding parameters. The effective dielectric constant can be approximately determined using transmission line calculation tools, with an initial length set to one-quarter wavelength at the center frequency. The other impedance values are arranged in descending order from high to low, and through a simple optimization process, and the impedance values for each stage can be obtained.

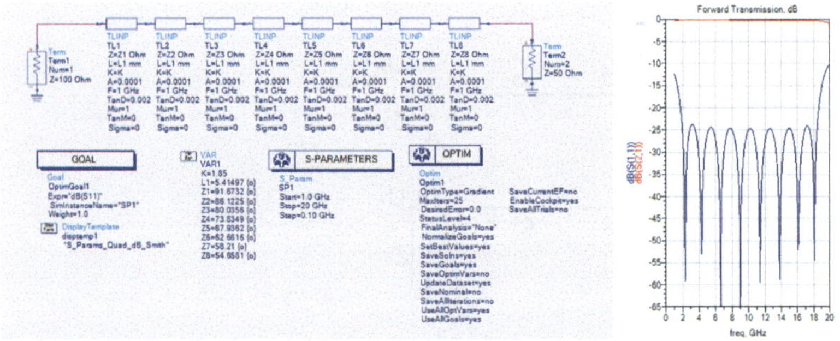

Fig. P2.4 Impedance transformer model and results in ADS

Using transmission line calculation tools such as the ones included in ADS or Polar SI9000, the actual widths of each impedance can be determined for a given substrate material, Rogers 5880, with a dielectric constant (ε_r) of 2.2, and a thickness

(t) of 0.254 mm. The results of the calculations are presented in the following table:

Table P2.1 Calculated values of impedance for different widths of substrate material rogers 5880

	Z_1	Z_2	Z_3	Z_4	Z_5	Z_6	Z_7	Z_8	Z_o
Impedance	91.7	86.1	80	73.8	67.9	62.6	58.2	54.6	50
Line width	0.23	0.27	0.32	0.37	0.44	0.51	0.58	0.64	0.74

- Simulation of isolation resistance: Based on the previous design results, a comprehensive theoretical model of the power divider was established in ADS. The transmission line parameters were fixed, and an isolation resistor was introduced. By providing an initial value and performing a two-port standing wave optimization, the numerical value of the isolation resistor can be conveniently obtained.

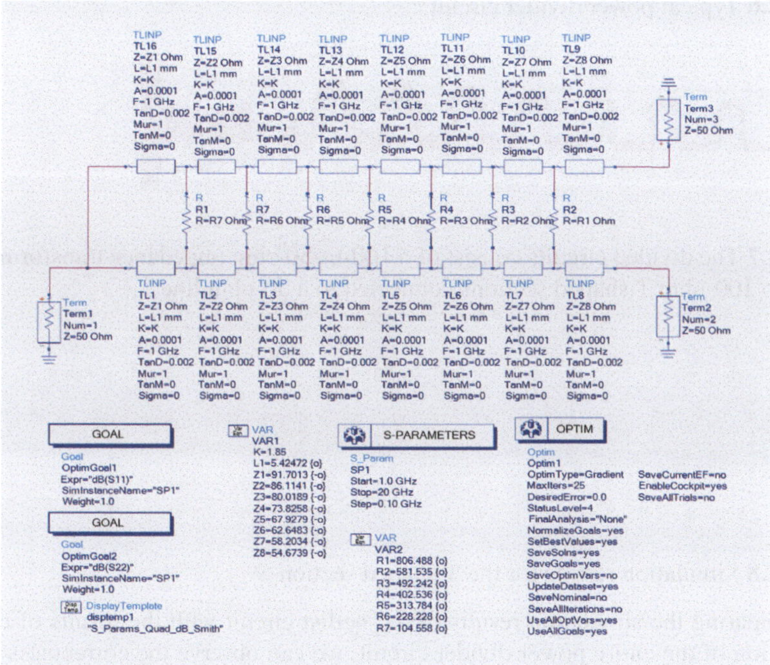

Fig. P2.5 Complete theoretical model of power divider

- Electromagnetic simulation of the impedance transformer and T-shaped section: Based on the calculated dimensions of the impedance transformer, a model for the impedance transformer was established, and simulations were conducted in Sonnet. Directly simulating the entire power divider as a single entity can be computationally intensive and time-consuming due to its large model size. To mitigate resource consumption, we divided the simulation into two separate parts and optimized them individually. Subsequently, a combined simulation was performed to verify and compare the results against those obtained without any optimization.

This comparison demonstrated the high accuracy of the parameters calculated in the ADS model.

- Overall simulation of the power divider model:
  The overall simulation of the power divider involves combining the impedance transformer and the T-shaped section into a complete power divider circuit. A netlist circuit is constructed in Sonnet for this purpose.

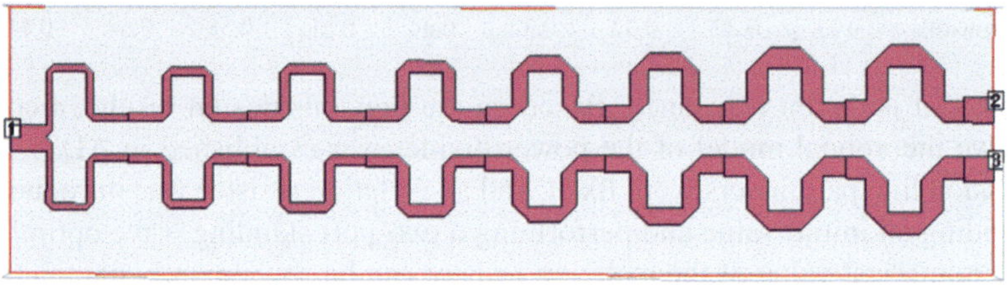

Fig. P2.6 Typical power divider circuit

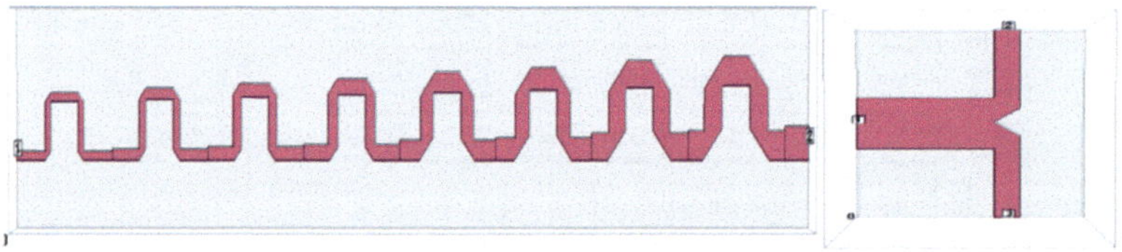

Fig. P2.7 The divided circuits consist of a 100-to-50 ohm impedance transformer and two 100-ohm T-shaped sections connected to a 50-ohm line

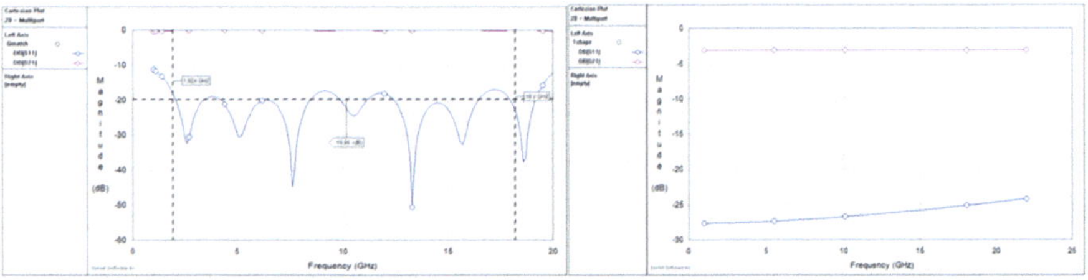

Fig. P2.8 Simulation results for the T-shaped section

By comparing the simulation results of the netlist circuit with the results of the simulation of the entire power divider circuit, we can observe the correctness of the circuit segmentation and the accuracy of the entire power divider simulation process. Finally, after incorporating the resistors, the overall simulation of the power divider is essentially completed.

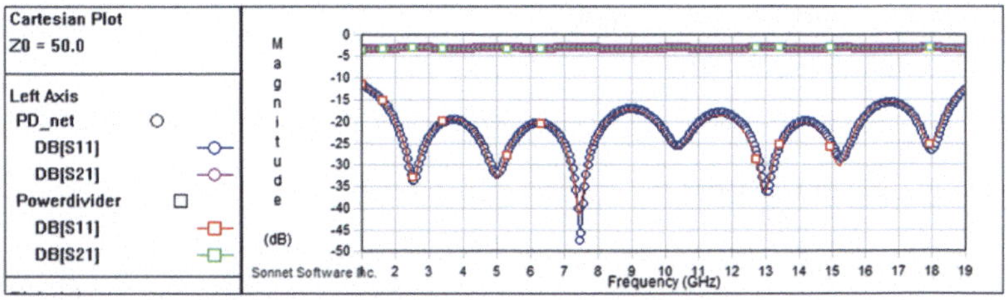

Fig. P2.9 Comparison of netlist circuit simulation and full circuit simulation results

P2.2

The solution provides the general operating steps for EMPIRE XPU.

- Start: Start Empire XPU
  Select "New Project," OK
  Select File—"Save As"
  Select/create storage folder, e.g., *tutorial*
  Enter file name, e.g., *patch*
  Save
- Group creation
  Right-click on group name #001
  Select *EditName*, enter *Substrate*
  Open Group
  Double-click on *Height*
  Set Height: z = 0...635
- Property setting
  Double-click "conductor" to change property
  Select Dielectric
  select Database tab
  Select Rogers
  Rogers-RT/Duroid5880
  OK to leave the property editor
- Create Box with values Click *Create Box*
  2 clicks at arbitrary points (to be edited later*)
  Long click to use group height
  Adjust coordinates
  OK
  Zoom extents
- Create Box with cursor
  Click *AddGroup*
  Enter *Patch* as name, recolor
  Double-click Height, set z = 635...645
  Keep conductor as property
  Create patch:
  Click *CreateBox*
  Left click at x = 20000, y = 20000

Left click at x = −20000, y = −20000
Long click to use group height
OK

- Port definition
  Create group, rename to *Port*
  Set Height: z = 0 …635
  Click *CreateSource*
  Select *PerpendicularPort*
  3 clicks at arbitrary points (to be edited later)
  Select Height *FromGroup*
  Adjust point values
  OK
  Redraw
- Simulation setup
  Click *SimulationSetup*
  EMSetup Tab:

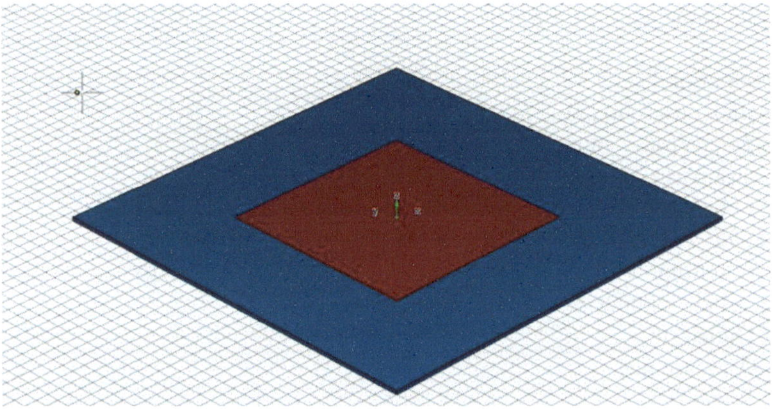

Fig. P2.10 The designed patch antenna in EMPIER XPU

Start Frequency: 1.25 GHz
Stop Frequency: 2.50 GHz
Target Frequency: 2.45 GHz
Boundary Tab:
zmin to Electric, keep zmax,
Change the others to *Absorbing*4(> 20 dB)
OK

- Far-field recording
  Click *CreateFieldMonitor*
  Select *EMFarField*, Boundary
  Select *ProcessingSetups*
  – Select Far Field Setup 1, Advanced Tab
  – Enable Boundary Mirroring at zmin
  Repeat-with Far-Field Setup 2–3
  Exit with OK

- Simulation Click *StartSimulation*
  "OK"
- Results
  Voltage ut1
  S-Parameter S_{11}
  Far field (Polar Lin)
- Near fields Return to 3D Design Tab
  Click Create Field Monitor

  Select *EMField*, *Plane*, OK
  Open Field Monitor, Set Height:635
  Repeat simulation *
  3D Results Tab, switch off far-field monitor
  Right-click *PlanarEMField* -Edit
  In *PlotOptions*** change settings (see next page)
  Click *Apply*

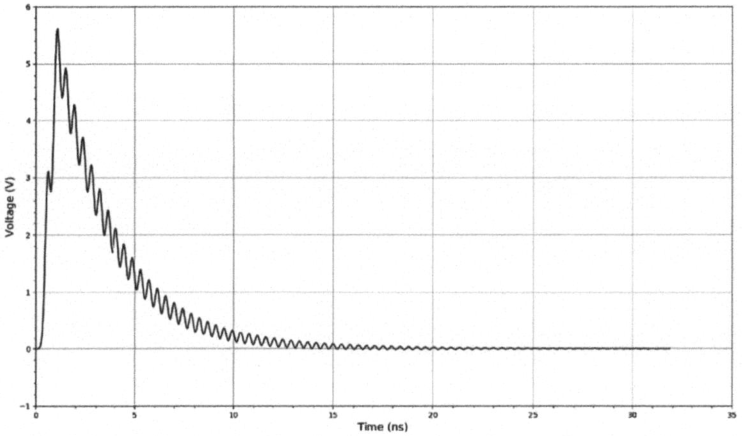

Fig. P2.11 Voltage simulation result

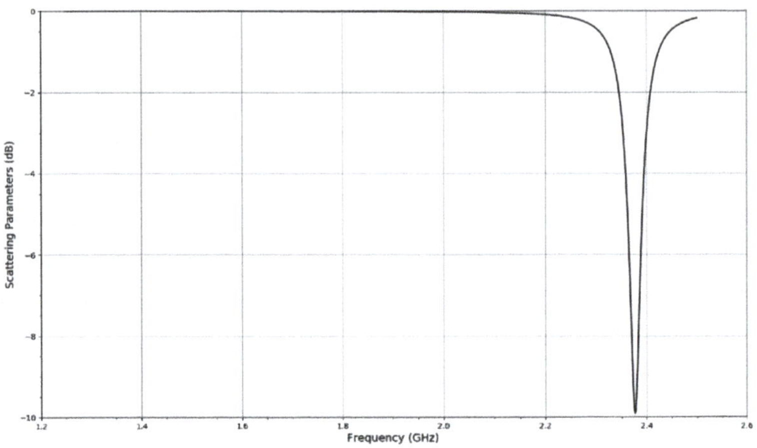

Fig. P2.12 S-Parameter simulation result

Optionally switch off groups to improve field display

- Plot Options Defaults, electric field
  Defaults + Field Plot Amplitude: 50000
  Defaults + Amplitude: 50000 + Contour Lines + Color: Black
  Defaults + Contour lines + Arrow Display-Scale Arrow

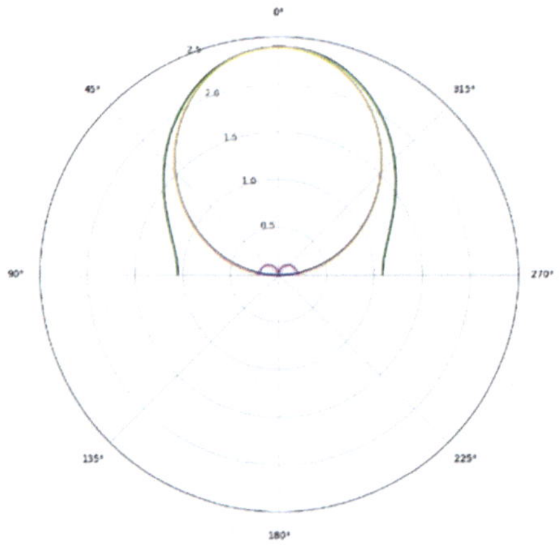

Fig. P2.13 Farfield simulation result

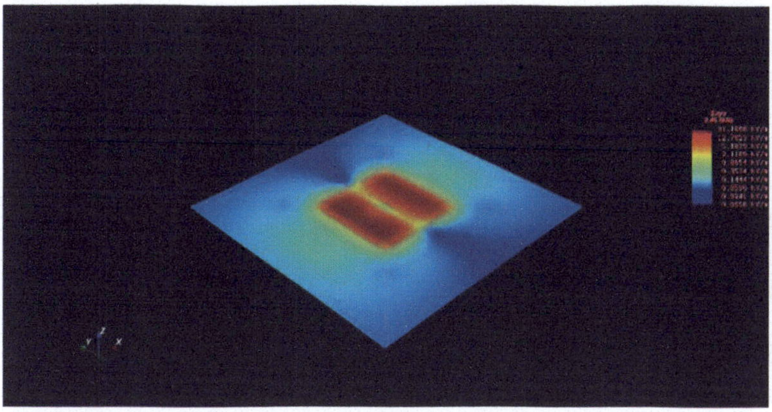

Fig. P2.14 Electric field distribution simulation result

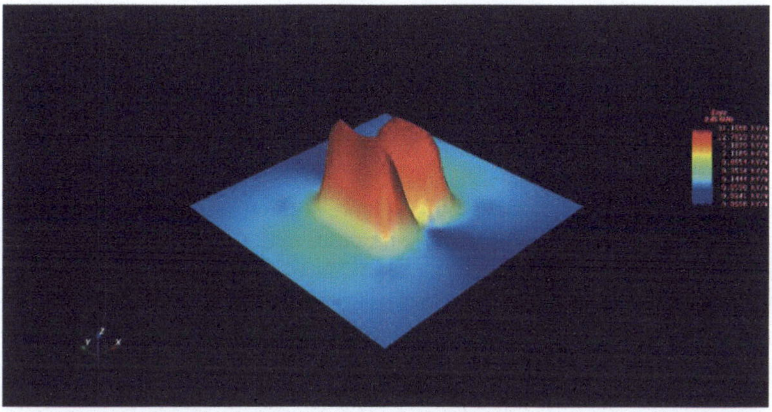

Fig. P2.15 Electric field with increased amplitude

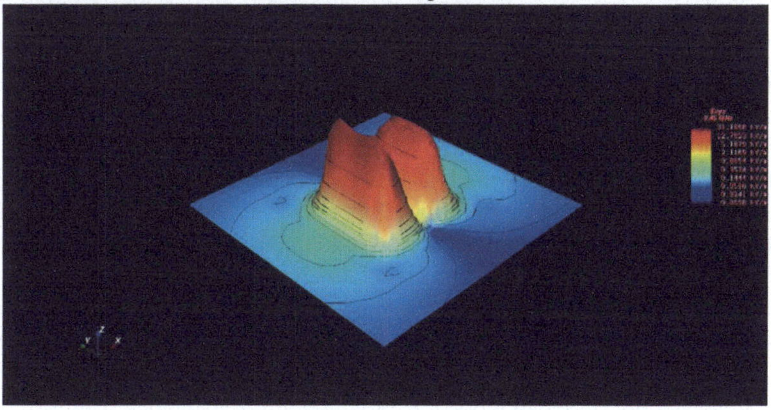

Fig. P2.16 Enhanced electric field visualization

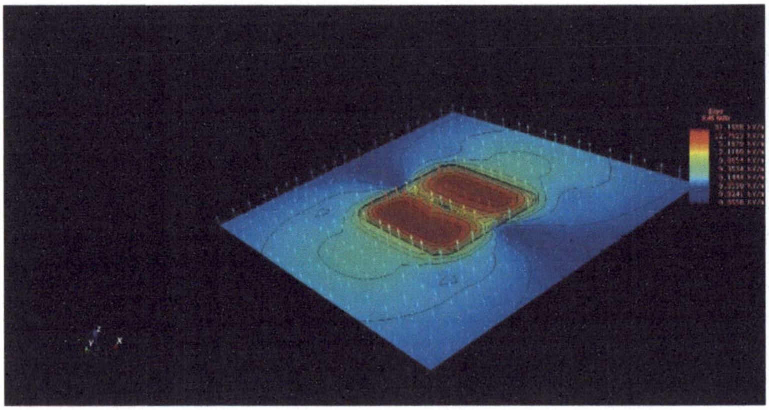

Fig. P2.17 Comprehensive electric field mapping with contour and vector arrows

Problems of Chapter 3

P3.1

In this filter design, the initial values of the filter are obtained through modeling using Genesys software. The filter has a wide bandwidth and requires high out-of-band suppression. It needs to mitigate interference from broadcasting stations near 110 MHz and strong signal interference at 1.86 GHz. For ease of design, the suppression at the frequency point of 1.86 GHz can be initially disregarded. The design process begins by first creating a filter that meets the other requirements, and subsequently adding a series resonator to the input port of the filter with a resonance frequency of 1.86 GHz, effectively creating a transmission zero at that frequency. When a strong signal at 1.86 GHz is present at the input, it is directed to the ground through this resonator. The entire design process is divided into the following three steps:

1. Obtain the initial values of the filter using Genesys.
   Specify the filter type, transmission zero, cut-off frequency, in-band ripple, out-of-band suppression, number of filter stages, and other parameters. The need to shield signals near 110 MHz, which are very close to the lower end of the pass-band, requires the adoption of an elliptic function structure to achieve better filter cut-off characteristics. Generally, at lower frequencies, implementing elliptic function filters using lumped elements is relatively straightforward. However, as the frequency increases, realizing elliptic function bandpass filters becomes challenging due to the significant impact of parasitic parameters. By continuously adjusting these parameters, a set of initial filter values that meet the requirements, and even exceed the specified targets, is obtained in the Response window. This is because when replacing the components in the Genesys-generated filter with packaged actual elements and constructing the practical circuit in ADS, the simulation results may experience some degradation. Therefore, when generating the initial filter values, it is important to consider necessary margins for the perfor-

mance indicators. Finally, the filter initial values generated by Genesys are shown in the figure below:

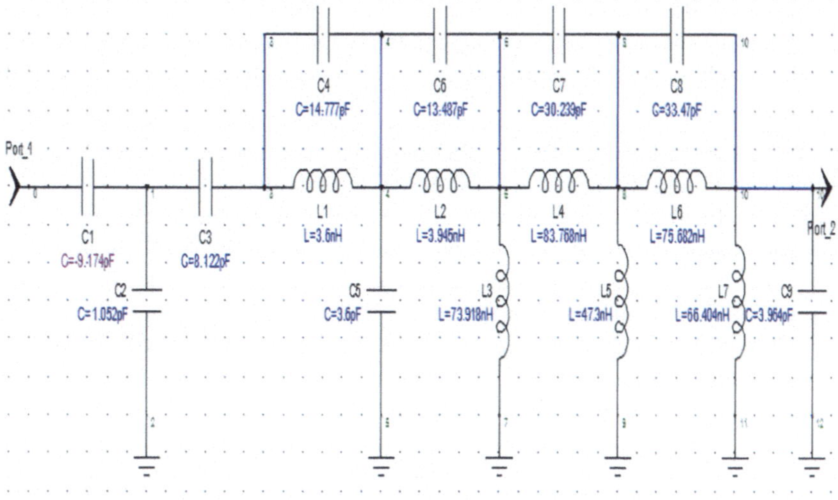

Fig. P3.1 Initial filter value generated by Genesys

2. Initial ADS schematic design.
   After obtaining the preliminary values from Genesys, the circuit is imported into ADS. Considering the requirement for suppression at the frequency point of 1.86 GHz, a series resonator is added at the input with a resonance frequency of 1.86 GHz. The ideal capacitors and inductors generated by Genesys are then replaced with packaged capacitors and inductors from the ADS component library. Care is taken to ensure that the component values selected in ADS are as close as possible to the ideal component values obtained from Genesys. The selection of packaged components should be kept reasonably small to minimize the introduction of significant parasitic effects, which may otherwise lead to unpredictable impacts on the filter characteristics. Additionally, consideration is given to the existing package sizes of components available in the laboratory to optimize time, effort, and design accuracy.
   Subsequently, the electrical connections in the schematic are replaced one by one with microstrip lines. The microstrip lines in the design are chosen to have a characteristic impedance of 50 Ωs. The width of the microstrip lines can be determined using the Line Calc tool in ADS. The electrical parameters and frequency of the selected substrate are input to calculate the line width for a 50-Ω impedance at the passband center frequency (since the variation of line width with frequency is not significant in this design).
3. Perform optimization on the circuit to achieve the final results.
   Set optimization objectives according to the design requirements. Due to the close resemblance of simulation results obtained from Genesys to the desired specifications, gradient algorithms can be employed for optimization to expedite

the process. During optimization, manual collaboration can be conducted using the Status Window, which provides an effective human-machine interaction interface. Through this window, the optimization process of the filter can be intuitively observed. Based on the prompts in the Status Window and the final optimization results, modifications can be made to the optimization scope, objective function, and weight factors to ensure that the simulation results ultimately meet the design requirements.

P3.2 The integration of ANSYS HFSS with National Instruments' AWR Microwave Office is achieved through the ANSYS AW Integration Link.

Here is the Microwave Office simulator window displaying the model created and designed using Microwave Office. Now, upon opening the "Em structures" folder, we can access the model options, which reveal that this model will be simulated using HFSS's three-dimensional finite element method solver. In the structure tool, we can observe the board, chip, and package components. In Microwave Office, we can right-click on the HFSS module and select "simulate," which automatically initiates the simulation in HFSS. Alternatively, we can directly open the model in the native editor, which will automatically launch HFSS. In HFSS, we can see the same layout and analysis settings, and we can also view different material and port configurations. HFSS offers a three-dimensional layout viewer to examine the circuit layout in greater detail. We have the capability to export the HFSS 3D layout model to HFSS, allowing full access to all of HFSS's functionalities, such as component drawing and adding coordinate systems.

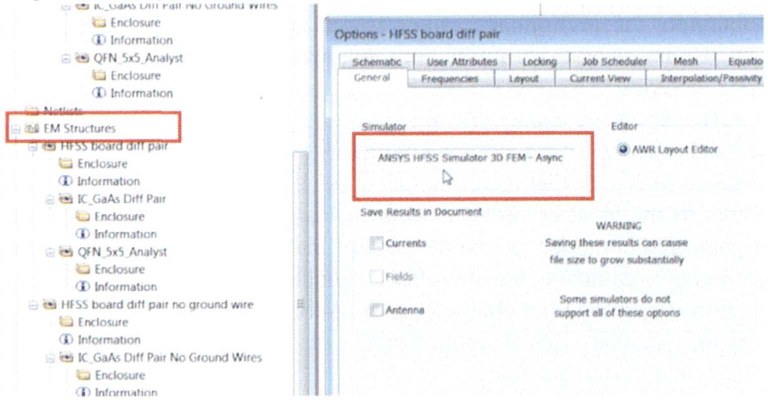

Fig. P3.2 The integration link of ANYSYS HFSS is Microwave Office

Problems of Chapter 4

P4.1

In CST, the dipole antenna model and S-parameter simulation results are shown in the Figs. P4.1 and P4.2.

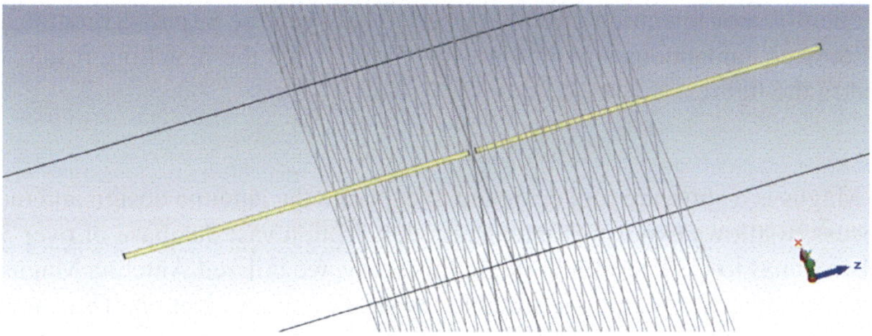

Fig. P4.1 Modeling of dipole antenna in CST

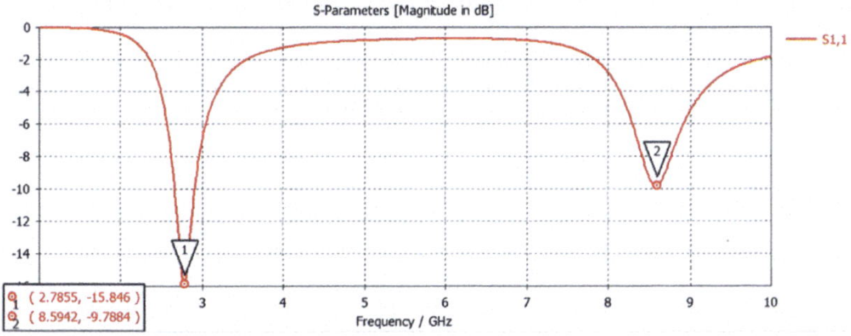

Fig. P4.2 S_{11} parameter simulation result

In the ADS, a data module is added through Data Items, and the S_{11} data "single dipole.s1p" is imported. In the ADS, a 50-Ω port and GND are included, along with the S-parameter sweep module. The S_{11} results are shown in the Fig. P4.3.

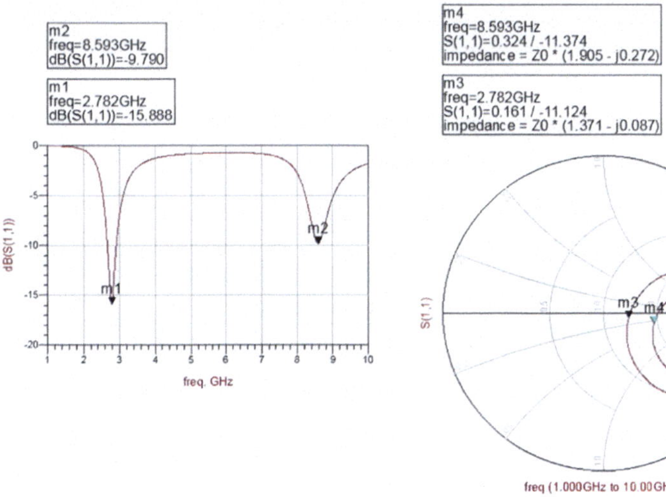

Fig. P4.3 S_{11} results of adding 50 ohm port and GND in ADS

Subsequently, a pi-matching circuit is inserted between the S1p data module and the 50-Ω port. Instantaneous simulation is performed, and the matching results are visualized in the figure.

P4.2

Antenna Magus is a software tool designed to expedite the antenna design and modeling process. It allows exporting antenna models from a vast database of over 350 validated antennas to CST Studio Suite. In this study, we utilized Antenna Magus to design a rectangular patch antenna suitable for Wi-Fi communication. The antenna operates at a center frequency of 2.45 GHz. Using the interface shown in the figure, we identified the rectangular patch antenna and adjusted the parameters accordingly. Finally, the antenna model was exported by selecting the "Model Export" option within the "Design Mode."

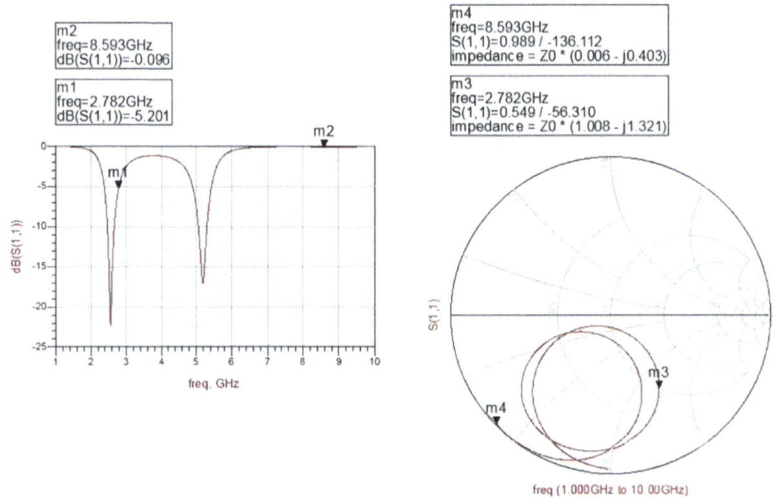

Fig. P4.4 S_{11} results of adding Pi type matching circuit in ADS

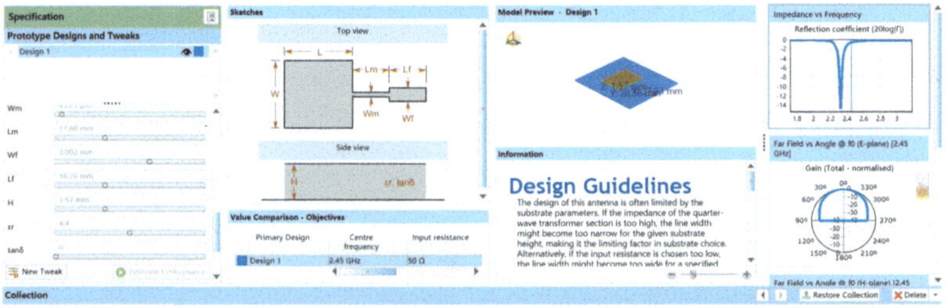

Fig. P4.5 Design guidelines and generated models for microstrip patch antennas in Antenna Magus

Problems of Chapter 5

P5.1 The following figure plots (Dk_{eff}) over frequency derived from (S_{21}) phase or time delay ((TD)); ($Dk_{\text{eff}} = (TD_{c0}/\text{length})^2$) from a Megtron-6 stripline case study. This method is different than IPC-TM-650 test method in that it determines (Dk_{eff}) from unwrapped phase delay rather than calculating (Dk/Df) from resonant peaks over the frequency range defined in the spec.

The blue plot is a simulated case based on core and prepreg (Dk) values from published (Dk/Df) tables at 12 GHz. When (Dk) is corrected due to roughness, using the equation, and resimulated, (Dk_{eff}) is shown in pink. Although the (Dk_{eff}) has improved, it still does not agree with the measured (Dk_{eff}) from the device under test (DUT), shown in red.

P5.2

1. Import PCB: The Ansys electromagnetic simulation platform allows seamless import of industry-standard PCB design files, such as Cadence, Zuken, ODB++, IPC-2581, and others, with a single click. The net list information remains consistent with the original design, eliminating the need for redundant settings and providing a convenient and efficient workflow.
2. S-parameter simulation:

   - After importing the PCB design file, port settings are configured in Ansys SIwave to analyze the network that generates electromagnetic interference (EMI) radiation.

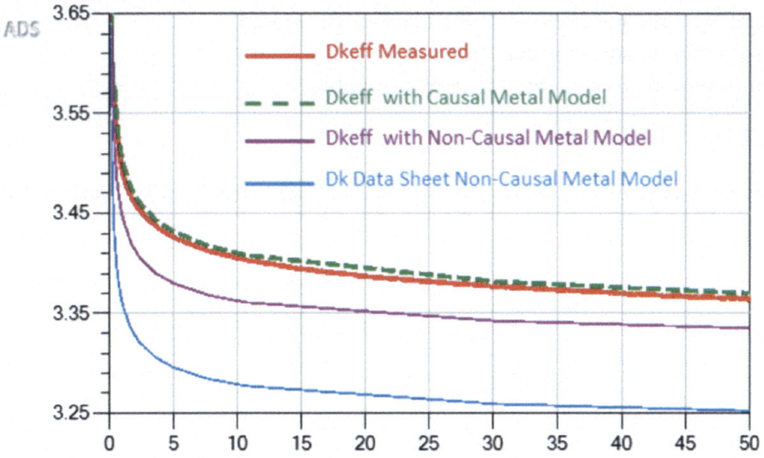

Fig. P5.1 Modeled with Simbeor and simulated with Keysight ADS

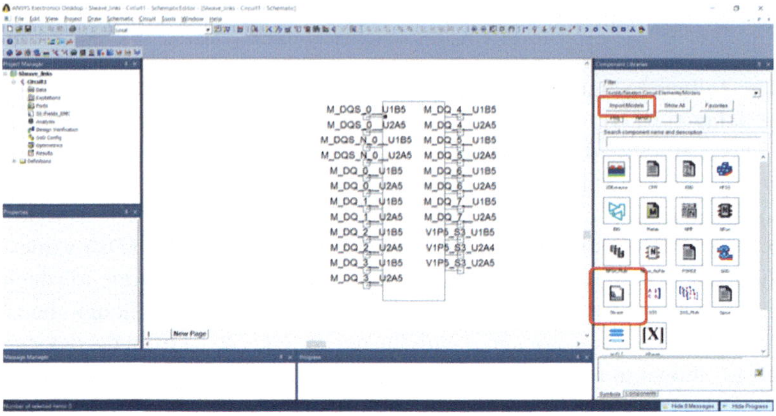

Fig. P5.2 Adding SIwave components in Circuit

- The S-parameter configuration is performed, followed by the simulation process.

3. Simulation combining fields and circuits—dynamic link simulation between Circuit and SIwave:

   - In the Circuit project, incorporate the completed SIwave project file from the aforementioned simulation.
   - Incorporate the chip's IBIS model into the Circuit project.
   - Complete the overall schematic design for the simulation.
   - Configure the parameters of the chip's IBIS model.
   - Perform transient simulation.
   - Output the results of the transient simulation.

4. PCB EMI radiation simulation:

   - Feed back the results of the simulation combining fields and circuits to SIwave.

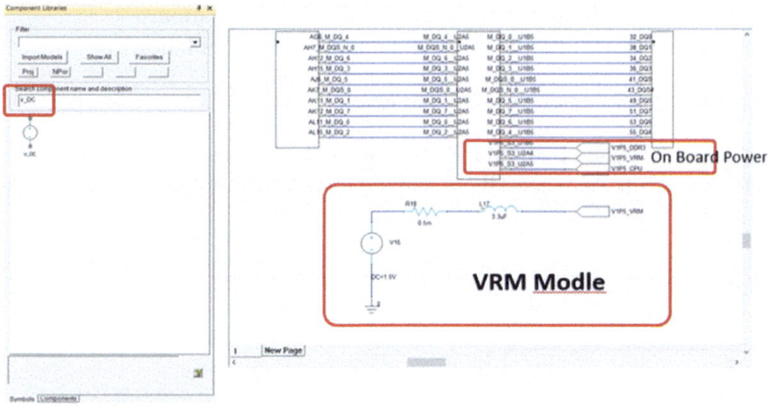

Fig. P5.3 Overall simulation schematic

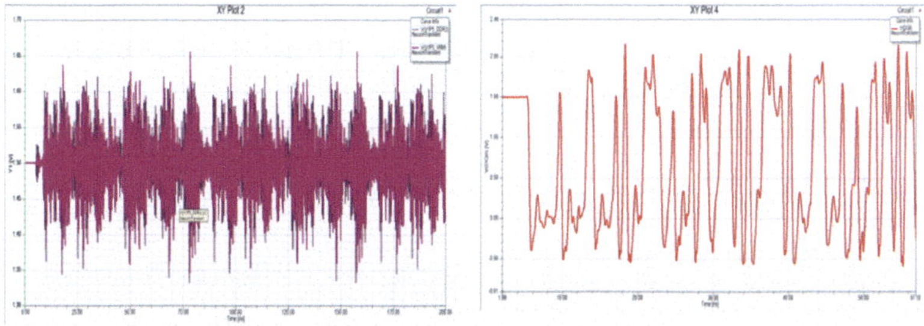

Fig. P5.4 Transient simulation results

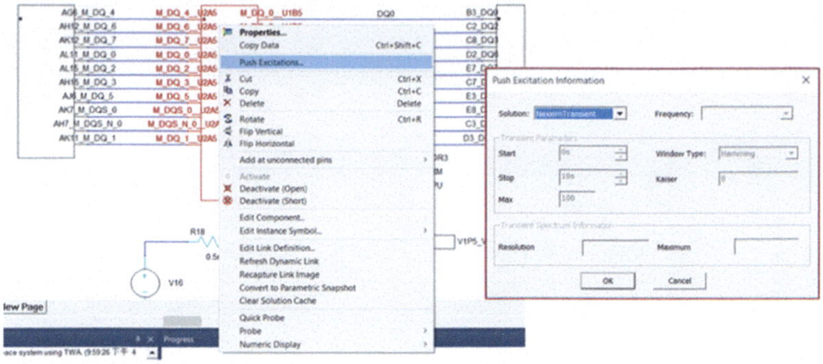

Fig. P5.5 Setting up push excitation information

- Conduct near-field and far-field analysis of PCB radiation.
- Output the results of PCB near-field and far-field simulations.

5. EMI analysis at the enclosure system level:

- Output the results of SIwave PCB near-field analysis.
- Establish a link between HFSS and the radiation source field.

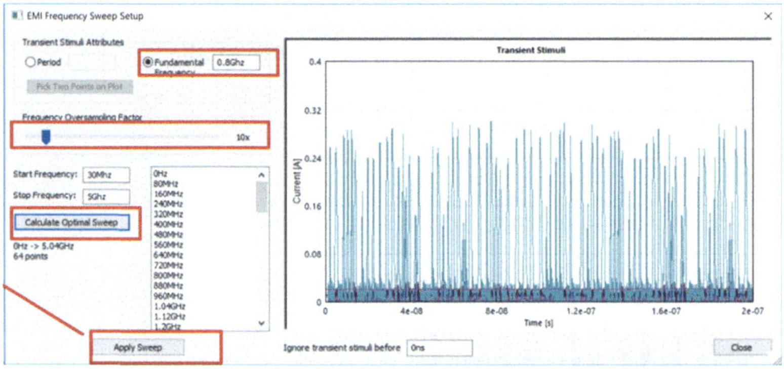

Fig. P5.6 PCB radiation near and far field analysis

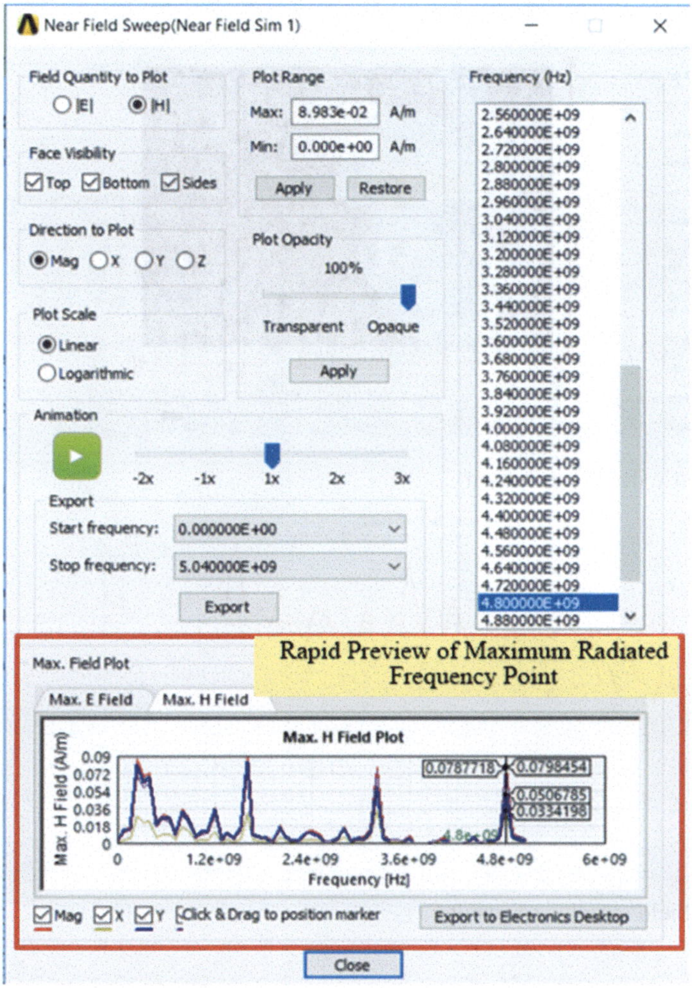

Fig. P5.7 PCB near field sweep results

- Import the enclosure model.
- Perform a full three-dimensional electromagnetic field simulation in HFSS.

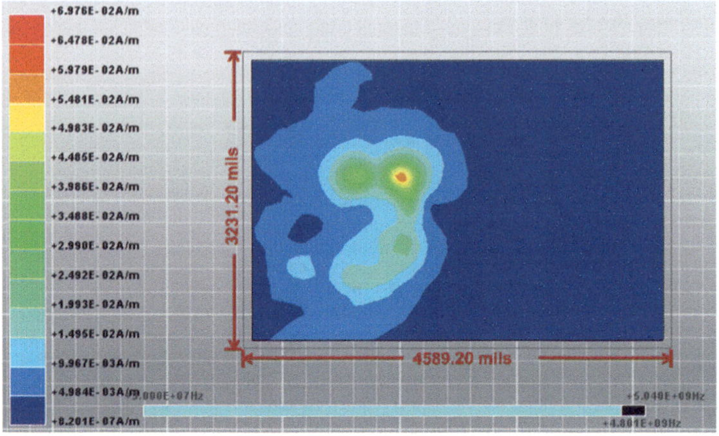

Fig. P5.8 Electric field distribution

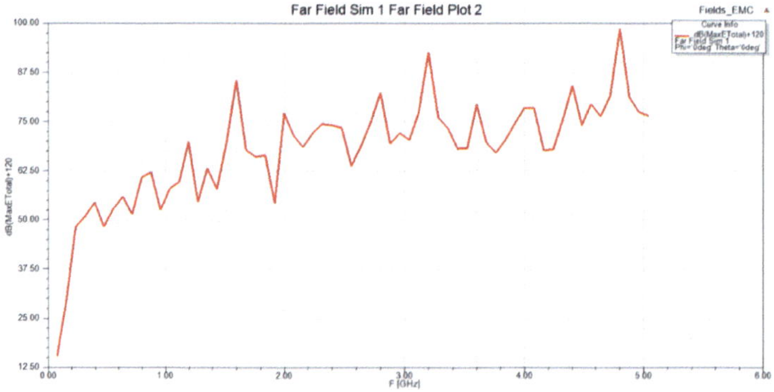

Fig. P5.9 Far Field simulation results

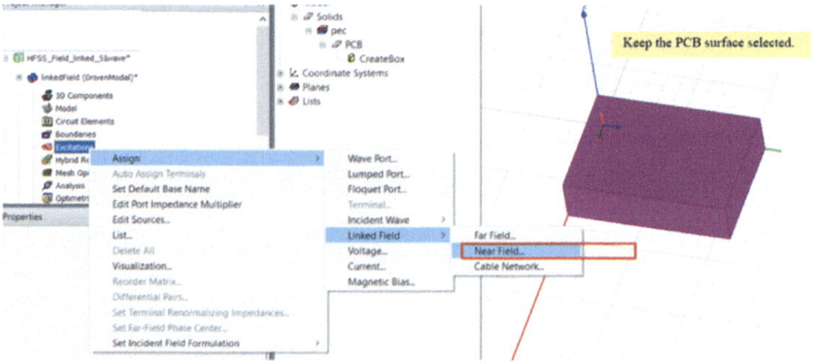

Fig. P5.10 Configuring near field simulation in HFSS software

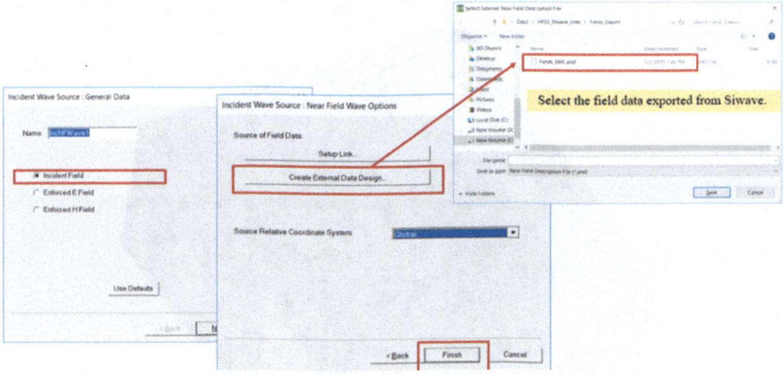

Fig. P5.11 Source setup for incident wave

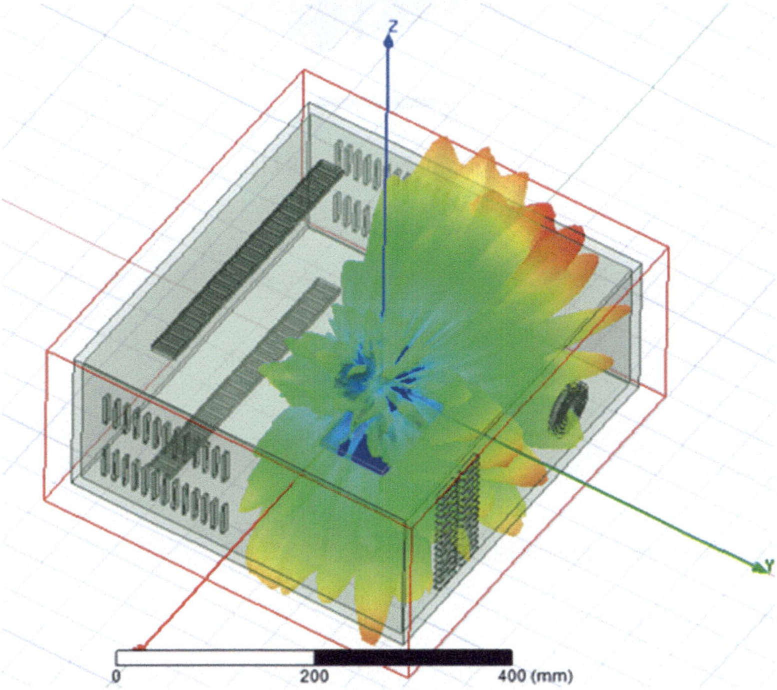

Fig. P5.12 Simulation for total far Field at 4.8 GH

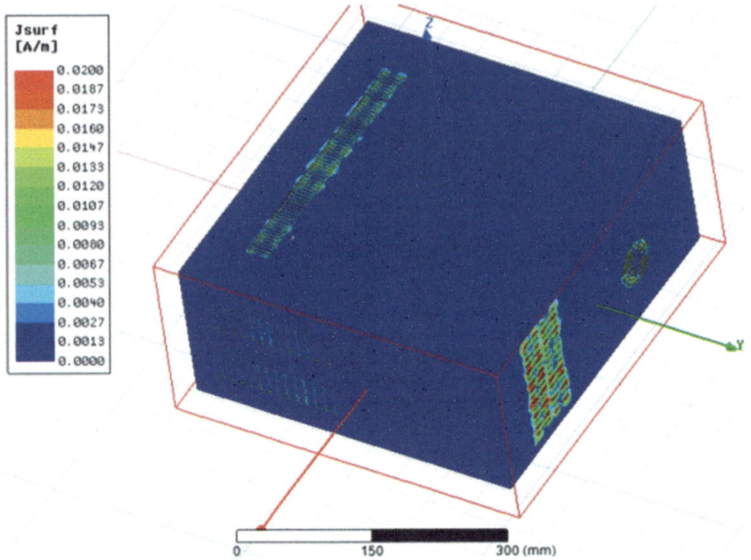

Fig. P5.13 Surface current density visualization at 4.8 GHz

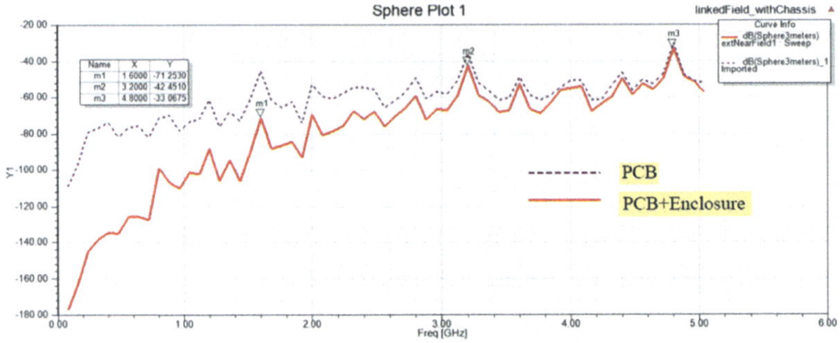

Fig. P5.14 Sphere result comparison

Problems of Chapter 6

P6.1 The main solution involves using ADS (advanced design system) to establish circuit and electromagnetic simulation models for the LNA-MMIC (low noise amplifier—monolithic microwave integrated circuit) chip, and using Flotherm to create a thermal simulation model for the chip's packaging. The ADS simulation modeling utilizes the built-in DemoKit-Non-Linear standard library, with a relative dielectric constant of 12.9 and a substrate thickness of H = 100 μm, using a QFN (quad flat no-lead) packaging structure. The Flotherm packaging model includes thermal layers, substrate layers, thermal vias, plastic packaging, bare chip circuit, wire bonds, solder paste, and solder layers, as illustrated in the figure. Flotherm software is utilized to perform the packaging modeling for the bare die of the low noise amplifier, obtaining chip temperature metrics.

Data transfer between the engineering disciplines is achieved through CSV (comma-separated values) files, enabling serial multi-sample simulations for the LNA-MMIC chip across the electrical, electromagnetic, and thermal domains.

P6.2 The main steps for performing electro-thermal one-way coupled simulation using HFSS and Icepak are as follows:

1. ANSYS HFSS setup and computation.

   Firstly, use HFSS to set up materials, boundaries, excitation conditions, and solver settings for the model. Compute the surface losses of the metal layers and the volume losses of the dielectric layers, which will be used as heat sources for Icepak. Volume power losses include Ohmic losses, dielectric losses, and magnetic losses of objects with linear material properties (requiring internal solving). After electromagnetic computation, the loss distribution in the metal and dielectric layers can be viewed through post-processing results, and the total loss can be calculated using field calculators.
2. Designmodeler setup and update.
   To perform coupled simulations between HFSS and Icepak, the HFSS model must be imported into geometry via the ANSYS Workbench platform. The model is then processed using Designmodeler to convert it into a format that Icepak can accept.
3. ANSYS Icepak setup and computation.
   Under the Workbench platform, import the HFSS model and the computed loss results into Icepak. Set up the computation region, cooling methods, air velocity, and model materials. Then, create a mesh and proceed with the computation. During the computation, residual curves and monitor point curves can be monitored in real time. Post-processing results can display temperature contour plots on cross sections, where temperature non-uniformity is mainly caused by uneven losses. In the Summary report interface, device losses can be quantified and compared between HFSS and Icepak computations. Consistent results between HFSS and Icepak demonstrate that ANSYS Workbench enables precise electromagnetic-thermal coupled simulations using HFSS and Icepak.

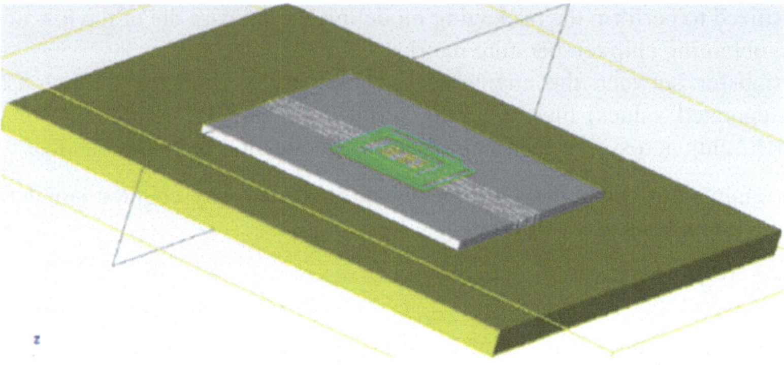

Fig. P6.1 Package model created by Flotherm

For performing bidirectional coupled simulations between HFSS and Icepak, the model's materials in HFSS need to be set as temperature dependent. In Icepak, the temperature feedback option should be selected, and temperature feedback is set up in the Workbench platform to perform electro-thermal bidirectional coupled simulation.

Problems of Chapter 7

P7.1 The question addresses the open-ended exploration of ADS's multi-physics simulation capabilities to evaluate power amplifier stability across diverse operational scenarios. This involves utilizing enhanced features of ADS, possibly in conjunction with RFPro's electromagnetic visualization tools, to identify the physical locations and frequencies that contribute to instability. By doing so, potential issues can be rectified prior to hardware manufacturing. The answer to this would depend on the specific simulation and design parameters, and the reader is encouraged to delve into ADS's documentation and resources for detailed guidance.

P7.2

- In EDT, solve a coupled Icepak-HFSS model.

  Simply transferring body temperature directly from the Icepak (Classic) Workbench application. This generates an initial set of temperatures, a fluid flow field, and a corresponding surface map of heat transfer coefficients.

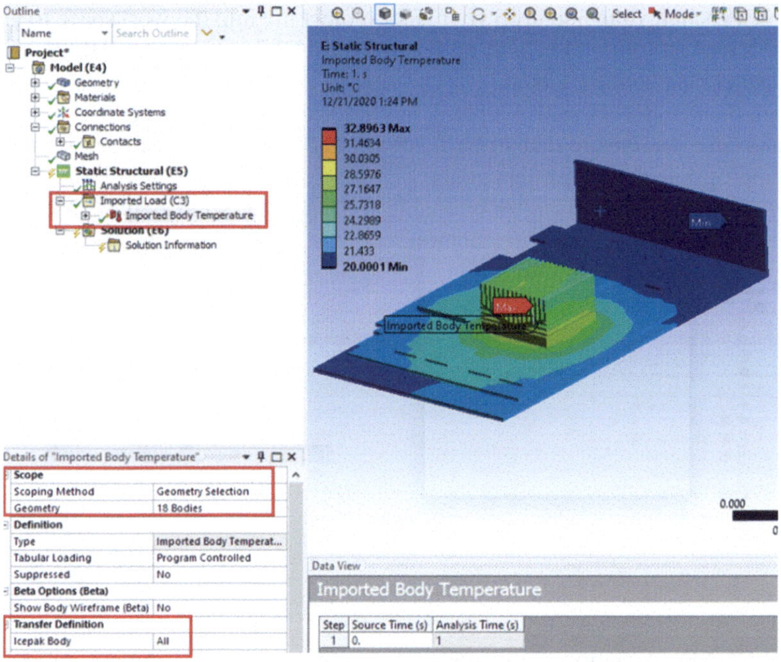

Fig. P7.1 Thermal analysis of imported body temperature in static structural simulation

- Connect multiphysics models in Workbench

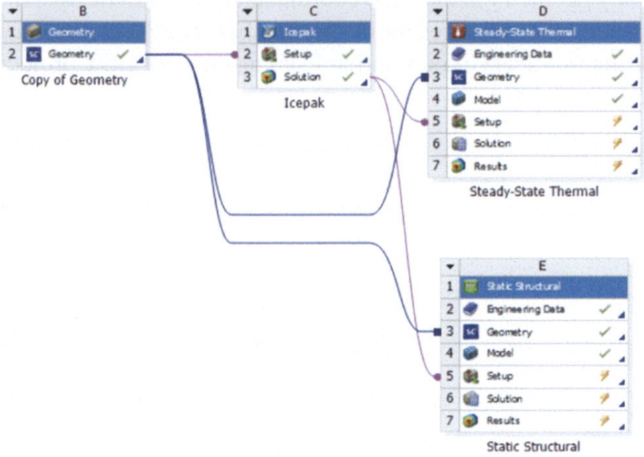

Fig. P7.2 Workflow integration for thermal and structural analysis

In Workbench, drag and drop your .aedt project file into the workspace to get an HFSS analysis type. This .aedt project file must ONLY contain an HFSS project, it cannot also contain Icepak.

- Open the new HFSS analysis block in your Workbench project
  Make sure you ensure that stress feedback and/or thermal feedback is enabled, as appropriate for the project goals.
  1. For stress/deformation, only enable feedback for solid bodies that will be present in the Mechanical simulation.

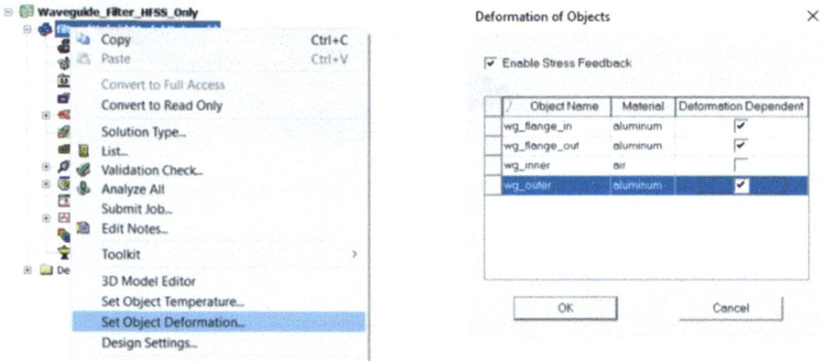

Fig. P7.3 Deformation of objects

  2. For temperature feedback, only bodies with an assigned material property that includes a thermal modifier are enabled. So, make sure the materials are correctly defined inside of EDT first.

- Match geometry and coordinate frames
  The same solid body geometry in the same coordinate frame must be used in both mechanical/thermal and HFSS cases for the bodies that are actually exchanging information. Use the Modeler–Export option from inside your EDT project and generate a .stp geometry file.

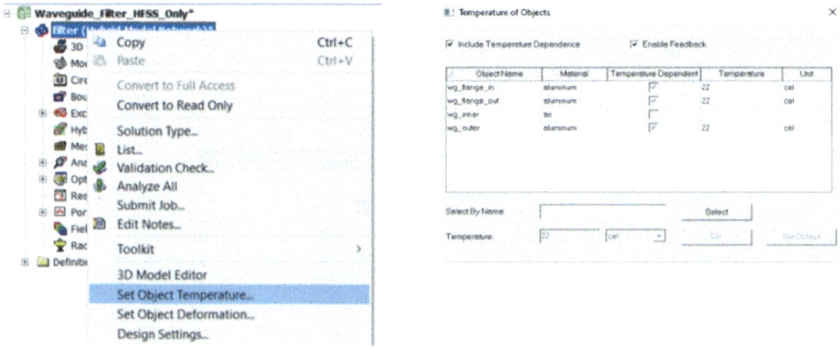

Fig. P7.4 Temperature of objects

- Create workflow in Workbench
  In the Workbench view with just your HFSS system present, build out the complete workflow like so:
  1. Drag a steady-state thermal system onto the "Solution" line of the HFSS system.
  2. Drag "Geometry" from HFSS onto "Engineering Data" in Steady-State Thermal.
  3. Drag a Static Structural system onto the "Solution" line of the HFSS system.
  4. Connect Engineering Data and Geometry from Steady-State Thermal to the corresponding items of static structural.
  5. Connect "Solution" from Steady-State Thermal to "Setup" of Static Structural.
  6. Connect 'Solution' from HFSS to "Setup" of Static Structural.
  7. Right-click on "Geometry" for Steady-State Thermal and use Import Geometry to bring in the .stp file from 4.
  8. Add an "External data" component system to the workflow, and drag "Setup" from External Data to "Setup" on Steady-State Thermal.
  9. The end result should look like below. It is important at this time that the "Model" of Steady-State Thermal and Static Structural not be attached, as we do not want/need the same mesh for both analysis types, and linking this can mess up the HFSS data transfer later on.
- Map Loads Open the "Setup" of External data, add the HTC map file (From step 1), then assign your data columns. You will need x/y/z point coordinates along with a heat transfer coefficient value and an ambient temperature value.

  Once the data file is added and columns assigned, go back to the project view and right-click Update the External Data "Setup," then $Updateupstreamcomponents$.
- Update the Thermal Model
  Right-click > Update on "Setup" for Steady-State Thermal. If HFSS was not already solved, this will solve the initial non-deformed isothermal case and enable future connection to the thermal/mechanical solvers. After the update is complete, open the Steady-State Thermal model:
  1. Assign materials to each solid body as needed.

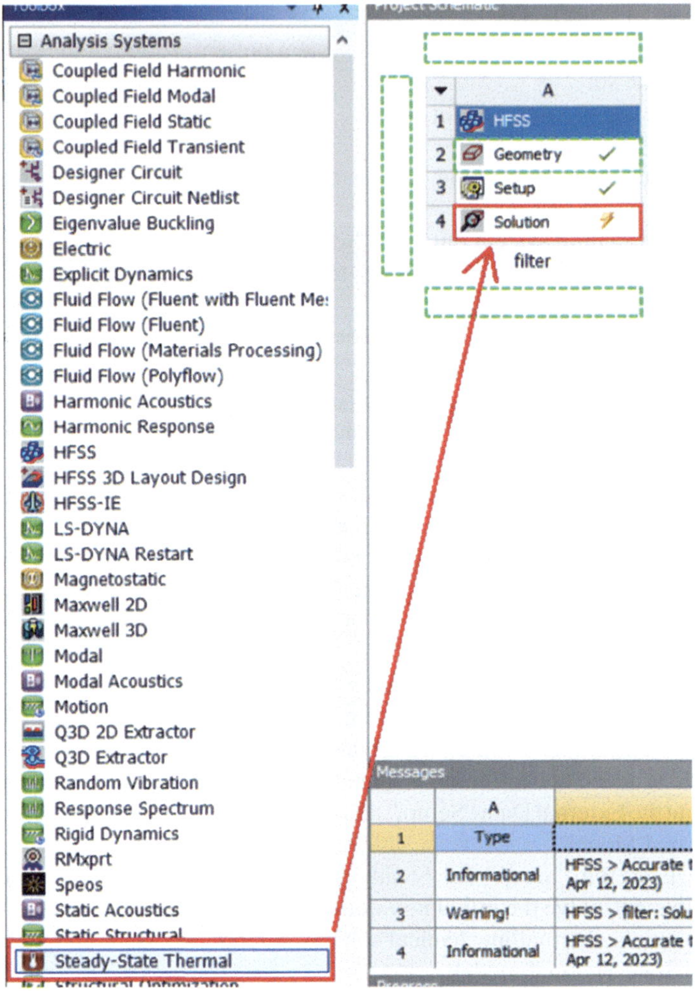

Fig. P7.5 Workflow of steady-state thermal simulation

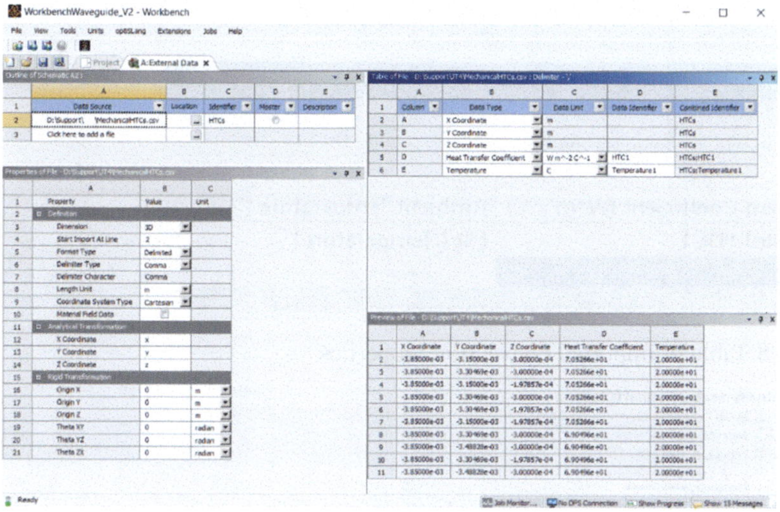

Fig. P7.6 Data import and property mapping interface

2. On the "Imported Load" for External data, right-click and insert a Convection Coefficient. Adjust the geometry scope so that it includes all the faces that were exported from the original Icepak thermal project. I suggest using the face selection filter and box select tool to grab everything.

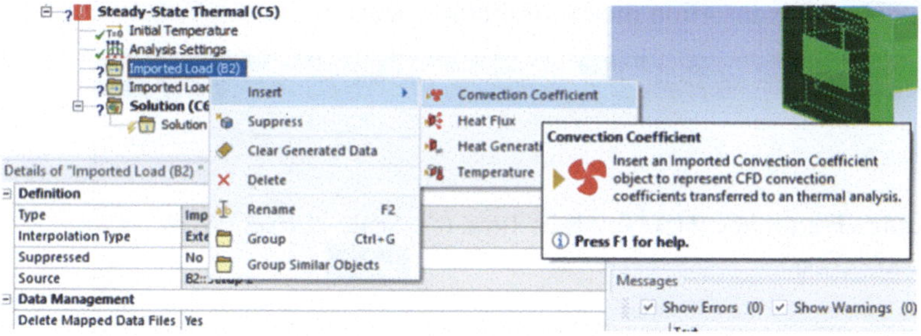

Fig. P7.7 Convection coefficient mapping for thermal analysis

If external data was loaded properly, the Film Coefficient and Ambient Temperature fields should automatically populate:

3. For the Imported Load from HFSS, add a heat flux and scope it to all relevant surfaces:

Note that you can Scale and Offset the imported load if desired:

4. With "Imported Load" from HFSS still selected, go into Details > Export Definition and change "Export After Solve" to Yes:

5. Add any result fields you want (Temperature, for example), and then solve the Steady-State Thermal problem. In the message section, there needs to be an "Info: Temperatures were successfully exported" message:

- Repeat the previous step, but for Static Structural.

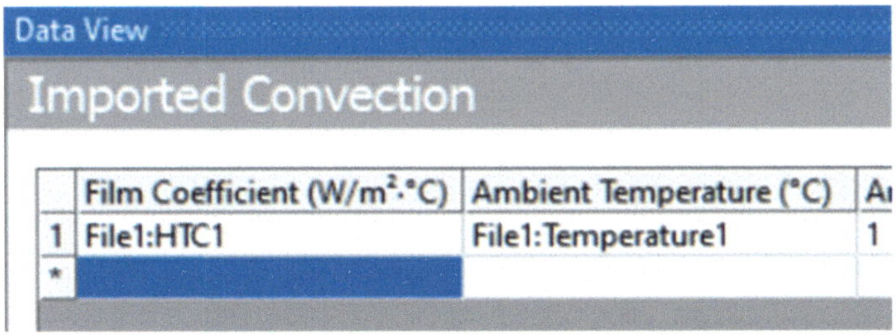

Fig. P7.8 Table of imported convection properties

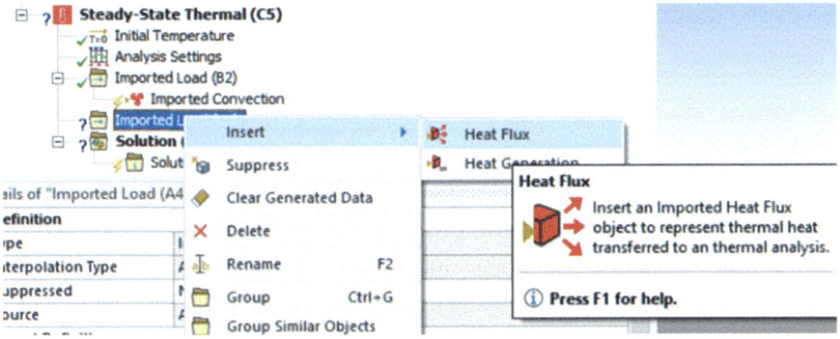

Fig. P7.9 Heat flux insertion menu for thermal analysis

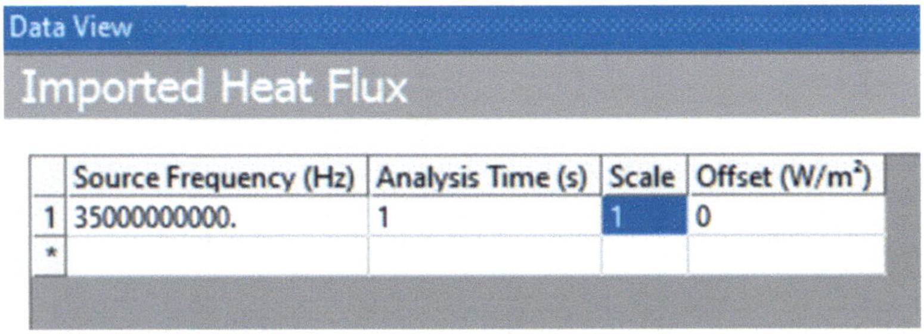

Fig. P7.10 Imported heat flux data parameters

For Imported Loads, there will be one from Steady-State Thermal and one from HFSS—the Steady-State Thermal connection should be fine by default. For HFSS, add either a body force or surface force with whatever object scoping is appropriate, and then make sure to enable "Export After Solve" as before. As always, Mechanical boundary conditions will need to be applied to properly constrain the problem. Additionally, care must be taken such that HFSS requirements are not violated—e.g., port excitation surfaces must remain planar. Solve the Static Structural system and look for the same "Info: Structural results were successfully exported" message.

- Enable HFSS Update.
  With Steady-State Thermal and Static structural each solved once, go back to the Workbench project view and right-click on "Solution" under HFSS—Enable Update:

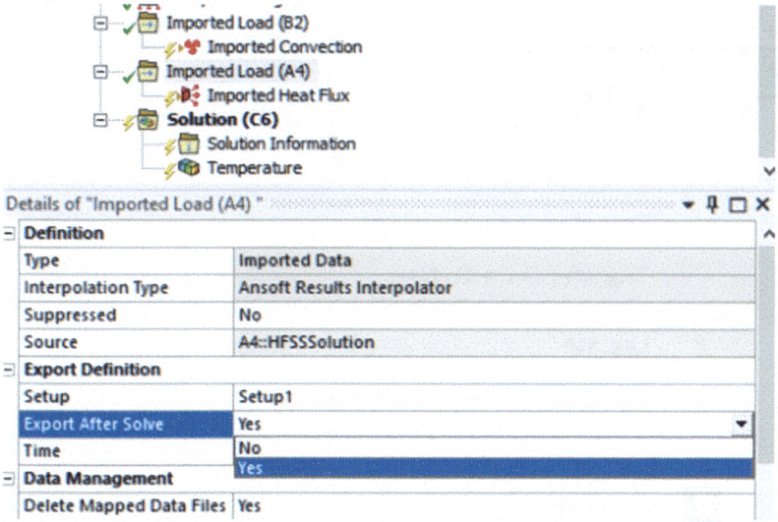

Fig. P7.11 Imported load configuration for thermal analysis

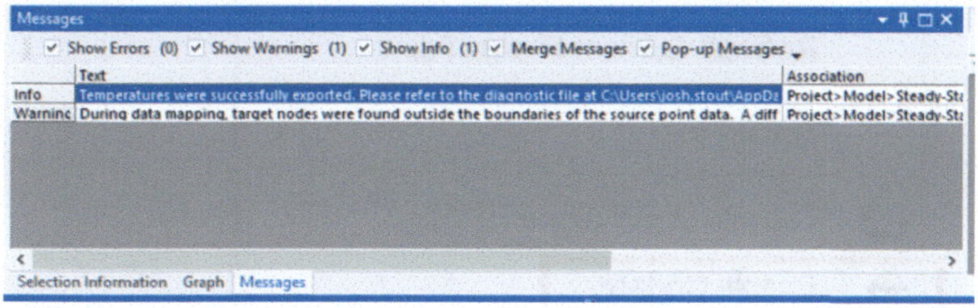

Fig. P7.12 The message section

- Update HFSS.
  Then, right-click "Update" on the HFSS solution. This solves HFSS again with the updated temperature field and structural deformation.
  You can now go back into the HFSS system and double-check the mapped temperatures, deformations, and general HFSS results via field plots.
- Repeat until your Multiphysics Simulation converges.
  This update process can continually be repeated to iterate through solutions until converged. Or, drag the "Feedback Iterator" component system onto the HFSS Setup line, set convergence criteria for delta temperature and/or displacement values and let it automatically loop through the process for you.

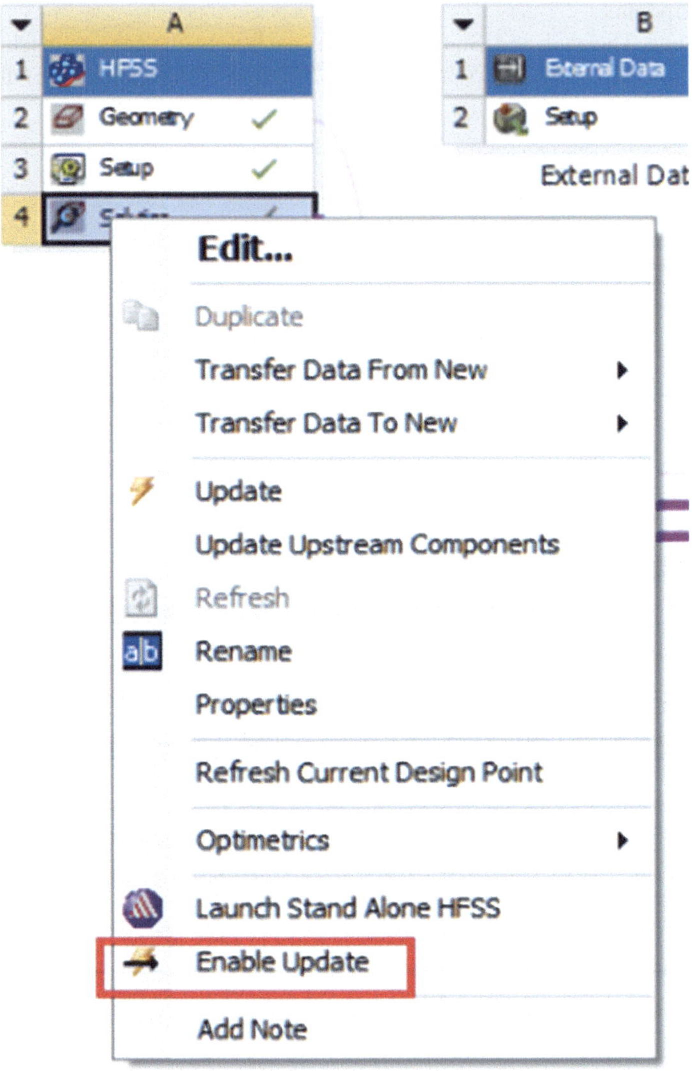

Fig. P7.13 Enable HFSS update in the workbench project

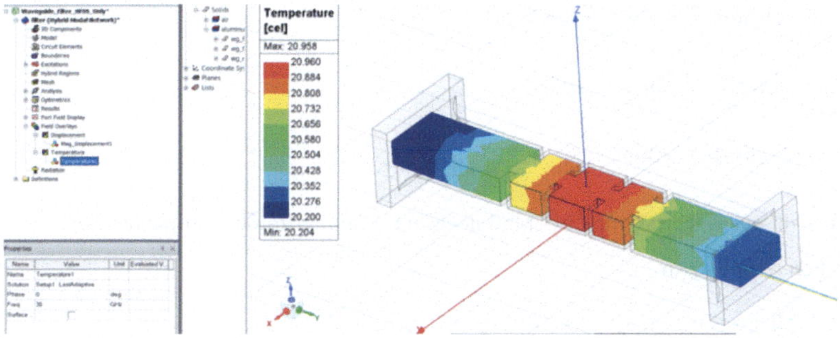

Fig. P7.14 The updated temperature field and structural deformation

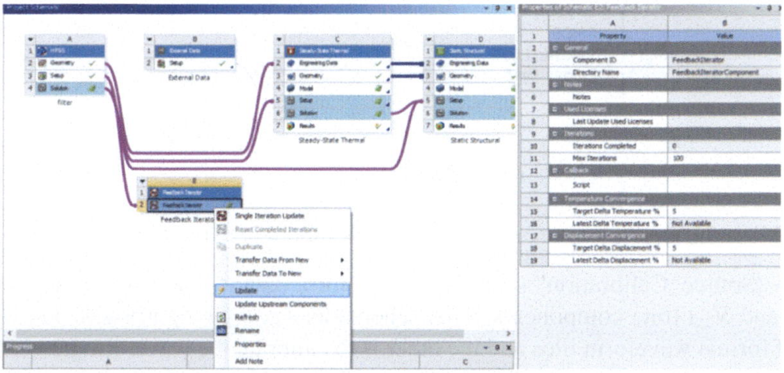

Fig. P7.15 Repeat until multiphysics simulation converges

Problems of Chapter 8

P8.1 The mat file is a binary format file used by MATLAB software for data storage. Standard ADS software does not natively support importing data in this format. To address this, data processing is performed using Datalink with the help of Python scripts and formulas. By decompressing and opening the provided official ADS project file "DL_mat2mdf_wrk," the MAT file to be imported is copied to the "data/python" directory. Then, the data display window "Mat2Mdf" is opened. The conversion function formula is selected, and by right-clicking the mouse, the "Calculate" option is chosen to execute the formula. A successful execution of the formula is indicated by the formula turning green, and the result showing "OK," which denotes a successful conversion. Upon successful execution, the "QPSK_200K.mdf" file will be generated in the "data/python" subdirectory of the project.

Fig. P8.1 Calling the conversion function formula

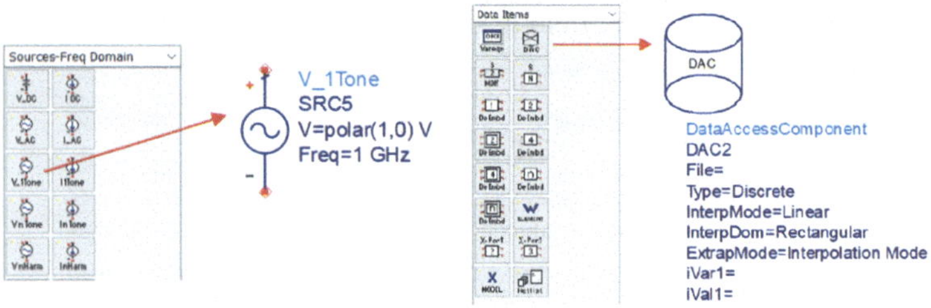

Fig. P8.2 Use the DAC component and V_1 Tone component to read in waveform data

In the "Source_Calibration" schematic, waveform data is read using DAC components and V_1Tone components. This schematic serves two purposes: one is to read mdf format waveform files and the other is to calibrate the waveform power for adjusting the waveform power to the required level in simulations. In the bottom-right corner, the .mdf filename, carrier frequency, waveform sampling rate, simulation end time, and other data are entered.

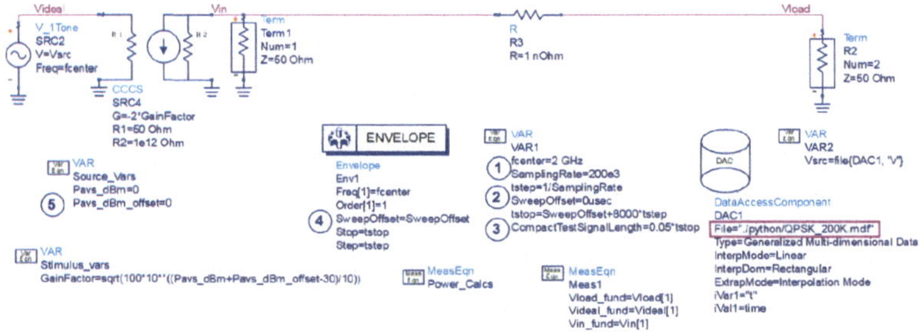

Fig. P8.3 Source_ Calibration schematic diagram

The simulation results include the spectrum of the input waveform, IQ constellation plot, waveform time amplitude, and CCDF curve. For more information on the data interface between ADS and Python in Datalink and debugging Python code, refer to the ADS help documentation.

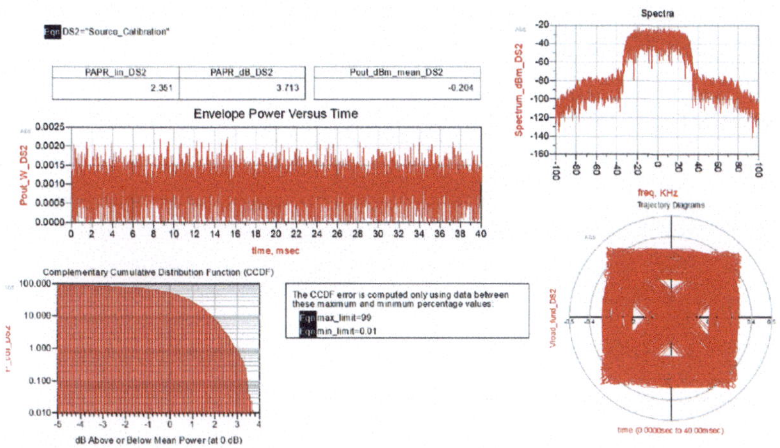

Fig. P8.4 The spectrum, IQ vector diagram, waveform time amplitude and CCDF curve of the input waveform in the simulation results

P8.2 In MATLAB, HFSS variables can be added using "hfssaddVar." When programming, the following code snippet can be used to utilize the relative coordinate system in HFSS:

```
for n=1:N
for m=1:M
i=N*(m−1)+n;
\%\%\%\%add local CS
CSName = ['CS', num2str(i)];
Origin = \{['(',num2str(m),' −(M+1)/2)*p'],
['((N+1)/2−',num2str(n),')*p'],'0'\};
XAxisVec = [1 0 0 ];
YAxisVec = \{0, 1, 0\};
hfssCreateRelativeCS(fid, CSName, Origin,
XAxisVec, YAxisVec, units);
hfssSetWCS(fid, CSName);
\%\%\%\%\%\%\%Array of metasrface(square loop)
squarebig\{i\} = ['squarebig\_',num2str(i)];
Start\_squareloop = \{'−te/2','−te/2','h'\};
hfssRectangle(fid, squarebig\{i\},
'Z', Start\_squareloop, 'te', 'te', units);
squaremedium\{i\} = ['squaremedium',num2str(i)];
Start\_squareloop1 = \{'−tc/2','−tc/2','h'\};
hfssRectangle(fid, squaremedium\{i\},
'Z', Start\_squareloop1, 'tc', 'tc', units);
hfssSubtract(fid, squarebig\{i\},
squaremedium\{i\});
cut\_square\{1,1\}= squarebig\{i\};
```

```
PEbig\{i\}=['PEbig\_',num2str(i)];
hfssAssignPE(fid, PEbig\{i\},cut\_square, false);
...
hfssSetWCS(fid, 'Global');
end
end
```

Index

M. S. Tong and X. Y. Li, *Co-simulations of Microwave Circuits and High-Frequency Electromagnetic Fields*, https://doi.org/10.1007/978-981-99-8307-0

H

I

K

L

X

Y

SPRINGER NATURE

GPSR Compliance

The European Union's (EU) General Product Safety Regulation (GPSR) is a set of rules that requires consumer products to be safe and our obligations to ensure this.

If you have any concerns about our products, you can contact us on ProductSafety@springernature.com

In case Publisher is established outside the EU, the EU authorized representative is:

Springer Nature Customer Service Center GmbH
Europaplatz 3
69115 Heidelberg, Germany

The manufacturer's authorised representative in the EU is Springer Nature Customer Service Centre GmbH, Europaplatz 3, 69115 Heidelberg, Germany. If you have any concerns regarding our products, please contact ProductSafety@springernature.com

Printed and bound by CPI Group (UK) Ltd, Croydon, CR0 4YY
01/10/2026
02225862-0014